Aging of the Genome

Aging of the Genome

The dual role of DNA in life and death

Jan Vijg

Buck Institute for Age Research, Novato, CA, USA

OXFORD
UNIVERSITY PRESS

OXFORD
UNIVERSITY PRESS

Great Clarendon Street, Oxford OX2 6DP

Oxford University Press is a department of the University of Oxford.
It furthers the University's objective of excellence in research, scholarship,
and education by publishing worldwide in

Oxford New York

Auckland Cape Town Dar es Salaam Hong Kong Karachi
Kuala Lumpur Madrid Melbourne Mexico City Nairobi
New Delhi Shanghai Taipei Toronto

With offices in

Argentina Austria Brazil Chile Czech Republic France Greece
Guatemala Hungary Italy Japan Poland Portugal Singapore
South Korea Switzerland Thailand Turkey Ukraine Vietnam

Oxford is a registered trade mark of Oxford University Press
in the UK and in certain other countries

Published in the United States
by Oxford University Press Inc., New York

British Library Cataloguing in Publication Data
Data available

Library of Congress Cataloging in Publication Data
Data available

Typeset by Newgen Imaging Systems (P) Ltd., Chennai, India
Printed in Great Britain
on acid-free paper by
Antony Rowe Ltd., Chippenham

ISBN 978–0–19–856922–0 978–0–19–856923–7 (Pbk.)

10 9 8 7 6 5 4 3 2 1

■ PREFACE

Science is a major force in the introduction of new ideas and information in society. This has not always been so. After a hesitant beginning in the high middle ages, science definitely took off during the late Renaissance as a competing force with religion and began to capture the hearts and minds of many people. While originally motivated by the desire to know life, how it originated, and how it could be extended, science was soon absorbed by the prosaicism of the Industrial Revolution in the eighteenth and nineteenth centuries. From then on science was subject to practical purposes such as industrial manufacture, environmental control, and fighting human disease. As such, science was generally accepted by the general population.

Meanwhile, the quest for the origin of life, of who we are, how we live, and how we die never expired and eventually resulted in a remarkably clear picture that is now generally adopted by the more enlightened in society. Ironically, this insight is highly controversial in society as a whole and not accepted at all by a large fraction (probably the vast majority) of the world population. Indeed, Darwin is as controversial now as in the nineteenth century. Meanwhile, biology has come to dominate the science of the twenty-first century and it is no wonder that again, as in the seventeenth century, it is the limit to life that takes hold of the minds of many of our best thinkers. To some extent we have come full circle. The question is again whether we can beat the aging process and disassemble the roadblocks to immortality, this time through the accomplishments of the new biology. Can modern science succeed where hermeticism failed?

To know whether it is possible to prevent or cure aging we need to know what it is that makes us lose our vigor, causes disease, and finally, inescapably, leads to death. This book is a recapitulation of one of the oldest and arguably the most consistent theories of how we age. First formulated in the 1950s, the somatic mutation theory explains aging as a gradual accumulation of random alterations in the DNA of the genome in the cells of our body. This theory has proved to be remarkably robust and is compatible with the other major theory of aging that does not die: the free-radical theory of aging. Whereas the latter provides a logical explanation for where most of life's wear and tear comes from, the somatic mutation theory explains how this can result in physiological decline and increased disease. Or does it?

Based on what we now know about the genome, ours as well as those of many other species, how the information it contains is maintained as part of its structural characteristics, and how this information is retrieved and translated into function, is it still reasonable to see this as the main cause of aging in a time when most of us are convinced that the process is multifactorial and must have many causes? Is it possible that the inherent instability of our genomes is not only responsible for the increased chance of getting cancer in old age, but also in some way has an adverse effect on cell function, results in reduced organ capacity, and causes a variety of physiological changes, as well as such diseases as cardiovascular disease, neurodegenerative disorders, and diabetes? Finally, what are the implications of such a stochastic, molecular basis of aging for all those strategies that are now being designed to keep us alive and healthy a bit longer and possibly forever? This and more will be discussed in this book.

I have not been shy to include many results obtained in my own laboratory, but a book of this kind depends heavily on other people's research and other people's writing. I have tried to acknowledge this great debt to others as much as I could and there are of course the references. Nevertheless, I am afraid that a substantial portion of what I read in some publication, website, or newspaper, not to mention elements picked up during scientific conferences or learned from some of my colleagues, is not properly acknowledged. I apologize for that in advance and would like to hear about it if at all possible.

I am heavily indebted to some of my colleagues for their critical comments on earlier drafts of the different chapters. More specifically, I would like to thank Judy Campisi (Berkeley, CA, USA) and Steve Austad (San Antonio, TX, USA) for their comments on Chapter 1, Steve Austad (San Antonio, TX, USA) and Gordon Lithgow (Novato, CA, USA) for their comments on Chapter 2, Tom Boyer (San Antonio, TX, USA) for comments on Chapter 3, Judy Campisi and Jan Hoeijmakers (Rotterdam, The Netherlands) for comments on Chapter 4, Paul Hasty (San Antonio, TX, USA) for comments on Chapter 5 (which is based on a joint publication), Peter Stambrook (Cincinnati, OH, USA) and Martijn Dollé (Bilthoven, The Netherlands) for comments on Chapter 6, George Martin (Seattle, WA, USA), Huber Warner (St. Paul, MN, USA), and my wife, Claudia Gravekamp (San Francisco, CA, USA), for their comments on Chapter 7, and Huber Warner and Aubrey de Grey (Cambridge, UK) for comments on Chapter 8.

I am extremely grateful to my friend and colleague, Yousin Suh (San Antonio, TX, USA), for critically reading the entire manuscript and her many useful comments. Thanks to her helpful input at a very early stage I have been able to find the right direction.

I thank the members of my laboratory, now and in the past, for sharing their results with me, for all their hard work and their flexibility in dealing with my often unreasonable demands. I am especially grateful to Jan Gossen and Martijn Dollé, perhaps the best scientists who came from my laboratory and superb scholars in their own right, and to Brent Calder for making many of the figures and for always being ready to help me out during the preparation of the manuscript.

Finally, I would like to thank the people of Oxford University Press, especially Nik Prowse for his careful editing and many useful suggestions for improvements, and Stefanie Gehrig and Ian Sherman for their frequent advice during the preparation of the manuscript. I am also grateful to the anonymous reviewers of the original book proposal for their many useful suggestions, and to Maria Gaczynska and Pawel Osmulski (University of Texas Health Science Center) for contributing the cover illustration.

And last, but not least, I thank my wife, Claudia Gravekamp, for her patience and non-abating support during the course of this work.

■ CONTENTS

Antonie van Leeuwenhoek observes protozoa, bacteria, and germ cells, providing the evidence that life begets life — 1677

1735 — Carolus Linnaeus publishes the first complete classification of living species

Matthias Schleiden and Theodor Schwann conclude that cells are the basic units of all life forms — 1830

1838 — Debate between Étienne Geoffroy Saint–Hilaire and Georges Cuvier on form and function

1858 — Charles Darwin and Alfred Wallace propose natural-selection theories of evolution

Gregor Mendel presents his basic laws of heredity — 1865

1893 — August Weismann recognizes the dichotomy between germ-line and somatic cells

August Weismann formulates the first non-adaptive theory of aging — 1902

1910 — Thomas Hunt Morgan establishes chromosomes as the location of Mendel's factors, now termed genes

Theodosius Dobzhansky links evolution to genetic mutation — 1937

1944 — Oswald Avery shows that DNA is the carrier of genetic information

Peter Medawar formulates the first evolutionary theory of aging — 1952

1953 — James Watson and Francis Crick propose a double-helical structure for DNA, explaining the perpetuation of genetic information

Denham Harman proposes that free radicals are the primary cause of aging — 1956

1958 — Leo Szilard formulates the first somatic mutation theory of aging

Peter Mitchell introduces the chemiosmotic hypothesis of energy production — 1961

Leslie Orgel proposes the error catastrophe theory of aging — 1963

1977 — Thomas Kirkwood proposes the disposable soma theory

Thomas Johnson provides the first evidence for single gene mutations that extend lifespan of an organism — 1984

2003 — The International Human Genome Sequencing Consortium publishes the complete draft of the human genome sequence

Aging of the Genome: timeline

1 Introduction: the coming of age of the genome

Science and technology extend life and improve the quality of life. Whereas in a sense this may have been true since the origin of *Homo sapiens*, it has never been more apparent than after the Industrial Revolution in the nineteenth century, when great strides in physics, chemistry and medicine significantly improved life for rich and poor alike. By 1900 most European countries had been liberated from the danger of recurrent famine. In addition, improved sanitary conditions, vaccination, and the widespread availability of antibiotics have been responsible for the dramatic increase in average lifespan over the last 200 years. Most of this increase in lifespan has been due to the rapid decrease in infant mortality, since the lives of babies and young children are especially precarious in times of hunger and disease, the latter usually following the former. However, evidence is now emerging that since the 1970s, possibly due to greater awareness of adverse lifestyle habits—such as smoking—and more effective medical care, mortality and morbidity of the elderly has been rapidly declining (at least in developed countries)[1,2]. In Sweden, a highly developed country with reliable demographic data on human lifespan since 1861, maximum age at death has risen from about 101 years during the 1860s to about 108 years during the 1990s, suggesting that the maximum lifespan of humans and possibly other animals is not immutable[3].

Whereas average lifespan is deduced from the age at death of all individuals of a population, including those who die very early, maximum lifespan is the maximum attainable duration of life for an individual of a given species. In principle, therefore, the maximum lifespan of our species is the age at death of the longest-lived human, which is 122 years. Jeanne Calment, a French woman who attained this respectable age, died in 1997. A better measure of the trend in achieved human lifespan is the change in upper percentiles of the age distribution of deaths, as was used in the study on maximum lifespan in Sweden cited above. In June 2006 the longest living human was Maria Esther Capovilla from Ecuador, who was then 116 years old. Before her, several human so-called supercentenarians died in quick succession around this age, underscoring the limitations of our species-specific genetic make-up in keeping us alive over extended periods of time. Further optimization in the way we live, even with the best possible medical care, will not appreciably change that situation. Under these ideal conditions, lifespan will likely continue to increase, but slowly and gradually. However, what will happen if science is able to alter the way we are, rather than the way we live? Will the recent dramatic developments in the biological sciences free us from the bonds, which, as in any other species, fix the time of our lives? Is biology crossing a threshold, from a strictly intellectual exercise in understanding life, to an orchestrated effort to halt its demise? Most importantly, can such an effort succeed or are there some inherent mechanistic limitations, which will ultimately prevent us from rapidly achieving, say, a doubling of human lifespan? As I will try to argue in this book, the answers to these questions may be hidden in the genome. The rapid rise of modern biology is very much the story of the coming of age of the genome, the complete set of genetic information of an organism. Genome research has not only provided us with our current basic understanding of the logic of life, but has also supplied the tools to practice a whole new form of biomedicine, now termed genomic medicine. It is the genome as a fluid entity that bears witness to the history of life as it has unfolded on our planet since the first replicators. It is the genome that carries the seeds of our development from fertilized egg into maturity. And it may be the genome, with its inherent instability, that will be responsible for our ultimate demise.

In this first chapter I will sketch the major developments in the science of biology, from the Renaissance to the genome revolution, in two parallel lines: one that explains how we gradually gained a mechanistic understanding of how life perpetuates itself through random alterations in DNA, with aging of its carriers as the inevitable by-product, and a much more complicated learning curve that thus far has merely provided the starting points of how we hope to gain a more complete understanding of how life forms are ordered at the molecular level and how this order turns into disorder during aging.

1.1 The age of biology

With physics and chemistry at their zenith in the nineteenth and twentieth centuries, biology, the study of life, is often considered the premier science of the century we have just entered, with the promise to revolutionize human existence. The information explosion in biology, which started relatively late, will soon reach a stage when, for the first time in human history, we might be able to extend and improve our life in a more fundamental way than through manipulation of our environment or lifestyle; that is, by intervening in our basic biological circuits in a way that will allow us to break the constraints of our species-specific genetic make-up. To reach this stage, biology has evolved from an originally descriptive science, through a period of hypothesis-driven experimental research, to the data-driven era, which we have now entered, with the prospect of rational interventions based on *in silico* models that can provide an integrated understanding of the processes that give and maintain human life.

At the dawn of modern biology two major, often intertwined, branches of knowledge-gathering sprung from the same source: the invention of the microscope in the new permissive era of the Renaissance, which allowed for the first time a detailed observation of the various manifestations of life. A dual quest began to discover life in all its splendid variability and to find out the details of its workings. Along these parallel paths of studying why life is and how it works, the science of aging emerged from the why and how of life's natural limitation, observed in so many of its individual representatives (see Timeline, p. xi).

1.1.1 THE LOGIC OF LIFE

The question of life's origin and its perpetuation in such a wide variety of forms appeared to be the most challenging of questions and was tackled in successive stages by a number of great minds from the seventeenth to the twentieth centuries. This quest culminated in Darwin's theory of evolution by natural selection and Watson and Crick's discovery of the molecular structure of DNA. The grand understanding of the logic of life would prove equally important for understanding its demise: the logic of aging.

Before the seventeenth century our state of knowledge was static and, in Western Europe, mainly based on a synthesis of the Greek–Roman heritage and the Christian Church. Following Aristotle (384–322 BC) the general consensus at the beginning of our modern era was that small animals like flies and worms originated spontaneously from putrefying matter. Antonie van Leeuwenhoek (1632–1723) was one of the first to discredit this popular notion of spontaneous generation, based on his direct observations of bacteria, protists, and living sperm cells with home-made microscopes—an early example of technology driving progress in biology. After examining and describing the spermatozoa from mollusks, fish, amphibians, birds, and mammals, he came to the novel conclusion that fertilization occurred when the spermatozoa penetrated the egg.

Having reached the consensus that life begets life an explanation was sought for the bewildering variation of life forms on earth. Aristotle had provided the world with a grand biological synthesis, including a classification of animals grouped together in genera and species. He was of the opinion that the current biological diversity had existed from the start, which was later adopted by the church in the form of the dogma that all creatures were created independently of one another by God and organized into a hierarchy. It was Carl Linnaeus (1707–1778) who provided us with a system for naming, ranking, and classifying organisms, still in wide use today, which would become the ultimate tool for recognizing the logic of a system of evolutionary descent. Initially, Linnaeus believed that species weres unchangeable, and he never abandoned the concept of a preordained diversity of life forms. But Linnaeus observed how different plant species could hybridize to create forms which looked like new species. He abandoned the concept that species were fixed and invariable, and suggested that some—perhaps most—species in a genus might have arisen after the creation of the world, through hybridization[4].

Alfred Wallace (1823–1913) and Charles Darwin (1809–1882), then, provided the now generally accepted explanation for the intriguing similarities among organisms, so beautifully organized by the system of Linnaeus. Whereas the different species had generally been assumed to be immutable and stable since the era of Plato and Aristotle, Darwin had begun to see life as fluid, and recognized that ample variation was present, even among individuals of the same population. Like several scientists before him, Darwin had come to believe that all life on Earth evolved (developed gradually) over millions of years from a few common ancestors. However, the primary mechanism of this process of evolutionary descent was unknown. Based on careful observations of many variations among plants and animals on the Galapagos Islands and South America during a British science expedition around the world, he proposed a process of natural selection to advance certain characteristics best adapted to environmental conditions. The results of this work were published as *On the Origin of Species by Means of Natural Selection, or the Preservation of Favoured Races in the Struggle for Life* (1859), commonly referred to as *The Origin of Species*[5].

Evolution by natural selection was controversial from the beginning and is still less generally accepted than, for example, Einstein's theories of relativity. This already indicates the sensitivity of society to new concepts in biology involving humans and our position in the living world. The original criticisms of evolutionary descent focused on the need to accept that current life, among which the human species was only one tip on a branching tree, extended back through ancestral species over a time period much longer than the biblical 6000 years. However, the most serious problem, still the main hindrance today for many people to accept Darwin's theory, is the lack of purpose and direction that speaks from his explanation of life. Natural selection makes use of existing, natural differences among individuals in a population of a species in their suitability to adapt to special problems in their local environment. We now know that such differences in heritable traits continually arise in our germ cells by random changes in the genes that control those traits. Individuals less fit in a given environment are eliminated, whereas those with the most favorable traits leave a disproportionately high number of offspring. As recognized by the great evolutionist Ernst Mayr (1904–2005), the process of adaptation to special problems of local environments gives rise to new species when fragments of a population become geographically and reproductively isolated; this is known as allopatric speciation. (Other, less well explored mechanisms of speciation may also operate.)

The concept of open-ended evolution, not necessarily governed by a Divine Plan and with no predetermined goal, is still unaccepted by many. Confusion and resistance to new scientific discoveries are not uncommon, as exemplified by popular reactions to Heisenberg's uncertainty principle and Freud's revelations of the subconscious at the beginning of the twentieth century. However, the alarm felt by many when confronted with the implications of Darwin's theory regarding the position of humans in life as a whole are quite unique. Indeed, the validity of the physical principles underlying the automobile, air travel, and the personal computer are never doubted by the general public. By contrast, equally solid principles in biology are often rejected out of hand by sizable segments of the educated public based on the strong intuitive appeal—often inspired by religion—of intelligent design and purposeful direction. Biology will continue to raise feelings of uneasiness in the years to come.

After Darwin, the next major development in biology was the emergence of the concept of the gene. A problem with Darwin's theory of natural selection as the mechanism of evolutionary change was the lack of knowledge as to how random variations in heritable traits could arise and how they could be perpetuated from parents to offspring. Ironically, the genetic principles governing this latter process had already been described in Darwin's lifetime by the Czech monk, Gregor Mendel (1822–1884). Working with different kinds of peas, Mendel demonstrated that the appearance of different hereditary traits followed specific laws, which could be understood by counting the diverse kinds of offspring produced from particular sets of parents. He established two principles of heredity that are now known as the law of segregation and the law of independent assortment, thereby proving the existence of paired elementary units of heredity (which he called factors) and establishing the statistical laws governing them. Mendel's findings on plant hybridization were ignored until they were confirmed independently in 1900 by three botanists.

After 1900, the physical basis for Mendel's laws was discovered in the form of the chromosomal basis for the transmission of genes from parents to offspring. Thomas Hunt Morgan (1866–1945) was the first to provide conclusive evidence that chromosomes are the location of Mendel's factors, termed genes by Wilhelm Johanssen in 1907 (in Greek meaning 'to give birth to'). Morgan chose the fruit fly, *Drosophila melanogaster*, as his experimental animal, which has remained a key experimental model system in genetics ever since. In 1910, he found a mutant male fly with white rather than the normal red eyes. Since all the female flies had red eyes with only some males having white eyes, Morgan realized that white eye color is not only a recessive trait but is also linked in some way to sex. This work led to the identification of four so-called linkage groups, which correlated nicely with the four pairs of chromosomes that *Drosophila* was known to possess. Their subsequent breeding experiments provided proof that the chromosomes are indeed the bearers of the genes, with different genes having specific locations along specific chromosomes. Traits on one particular chromosome naturally tended to segregate together. However, Morgan noted that these 'linked' traits would separate, from which he inferred the process of chromosome recombination: two paired chromosomes could exchange genetic material between each other, an event termed crossover. The frequency of recombination appeared to be a function of the distance between genes on the chromosome. The smaller that distance, the greater their chance of being inherited together, whereas the farther away they are from each other, the more chance of their being separated by the process of crossing over. The Morgan is now the unit of measurement of distances along all chromosomes in fly, mouse, and human.

In the meantime, cytologists had described the processes of mitosis and meiosis at the end of the nineteenth century. The chromosomes, thread-shaped structures under the microscope, were known to be located in the nucleus of a cell, but nobody knew their function. By correlating their breeding results with cytological observations of chromosomes, Morgan's group provided the physical reality for Mendel's hypothetical factors. It was recognized that chromosomes, which could be distinguished, quantified, and observed to occur in pairs, except in germ cells, housed the genetic material. Germ cells were demonstrated to have only one copy of each chromosome pair, with fusion of the germ-cell nuclei restoring a complete set of chromosomes, half from the father and half from the mother. A late highlight in this development was the work of Cyril Darlington (1903–1981), who made the connection between the structural behavior of chromosomes, including the mechanics of chromosomal recombination, and the functional consequences in terms of heredity[6]. The chromosomal theory of inheritance, with its distinction between somatic and germ cells, ended speculation by Darwin, Jean-Baptiste Lamarck (1744–1829), and others that offspring were a mere blending of the parents and that acquired traits could be inherited.

It was also around this time that the terms phenotype and genotype began to be distinguished. The phenotype of an individual organism comprises its observable traits (such as size or eye color) whereas the genotype is the genetic endowment underlying the phenotype. Of note, in those early days the genotype could only be determined on the basis of the phenotype because the nature of the genetic material was still unknown. Therefore, inheritance patterns could only be checked by breeding experiments. Based on the early separation between somatic and germ cells, August Weismann (1834–1914) first formulated the unidirectional theory that the phenotype cannot affect the genotype[7]. The distinction of germ line and soma would profoundly influence our ideas about aging. Weismann recognized that the germ cells are not affected by any variation that might occur in an individual. This is especially relevant for somatic changes in the structure of deoxyribonucleic acid (DNA), which we now know is the carrier of the genetic information. Such changes, termed mutations, in a somatic cell may damage the cell, kill it, or turn it into a cancer cell. But, whatever its effect, a somatic mutation is doomed to disappear when the cell in which it occurred or its owner dies. By contrast, germ-line mutations such as the one that gave rise to Morgan's white-eye trait, will be found in every cell descended from the zygote to which that mutant gamete contributed. If an adult is successfully produced, every one of its cells will contain the mutation. Included among these will be the next generation of gametes, so if the owner is able to become a parent, that mutation will pass down to yet another generation. Mutations in somatic cells may be expressed, but are not passed on to further generations. Mutations in germ cells can be both expressed and transmitted to descendents.

The distinction between germ line and soma exists only in animals. In plants, cells destined to become gametes can arise from somatic tissues. In organisms without sexual reproduction, such as many unicellular organisms, there is no distinction between germ and soma. In Weismann's view, the soma simply provides the housing for the germ line, seeing to it that the germ cells are protected, nourished, and combined with the germ cells of the opposite sex to create the next generation. This provided the logical basis for rejecting the ideas of Lamarck and others that characters acquired during lifetime could be inherited by the next generation. Weismann's views foreshadowed the concept by Richard Dawkins (Oxford, UK) of the gene as the fundamental unit of selection, instead of species, group, or individual[8], as well as the disposable soma theory of aging by Tom Kirkwood (Newcastle upon Tyne, UK)[9].

Weismann was also the first to explain the aging of metazoa in evolutionary terms. In the first instance he proposed that aging was an evolutionary adaptation to avoid the need for offspring to compete with their parents for scarce resources. The idea that old individuals die as an act of altruism to the rest of the group or species is now generally considered as naive and incompatible with the negligible impact of aging on animals in the wild (few animals survive long enough to experience old age). However, Weismann also presented the case for aging as a non-adaptive trait, which would again foreshadow modern thinking about why we age. In this case, he argued that characters that have become useless to an organism, such as eyesight in animals that never see the light, are not subject to natural selection. Applied to the 'useless period of life following the completion of reproductive duty' this theory would predict a weakening of selection against characters with adverse effects later in life. Moreover, it predicts the positive selection of such traits if there is some benefit in the earlier years of life[7].

In the 1940s, Weismann's neodarwinism was integrated with new findings in laboratory genetics and fieldwork on animal populations. This so-called evolutionary synthesis, in a sense the grand finale of the work begun by Darwin and his predecessors, started with T.H. Morgan, mentioned above, and reached a new height during the first decades of the twentieth century with the work of the great mathematical population geneticists, Ronald Fisher (1890–1962), Sewall Wright (1889–1988), and J.B.S. Haldane (1892–1964). They developed quantitative genetics as a synthesis of statistics, Mendelian principles, and evolutionary biology. They demonstrated that the same principles that applied to discrete traits (such as eye color) were also valid for quantitative traits, such as height and certain behavioral characteristics, which display continuous variation in the population. These concepts were later combined with explanations for the origin of biodiversity by Theodosius Dobzhansky (1900–1975), the previously mentioned Ernst Mayr, and others, resulting in the integration of Mendel's theory of heredity with Darwin's theory of evolution and natural selection.

The unification of genetics and evolution by natural selection also gave rise to the first discussions—in the new, mathematical language of the modern synthesis—of the evolutionary basis of aging. It was Fisher who noticed, probably for the first time, that the chance of individuals to contribute to the future ancestry of their population declines with age[10]. Later, this would lead Peter Medawar (1915–1987), a Nobel laureate and better known for his work on transplantation immunology, to propose that aging, at least in sexually reproducing organisms with a difference between the soma and the germ line, is a result of the declining force of natural selection with age (see Chapter 2 in this volume).

What was still not clear at the time was the nature of a gene and the mechanism of Mendel's transmission of heritable traits through the germ line. It was only in 1944 that Oswald Avery (1877–1955) and collaborators made a convincing case for DNA as the carrier of the genes[11]. They were studying a substance that could turn non-pathogenic variants (R cells) of *Streptococcus pneumoniae*, a bacterium that causes pneumonia, into pathogenic ones (S cells). This so-called transforming principle, which had a high molecular mass, was resistant to heat or enzymes that destroy proteins and lipids, and it could be precipitated by ethanol. Hence, it was most likely DNA, a substance already described by Johann Friedrich Miescher (1844–1895) in 1869 as occurring in human white blood cells and in the sperm of trout. However, the nature of the genetic code and a mechanism for how DNA was able to transfer this information from cell to cell and how it could convert this information into cellular function was still unknown. James Watson (Cold Spring Harbor Laboratory, NY, USA) and Francis Crick (1916–2004) provided the answer in 1953 in the form of the molecular structure of DNA: two helical strands of alternating sugar-phosphate sequences, each coiled round the same axis, held together by adenine–thymine- and cytosine–guanine-specific base pairing. The base pairing properties of DNA dictate the mechanism of gene replication[12].

Hence, it was now known that the complete set of genetic information of an organism, the genome, was written in its DNA. Genomes, which can vary widely in size, from 600 000 bp in a small bacterium to 3 billion in a mammal, were subsequently demonstrated to be the repository of the genes, the basic physical and functional units of heredity. The years immediately after Watson and Crick are now known as the classical period of molecular biology. First, Matthew Meselson (Cambridge, MA, USA) and Franklin Stahl (Eugene, OR, USA) experimentally confirmed[13] the process of semiconservative DNA replication predicted by the double-helical, base-paired model proposed by Watson and Crick. DNA isolated from *Escherichia coli* after growth in medium containing heavy or light isotopes of nitrogen showed a distinct density distribution in CsCl gradients. After switching medium, DNA of an intermediate density was obtained, which is expected if the newly replicated DNA is a hybrid molecule consisting of one parental and one newly synthesized strand.

Then, following the prediction by François Jacob (Paris, France) and Jacques Monod (1910–1976) that messenger ribonucleic acid (mRNA) transcribed from the DNA of a gene in the form of a single-strand complementary copy was the template for protein synthesis[14], Crick, Sydney Brenner (San Diego, CA, USA) and colleagues[15] demonstrated in 1961, by deleting bases one by one from DNA of the bacteriophage T4, that the genetic code was a triplet of bases. A string of triplets specifies the full sequence of amino acids in a protein chain. Using a cell-free translation system and synthetic homopolymers, Marshall Nirenberg (Bethesda, MD, USA)[16] and Har Gobind Khorana (Cambridge, MA, USA)[17] identified which codons corresponded to which amino acids. Meanwhile, the laboratories of Mahlon Hoagland (Worcestor, MA, USA)[18], Robert Holley (1922–1993)[19], and others had discovered transfer RNA (tRNA), predicted by Crick in his adaptor hypothesis as the entity that recognized triplets of bases on the mRNA. Adaptor enzymes link each kind of amino acid to the appropriate carrier, tRNA. Protein synthesis or translation is carried out by bringing the mRNA and the set of tRNAs charged with the appropriate amino acids to the ribosomes, discovered earlier as the protein-making apparatus in the cytoplasm.

The guiding role of Francis Crick in bringing this classical period to its zenith is now well recognized. Crick's predictions that the genetic code was universal to all forms of life and that genetic information can go only one way—that is, from DNA via RNA to protein—proved correct with minor exceptions. This so-called central dogma of molecular biology is another way of saying that acquired characteristics cannot be inherited.

With the discovery of the structure of DNA and the genetic code, the origin of Darwin's existing natural differences in heritable traits had also become clear. DNA in the living cell is not completely stable, but can undergo alterations in its base pair composition through errors during replication or the repair of chemical damage. Hermann Joseph Muller (1890–1967), a student of T.H. Morgan, had already demonstrated in 1927[20] that mutations could be induced by radiation. He identified mutations mainly by the observed effect on an organism, but was able to show that mutations can result from breakages in chromosomes and changes in individual genes. He also realized that the majority of such random mutational changes are deleterious, although an occasional mutation is beneficial, for example, by giving rise to a better-functioning protein. However, as we now know, the genetic code is tolerant of certain mutations. This degeneration of the code is due to the fact that there are three times as many codons as there are amino acids, hence the tolerance of some amino acids for a mismatch at the third position of each triplet.

With the evolutionary synthesis and the new understanding of its underlying molecular principles, the pursuit of the origin and perpetuation of life was essentially over. From now on, biology could fully focus on unraveling the structure and function of life's various manifestations.

1.1.2 SEARCHING FOR STRUCTURE AND FUNCTION

The desire to know all structural, organizational, and functional facets of life sprung from the same source as the theory of evolution and modern genetics: the careful observations made by the pioneers of science in the seventeenth century, with microscopy as their main tool. Then, as now, there was a significant relationship between the ability of craftsmen to provide good instrumentation and the direction of scientific investigation. Most notable among these early scientists, apart from the above-mentioned Antonie van Leeuwenhoek, was his contemporary, Marcello Malpighi (1628–1694). Malpighi was probably the first scientist to use model organisms—frogs and turtles—to obtain structural information on human organs, thereby inventing comparative anatomy. Following the early work of William Harvey (1578–1657) on human blood circulation, Malpighi discovered blood flow through capillaries in the lungs, opening the way to understanding the function of this organ in respiration. He conducted a famous comparative study of the liver, from snails through fishes and reptiles, right up to humans, and he was the first to give an adequate description of the development of the chick in the egg[21].

At this time, it had begun to dawn from Leeuwenhoek's work, as well as from microscopic observations by the great British natural philosopher Robert Hooke (1635–1703), that life was organized around a basic unit, termed a cell by Hooke. However, it took until 1839 before Mathias Schleiden (1804–1881) and Theodor Schwann (1810–1882) could make the conclusion that cells were the basic units of life. In animals, cells were progressively organized into tissues, organs, systems, and, finally, the whole body. The adult human body is an aggregate of more than 75 trillion cells. With the birth of modern cell theory, anatomists had widened their scope and new disciplines emerged, such as embryology, cytology, and physiology, all focused on understanding the mechanisms of life in all its facets, and how this unfolds from a fertilized egg to an adult organism. Meanwhile, in studying various life forms, the early scientific community was struggling with the question of whether organisms were integrated wholes, as advocated by Georges Cuvier (1769–1832), or whether morphology could be changed and affected by environmental conditions, as proposed by Étienne Geoffroy Saint-Hilaire (1772–1844). In other words, does function strictly dictate form with no modification possible, or do body plans constrain how organ functions are manifested? These positions, which were later synthesized, remain a leitmotiv for modern systems biology and functional genomics.

The dramatic increase in our understanding of how structure follows function was a result of the application of new insights in chemistry, most notably organic chemistry, to study different cellular components. This would first lead to biochemistry, the science dealing with the chemistry of living matter, and ultimately to molecular biology, the branch of biology dealing with the nature of biological phenomena at the molecular level through the study of DNA, RNA, proteins, and other macromolecules involved in genetic information and cell function. The undisputed highlight of this development was our ultimate understanding of how cells harvest the energy of food through the conversion of adenosine diphosphate (ADP) into the energy-carrying compound adenosine triphosphate (ATP) in subcellular structures called mitochondria. In his 1961 paper[22], Peter Mitchell (1920–1992) introduced the chemiosmotic hypothesis, connecting the electron-transport chain, through a proton (H^+) gradient across the inner mitochondrial membrane, with oxidative phosphorylation and the synthesis of ATP. Critically important to all biology and shaping our understanding of the fundamental mechanisms of this most important of all cellular activities, the elegance of the chemiosmotic model in correlating structure and function would have been appreciated by Cuvier.

The universality of the process of oxidative phosphorylation suggests its importance as a factor in aging. Ironically, even before Mitchell's landmark paper, another chemist, Denham Harman (Omaha, NE, USA), proposed that free radicals, the adverse by-products of oxidative phosphorylation, were a ubiquitous cause of aging[23]. This hypothesis is known as the free radical theory of aging and has been with us ever since. Free radicals are now generally considered as a most likely explanation for the damage that ultimately leads to our demise. It also drew attention to the mitochondria and their own independent genome, so close to the origin and main source of free radicals. Distinct from the far larger nuclear genome, the mitochondrial genome is now considered a major target for spontaneous mutagenesis. In turn, this may adversely affect the process of oxidative phosphorylation itself, thereby accelerating formation of free radicals. This is described in detail in Chapter 6.

As we have seen, molecular biology provided the insight that proteins were the work-horses of biological systems, and DNA the carrier of genetic information, organized in the form of a genome. Genes were shown to be specific sequences of base pairs that contain the instructions, in the form of a triplet code, for making proteins. Interestingly, not long after the discovery of the fundamental mechanism of protein biosynthesis, Leslie E. Orgel (San Diego CA, USA) proposed in 1963 that cellular aging involves the accumulation of defective proteins as a result of an inherent inaccuracy of the translational machinery. This is generally known as the error catastrophe theory of aging and longevity, based on Orgel's realization that the faulty RNA and DNA polymerases, also resulting from translational errors, could lead to an exponential increase of defects in protein, RNA, and DNA, causing the collapse of the cellular machinery for information transfer. This idea is not supported by experimental evidence, but it can be argued that errors are random, with each cell acquiring a unique set of errors. Since current technology is geared towards analyzing mixtures of cells rather than individual cells, we may simply be unable to detect error catastrophes.

In the decades following the discovery of the double helix, and especially after the development of recombinant DNA technology, molecular biology became a premier discipline in biology, always at the cutting edge of new developments. Initially, molecular biology remained separate from more traditional disciplines, such as physiology. However, gradually these other disciplines would include molecular biology as an aide in support of their own research endeavors. Meanwhile, the realization of the extreme complexity of the gene–phenotype relationship necessitated a whole new approach, which coincided with the informatics explosion, bringing powerful new computers and the internet. Eventually this would lead to a departure from the original reductionist approaches to holistic strategies, providing a more comprehensive understanding of life, and the emergence of functional genomics and systems biology.

1.2   **From genetics to genomics**

In the heydays of molecular biology it seemed natural to begin our effort of understanding the structure and function of various life forms with understanding individual genes and their activities in different organisms. Indeed, after Watson and Crick, the central dogma may have clarified the mechanisms underlying Mendel's laws, but virtually all known genes were still identified only by mutations and their phenotypic consequences. Genetics was a matter of studying inherited phenotypes, rather than genes, none of which had been isolated before 1973, when Stanley N. Cohen of Stanford University and Herbert W. Boyer of the University of California, San Francisco, developed the laboratory process to take DNA from one organism and propagate it in a bacterium. This process, called recombinant DNA technology, was used in 1977 for the production of the first human protein manufactured in a bacterium: somatostatin, a human growth hormone-releasing inhibitory factor. For the first time, a synthetic, recombinant gene was cloned and used to produce a protein[24]. The following decade saw a surge in the study of genes and their function, for which Tom Roderick (Bar Harbor, ME, USA) in 1986 coined the term genomics. Genomics was highly technology-driven, as exemplified by the rapid emergence of a host of new techniques and instruments. The undisputed highlight of this development was the discovery, by Kari Mullis, then at the Cetus Corporation, of the polymerase chain reaction (PCR), a technique for amplifying DNA sequences *in vitro* by separating the DNA into two strands and incubating them with oligonucleotide primers and DNA polymerase (Fig. 1.1). PCR can amplify a specific DNA sequence as many as one billion times, and quickly became essential in biotechnology, forensics, medicine, and genetic research as probably no method before.

Initially, genomics was not different from standard, investigator-initiated research and was entirely hypothesis-driven. This would change with the conception of the Human Genome Project (HGP), the international research effort that determined the DNA sequence of the entire human genome. The rationale behind the HGP was that by sequencing a complex genome, the amino acid sequences of all proteins as well as all sequence-encoded regulatory and structural characteristics of that genome would be immediately available, obviating the need to purify and characterize each feature separately. Cloning genes into expression vectors allowed the production of proteins, but also allowed their engineering, for example, for studying their phenotypic characteristics in cell cultures or experimental animals. Indeed, it was around this time—in the 1980s—that the methods to make transgenic mice were developed by Jon Gordon (New York, NY, USA)[25],

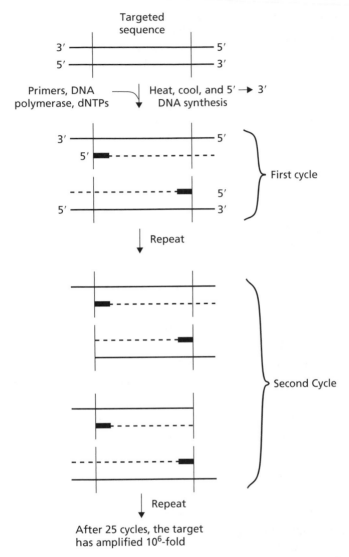

Fig. 1.1 In the PCR small single-stranded DNA fragments, complementary to known sequences that flank a nucleic acid sequence, are used as primers (black rectangles) to amplify this sequence millions of times through 25 or more cycles of *in vitro* enzymatic synthesis. dNTPs are the four deoxynucleotide Triphosphates.

Ralph Brinster (Philadelphia, PA, USA) and Richard Palmiter (Seattle, WA, USA)[26], and their co-workers. The first use of transgenic mice was to study gene function in the whole animal, in particular how and why a specific gene is turned on in some tissues and turned off in others. This diversity of gene expression that produces the distinct cell types and tissues of the body, making a muscle cell different from a liver cell, had quickly become of central interest in molecular biology. Access to comprehensive genome sequences of different species allowed scientists to systematically address this question.

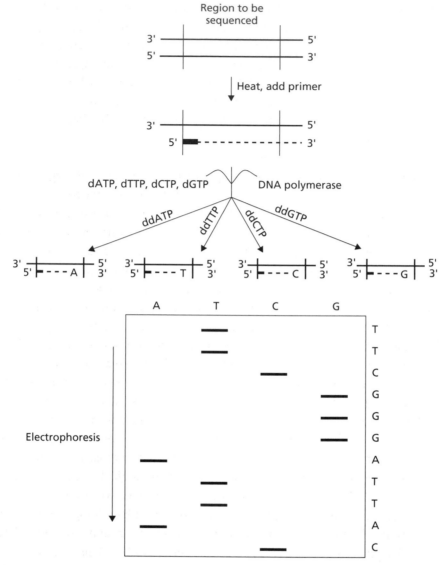

Fig. 1.2 Principle of nucleotide sequencing by the Sanger method. In this case a single primer (black rectangle) is used to generate a set of fragments with a common 5′ origin through base-specific interruption of *in vitro* enzymatic synthesis. A, adenosine; C, cytidine; G, guanosine; T, thymidine; dATP, dCTP, dGTP, dTTP, deoxy-adenosine, -cytidine, -guanosine, and -thymidine triphosphate, respectively. ddATP, ddCTP, ddGTP, ddTTP, dideoxy-adenosine, -cytidine, -guanosine, and -thymidine triphosphate respectively.

It was realized early on that the average research laboratory was too small to contribute significantly to such a project and that methods of scale were needed. This resulted in most of the work being done by large genome centers. Contributors to the HGP included the US National Institutes of Health and the US Department of Energy (where discussions of the HGP began as early as 1984[27]), numerous universities throughout the USA, and international partners in the UK, France, Germany, Japan, and China. A separate, commercial project to sequence the human and other genomes was initiated in 1998 by the Celera Genomics Corporation[28]. In the course of completing the sequence, two separate interim working drafts of the human genome were produced in 2001 with much publicity by both the public consortium[29] and Celera[30]. However, the major aim of the HGP was to obtain a comprehensive, 'finished' sequence of the entire 3×10^9-base haploid human genome, which was eventually published in 2004 and contained 2.85 billion nucleotides, covering 99% of the euchromatic genome. The paper in *Nature* reporting this accomplishment had over 2800 authors, an example of the emergence of large consortia of researchers at the expense of the more classical investigator-initiated approach of the past[31]. To the surprise of many, the complete sequence of the human genome did not immediately tell us how many human genes there are. This uncertainty is likely to last for a while because of the limitations in gene-prediction software, which thus far have precluded an accurate assessment of the total number of human protein-coding genes. It is generally thought that this number is somewhere around 30 000.

In hindsight, the achievement of its primary goal may have been less important than the impact of the HGP on the way biological research was conducted. Its legacy will probably always be associated with the transformation of biology from an almost exclusively solitary, hypothesis-driven science into an information science. This was based on the use of high-throughput methods and the increasing need to organize research as large collaborative efforts of multiple investigators from various disciplines. However, rather than abandoning individual, hypothesis-driven research, this development is more likely to eventually lead to the iterative and integrative approach of global analyses driven by hypothetical models, now known as a systems approach. I will discuss this extensively below.

The main driving force behind the globalization of biological research was the additional goal of the HGP to develop novel technologies and improve existing ones. The success of the HGP in converting regular methods in molecular biology into methods of scale is exemplified by the great improvements in the sequencing method first described by Frederick Sanger (Cambridge, UK) and co-workers in 1977[32]. This method is based on the use of a DNA polymerase to extend, in four separate reactions, an oligonucleotide primer from its annealing site at the beginning of the target sequence over a length of 500–1000 bp. Each reaction contains all four deoxynucleotide triphosphates plus a limiting amount of either adenine, thymine, cytosine, or guanine dideoxynucleotide triphosphate, which terminate the reaction upon incorporation. After electrophoretic separation of each fragment mixture at a resolution of 1 bp, the sequence of the target can be read directly from the resulting banding pattern (Fig. 1.2).

Initially, radioactive labeling was used to detect the fragments after size separation. Later, fluorescent labels were developed, which allowed automated detection of the electrophoretically separated fragments using a laser[33]. From this first partial automation of the Sanger dideoxy sequencing principle to the current, almost fully automated 384-capillary electrophoresis systems there has been an approximately 2000-fold increase in throughput. Ironically, this has not been achieved by new technology, as originally anticipated, but almost exclusively by the introduction of methods of scale. Although such improvements have now made it relatively easy to obtain the consensus sequence of a genome quickly, especially that of a small microbe, conventional sequencing is still not cost-effective enough for routine application, for example, in large-scale genetic epidemiology or clinical diagnosis. It is anticipated that novel sequencing principles, including single-molecule sequencing[34], will successfully address remaining limitations in cost, speed, and sensitivity. In this respect, it has been predicted that about 10 years from now it will be possible to sequence an entire human genome in 30 min for about $1000[35].

In addition to the human genome, hundreds of genomes of other species, from simple microorganisms, such as *E. coli*[36], to the mouse, rat, and chimpanzee[37], have now been sequenced completely. The information that can be derived from these sequences is vast. Large-scale sequencing totally transformed certain disciplines as genetics and physiology and created new ones, such as comparative genomics, the most powerful way to elucidate the roles of many related genes. Although the practice of comparing gene or protein sequences with each other, in the hope of elucidating functional and evolutionary significance, is well established, its application to complete genomes greatly expands its utility and implications. For example, phylogenetic trees can be built not from the sequences of a single gene (usually ribosomal RNA (rRNA) genes) but from multiple gene sequences as well as from non-sequence information, such as similarities in gene repertoire and gene order[38]. This requires the rational classification of genes and proteins, which is usually done in the form of a system of orthologous gene sets. (Orthologs are homologous genes that evolved from a single ancestral gene in the last common ancestor of the compared genomes; paralogs are genes related via duplication within a genome.) Major applications of cross-species genome comparisons are the identification of functionally important genomic elements, e.g. protein-coding and regulatory sequences, on the basis of homology[39]. This is based on the assumption that functionally important regions tend to have a lower mutation rate than non-functional regions. The rapid increase in whole-genome sequences from different mammals and the development of better tools for their comparison should lead to increased insight into the functional constraints of the human genome.

The HGP has also been the starting point for several new, large-scale initiatives in genomics. For example, the establishment of a catalog of all common sequence variants (single nucleotide polymorphisms or SNPs) in the human genome with their patterns of linkage disequilibrium (the HapMap project) has been initiated to facilitate the identification of genetic risk factors in disease susceptibility and other phenotypes, whereas the Encyclopedia of DNA Elements (ENCODE) project aims to identify different functional elements in the human genome[40]. The latter is very much based on the realization that gene function can only be understood in the context of the genome as a whole, with its multiple overlapping networks of regulatory sequences. All these large-scale genome projects are part of a general development to collect biological information in a systematic way and make it publicly available. The rapid growth of the Internet around the same time greatly facilitated the use of such shared resources, which now play a crucial role in conducting biological research and have become the basis for functional genomics and systems biology.

1.3 **A return to function**

Complete DNA sequence information is not the end, but merely the beginning of our quest to understand how genomes—and therefore organisms—function and how time, both evolutionary time and the lifetime of an individual, can affect such function. For this purpose, genes need to be identified; the function of their products (RNAs and proteins) must be elucidated and the role of non-coding regulatory sequences needs to be understood. Since the landmark completion of the HGP, the type of biology focused on the identification and functional analysis of genes, coding regions, and other functional elements of entire genomes on a high-throughput basis has been termed functional genomics. Whereas genomics implies the study of genes and their function, functional genomics attempts to integrate all genes, their products, and their resultant phenotypes into dynamic networks of molecular pathways that ultimately determine our physiology (Fig. 1.3). Such networks of interaction have now all but replaced the original one-gene/one-protein way of thinking.

   If the advances of molecular biology and genomics had made anything clear, it was the stupendous complexity of living cells and their interactions to generate complete functioning organisms. A discrete biological function can only rarely be attributed to one individual protein encoded by one gene. In reality, biological characteristics involve complex interactions among many components in cells, such as DNA, RNAs, proteins, and small molecules. Emphasizing this integrative nature of biological function, Hartwell and Hopfield coined the term functional modules[41]. In functional genomics, a distinction is made between the different levels of organization in the cell. The genome, as we have seen, denotes the totality of all genes on all chromosomes in the nucleus of a cell. The complete set of mRNAs, the next hierarchical level below the genome, is called the transcriptome. Next, there is the proteome, which is the set of all expressed proteins for a given organism. This is followed by the metabolome, a biochemical snapshot of the small molecules produced during cellular metabolism, such as glucose, cholesterol, and ATP, and

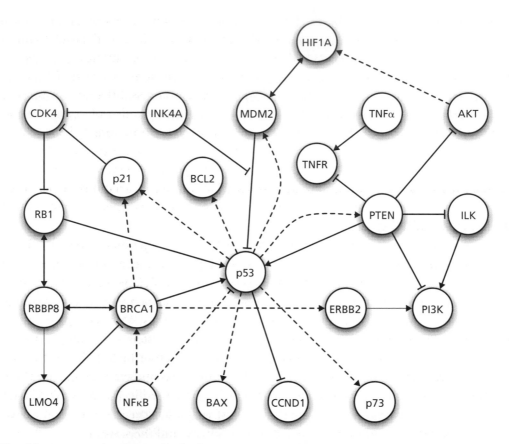

Fig. 1.3 An example of a network showing interactions between the TP53 protein and other gene products. Solid lines indicate protein–protein interactions whereas dashed lines show activation or repression at the transcriptional level.

several other comprehensive sets of biological information, such as the secretome (total of secreted molecules) and the interactome (a complete set of macromolecular interactions, such as protein–protein interactions).

Functional genomics was driven by the need to understand the formal relationships between genes and all the -omes, including the rules that control transition between these levels and from them to complex, functional systems, such as oxidative phosphorylation, genome maintenance, and the immune system. A key aim now became to systematically catalog all molecules and their interactions in a living cell. To do this, high-throughput methods for genome-wide data collection have become indispensable. Since the discovery of genes has outpaced our capacity to understand their biology, high-throughput methods to assess genotype–phenotype relationships are rapidly being developed and applied.

The most popular vehicle for high-throughput analysis in biology has become the microarray chip. In its first successful manifestation, hundreds to thousands or tens of thousands of cDNAs or oligonucleotides, complementary to parts of individual mRNAs, were attached to a glass slide. Hybridization of such slides with labeled probes obtained from reverse-transcribed RNA from tissues or cells of interest permits the analysis of changes in expression of a large number of genes simultaneously. This technology has the ability to reveal patterns of gene expression across different samples. For this purpose, genes are grouped into classes with similar profiles of activity, in an approach called cluster analysis. Such genes may have related functions or be regulated by common mechanisms. The structured gene-functional-categorization database, Gene Ontology or GO, provides the opportunity to partition genes into functional classes. Microarrays are now also used to study DNA sequence variations (SNPs), proteins, protein–protein and protein–DNA interactions, and various other structural characteristics of the cell or tissue.

Another tool that can be applied in a microarray format on a genome-wide scale, RNA-mediated interference (RNAi), proved to be of critical importance in bridging the gap between genotype and phenotype that had opened up since T.H. Morgan. First demonstrated in 1998 in *Caenorhabditis elegans*[42], RNAi allows the sequence-specific silencing of genes using synthetic double-stranded RNAs. Such exogenous RNAs co-opt a ubiquitously expressed, evolutionarily conserved gene-regulatory system consisting of micro-RNAs (miRNAs; Chapter 3). Endogenous miRNAs are transcribed as single-stranded precursors up to 2000 bp in length and exhibit significant secondary structure, resulting in stems and loops. Such primary transcripts are first processed in the nucleus and after entering the cytoplasm converted by the RNase III enzyme Dicer into double-stranded 21–23-nucleotide-long mature RNAs. Synthetic forms of miRNAs, so-called short hairpin RNAs (shRNAs), can be used experimentally to mimic their natural equivalents. Alternatively, it is possible to use short interfering RNAs (siRNAs), synthetic double-strand RNAs of less than 30 bp. Such siRNAs bypass cleavage by Dicer. All the small double-stranded RNAs need to associate with the RNA-induced silencing complex (RISC), which unwinds them and associates stably with the strand that is complementary to the target mRNA. Depending on the degree of homology, the complexes inhibit gene activity either by translational repression or triggering mRNA degradation. Apart from using synthetic variants of these RNAs, it is also possible to express them using a plasmid to silence gene expression for longer periods of time[43]. RNAi is a typical example of reverse genetic technology and can conveniently be applied in a microarray format. RNAi at such a genome-wide scale was applied early on in the science of aging to screen for genes regulating lifespan (Chapter 2).

In general, variation greatly increases from DNA to RNA to protein. For example, while the entire human genome may contain no more than 30 000 genes, there may be three times that many proteins, due to alternative splicing; this is the production of more than one transcript by including or excluding specific exons (the DNA segments of a gene that are protein-coding; see Chapter 3) or altering the length of a specific exon. This is without taking into account posttranslational modification, such as the attachment of phosphate, acetate, lipid, or sugar groups. To systematically describe the proteome and its different patterns of interaction in a complex organism is therefore more difficult than making an inventory of all genes. This is especially true because microarray technology for proteins is less well developed as it is for genes. Nevertheless, progress in this field is now also rapid, resulting in ever larger sets of proteins often subdivided according to their specific modification. For example, protein phosphorylation is estimated to affect 30% of the proteome and is a major regulatory mechanism that controls many basic cellular processes. Using microarray technology, the *in vitro* substrates recognized by most yeast protein kinases were recently identified, involving over 4000 phosphorylation events and 1325 different proteins. This collection of data was called the phosphorylome[44].

With phenotypic variation much more extensive than genotypic variation and an increasing number of global data-sets emerging at ever-shorter time intervals, the resulting deluge of data is truly transforming molecular biology, from the focused analysis of single genes and proteins to the systematic analysis of entire networks of coupled biochemical reactions and feedback signals. In this respect, the HGP has taught us to see the study of genomes as information science that requires support by advanced computational biology tools and databases. The new discipline of bioinformatics plays a critical role in implementing this endeavor. Bioinformatics uses information technology to organize, visualize, interpret, and distribute biological information to answer complex biological questions. It allows workers in functional genomics to cope with the flood of data and address biological questions in a fraction of the time it would take using traditional analysis techniques. It is bioinformatics that enables functional genomics to bring order out of a vast number of data points.

A central component of functional genomics, driven by high-throughput methods and information science, is the ability to standardize extensive sets of disparate data. A key attribute in standardizing biological databases, which makes them computationally accessible, is an ontology. An ontology formally defines a common set of terms that are used to describe and represent a domain. Such vocabularies of terms specify the concepts in a given field and avoid semantic confusion. An example is the Gene Ontology, which can be used to describe the biological process, molecular function, and cellular location of any gene product. Ontologies have also become important in systematically collecting phenotypic information. As mentioned above, the term phenotype refers to observable traits and can be applied to any morphologic, biochemical, physiologic, or behavioral characteristic of an organism. The complete phenotypic representation of a species is now known as its phenome. The Mouse Phenome Project is a consortium of academic and industrial participants that promotes the quantitative phenotypic characterization of a defined set of mouse strains under standardized conditions. Such coordinated efforts in obtaining phenomic databases are now replacing phenotypic investigations carried out by thousands of independent investigators throughout the world, most of whom have no communication with each other. Standardized and comprehensive, such databases are critically important in the unraveling of molecular networks in the context of a functional unit, such as an organ or an organism.

An important condition in using large, standardized data-sets in an efficient manner for testing hypotheses and generating new ones is their integration. A major challenge in bioinformatics is to create single platforms for the integrated analysis of multiple, distributed data sources; for example phenotypic data with protein-interaction data and gene-expression data. In other words, the different components of the biological landscape in the form of the different -omics levels are pulled together to gain an understanding of biology at a higher level. This convergence will happen in an approach known as systems biology (Fig. 1.4). Systems biology studies the interrelationships of all the elements in a

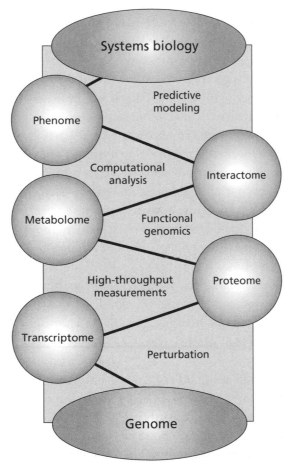

Fig. 1.4 The holistic science of systems biology attempts to define biological realities on the basis of global responses of cells, organs, or entire organisms to environmental or genetic perturbations.

system rather than studying them one at a time, in an effort to uncover hidden rules governing the ensemble of biomolecules working concertedly to perform certain functions in the cell. It aspires to use comprehensive data-sets, including such specifics as the experimental conditions under which the data were obtained, for building predictive models. In a sense, systems biology is the antithesis of the reductionist approach to biology, which has been so successful in the past in providing insights into the molecular machinery of many living systems and will continue to do so in further unraveling gene function in functional genomics approaches. However, it did not and cannot provide an understanding as to how molecular processes are integrated to provide function and how molecular function is regulated in living cells so as to give rise to dynamic cell, tissue, and organismal phenotypes.

General systems theory was conceived in the 1930s by the Austrian biologist Ludwig Von Bertalanffy (1901–1972), whose ambition was to create a 'universal science of organization'. His legacy is to have started systems thinking, thinking about a system as the emergent property of the interaction among all the components of the system and not as mere aggregates of its parts. System thinking in biology is not really new. Indeed, ever since the early microscopists it was always realized that the sum is more than the individual parts. In physiology, Claude Bernard (1813–1878) had already realized the common purpose of diverse physiological mechanisms to maintain homeostasis[45]. In molecular biology, however, such an integrative approach only became possible with the emergence of high-throughput technologies for measuring the large numbers of functionally diverse sets of elements in a cell with their patterns of selective interactions. Systems biology has now become an integrated approach to modeling biological systems in their entirety and to simulate their activity. For example, the human physiome project is to provide the framework for modeling the human body using computational methods[46]. The key challenge in these approaches is to distill the results of data-collection efforts into an interpretable computational form as the basis of a predictive model. A systems approach, then, involves repeated cycles of data collection and modeling. While systems biology is very far from computing the behavior of even a single cell, significant progress has been made. For example, models of heart function have already reached astonishing levels of detail, accuracy, and predictive power, as illustrated by realistic simulations of the beating of normal and abnormal hearts[47].

Studying the living world has brought us from the earliest microscopical observations of cells through the unraveling of the DNA sequence of hundreds of species to increasingly extensive collections of cellular constituents. The question then became how to convert this information into knowledge about the organism. Functional genomics and systems biology show great promise in becoming the centerpoints for exploring functionality in a quantitative manner, from the level of the genome and transcriptome to the physiology of organs and whole organisms. Can we use these same approaches to unlock the secrets of aging? And will that give us the means to develop interventions to ultimately halt its devastating effects?

1.4 **The causes of aging: a random affair**

What is aging? For practical reasons aging can be defined as a series of time-related processes occurring in the adult individual that ultimately bring life to a close. Aging is the most complex phenotype currently known and the only example of generalized biological dysfunction. Its effects become manifest in all organs and tissues. Aging influences an organism's entire physiology, impacts function at all levels, and increases susceptibility to all major chronic diseases. Organ systems communicate with each other in order to maintain homeostasis. We need to decipher how these communications change over the life course and which cells and biological macromolecules are involved in such changes. It is therefore obvious that a systems approach is required to address the core problem of biological aging: the loss of homeostasis. Indeed, a comprehensive explanation of how we age requires an understanding at all levels of the decline of the many complex functionally interacting subsystems of an organism. Such insight should provide us with a rational basis for tracking aging processes from their downstream manifestations to the primary causes. Depending on what those causes are, this may in turn permit the identification of novel molecular and cellular targets for prevention and treatment of aging-related illnesses through pharmacological means.

As we have seen, in the history of biology the discovery of the logic of life was followed by an understanding of the logic of aging. Following Weismann's original non-adaptive concepts of explaining aging, most researchers now accept that aging is ultimately due to the greater relative weight placed by natural selection on early survival or reproduction than on maintaining vigor at later ages. This decline in the force of natural selection with age is largely due to the scarcity of older individuals in natural populations owing to mortality caused by extrinsic hazards (Chapter 2). By contrast, our understanding of the proximal causes of aging is limited. One can argue that this is due in large measure to our inability in the past to study aging systems. Instead, ample information has been gathered about individual cellular components at various ages, but this has not allowed a clear understanding of the integrated genomic circuits that control mechanisms of aging, survival, and responses to endogenous and environmental challenges. With the emergence of functional genomics and systems biology, we finally have the opportunity to study aging in a comprehensive manner, as a function of the dynamic network of genes that determines the physiology of an individual organism over time[48].

Increased technological prowess increases confidence levels. Our increased capacity to handle complex problems in biology whetted appetites for knowing the pathways that control the gradual changes in the structure and function of humans and animals that occur with the passage of time and their relationship with functional decline, disease, and death. In the past, aging was not always considered a serious biological problem. In contrast to a disease, aging was thought to be inevitable, with attempts to intervene in its many adverse effects better left to charlatans trying to interest the public in anti-aging products that often lack any rational basis. Perhaps in part because of unusual difficulties in studying aging, its science was seen by many as less rigorous and mainly phenomenological. Nevertheless, it was in the days when aging research had a low profile that some major scientific minds laid the groundwork for our current understanding of why and how we age. These individuals, true giants who laid the foundations for the science of aging, include George Sacher (1917–1981), Nathan Shock (1906–1989), Bernard Strehler (1925–2001), Alex Comfort (1920–2000), John Maynard Smith (1920–2004), Zhores Medvedev (London, UK), Paola Timiras (Berkeley, CA, USA), Leonard Hayflick (San Francisco, CA, USA), George Martin (Seattle, WA, USA), the previously mentioned Denham Harman, and several others[49].

Whereas charlatans and their anti-aging products have by no means disappeared, studying longevity and aging has now become respected and its accomplishments frequently evoke great enthusiasm, as exemplified by an increasing number of high-profile publications and abundant interest from the respected lay press. This development was greatly aided by the discovery, originally in the laboratory of Tom Johnson (Boulder, CO, USA) in the early 1980s, that a single gene mutation, called *age-1*, dramatically extended lifespan of the nematode worm *C. elegans*[50]. This was important because at that time it was generally believed that aging was too complex to be significantly delayed by altering a single gene. This lack of single genes affecting the normal aging process made the field unattractive for scientists who were used to relying on consistent effects of a limited number of genes involved in specific mechanisms, related to developmental or disease processes.

With the discovery of the first gene affecting lifespan, aging had taken its place as a problem that could be addressed by studying the coordinated action of the products of multiple genes, similar to differentiation, development, and disease. In other words, aging had become a worthy object of study and began to attract highly reputed, well-funded scientists. At this point in time, there are hundreds of mutant genes in a variety of organisms, from yeast and fruit flies to mice, which increase longevity by dampening growth, reproduction, energy metabolism, or nutrient sensing. There are also mutant genes that cause accelerated aging, but these are not always generally accepted due to the difficulties in demonstrating that reduced longevity is genuinely due to accelerated aging or merely a result of a disease or developmental defect. This is discussed extensively in Chapter 5.

Nomenclature for genes affecting aging is confusing. Analogous to disease genes—genes that cause disease when mutated—genes with mutations causing increased longevity are sometimes called longevity genes. Since others speak of gerontogenes and yet others of aging genes, this leads to difficulties in understanding the normal function of the pathways in which these genes act. Since the pathway in which *age-1* and other longevity mutants in the nematode acts is really pro-aging, I will consistently call such genes aging genes. Longevity genes are genes encoding cellular processes that protect the organism against toxic insult. I realize that this departs from the disease gene nomenclature, but since this was wrong in the first place there is no need to also make the same confusing mistake for aging.

As described in the next chapter, the novel approaches of functional genomics and systems biology greatly facilitate the further unraveling of the functional modules of longevity control. Along these lines rapid progress is now being made in understanding the pro-longevity responses that result from dampening the activities of aging genes. It is likely that very soon the mechanisms of action of such genes and the network of interactions that gives rise to increased longevity will be resolved. The problem of aging, however, has two faces and the specific programmatic responses extending lifespan is only one of them.

Behind the other face of aging is stochasticity, an important aspect of biology in general, but often ignored since random variation is difficult to capture even with our current, highly sophisticated, biological tool set. While programmatic responses may defend us against toxic insults, it may be stochasticity that is behind the proximal cause of aging. For some time, evidence has been emerging that aging is caused by the accumulation of cellular damage, a random process. The programmatic component of aging may merely control the rate and extent of damage accumulation through dampening growth and reproductive processes with damage production as their inevitable side effects, or by upregulating processes of cellular defense. As we shall see in the next chapter, organisms are able to manipulate the allocation of their resources and balance reproductive efforts against somatic maintenance and repair. For some species, such as C. elegans, this flexibility is so high that interference in pathways of growth and reproduction, for example through single-gene mutations, can lead to 6-fold increases in lifespan. However, such dramatic effects are unlikely to occur in mammals due to their much greater complexity.

Hence, whereas functional-genomics approaches can help us to more fully understand how lifespan is controlled in a variety of organisms, for studying aging it will be necessary to focus on its proximal cause: the accumulation of somatic damage. In this respect, there is now ample evidence that damage to the genome can explain many of the most important phenotypes of aging. This book is focused on the possibility that the genome is both the creative engine behind longevity, as this emerged during evolution in ever more robust manifestations, and the main target of the somatic damage that ultimately limits life.

Since the original emergence of a genome 3–4 billion years ago, there has been a divergence into the current estimate of 30 million genomes, each representing a unique species. Evolution by natural selection requires the occurrence of mutations. If beneficial, such mutations are perpetuated. It is through mutations that Darwinian selection could lead to increasingly complex genomes and the adaptation of their hosts to the various challenges of a continuously changing environment. Because they occur at random, most mutations have adverse effects. During evolution, such genomes fall by the wayside as a consequence of natural selection.

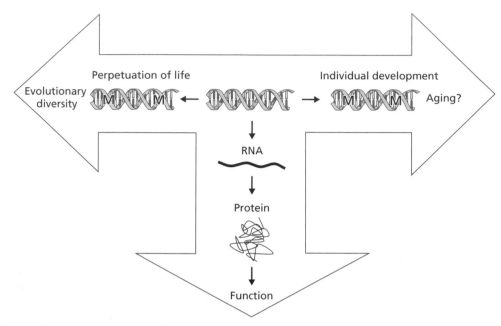

Fig. 1.5 The genome harbors all the instructions for providing function to the somatic cells of an organism through RNA and protein. Random alterations in its information content in the germ line drive evolutionary change, whereas similar changes in the somatic cells could be the cause of aging. M=mutation.

Genomes are not only subject to variation in the germ cells transmitted by parents to offspring. New variation accumulates also in the soma through mutation. Such instability of the somatic genome is much more extensive than originally thought in the immediate aftermath of the double helix. As we shall see later, organisms gradually move from having cells with very similar genomes towards mosaics of cells each with their own genotype. It is the thesis of this book that the same time-dependent instability of genomes that gives rise to evolutionary diversity also leads to aging at the somatic level. In both cases it can do that by randomly influencing the mechanisms by which genomes provide function (Fig. 1.5).

In the following chapters I will describe the logic of random genome damage as a major, highly conserved proximate cause of aging, discuss the ongoing efforts to unravel the many factors that determine this gradual loss of genome integrity of somatic cells, and critically review the evidence that genome alterations cause aging and its associated phenotypes, in the context of possible alternative explanations. In the final chapter I will discuss the impact of genome deterioration on our options to delay, halt, or even reverse the process of aging in mammals.

2   The logic of aging

In the eyes of many, the science of aging must look like a quest for the Holy Grail, a confusing series of contradictory approaches towards some elusive object. Today the field can roughly be divided into three main branches (Fig. 2.1): the biometric branch, seeing aging as infinitely complex and hardly amenable to intervention; the inductive branch, which attempts to explain aging in terms of few relatively simple, universal mechanisms; and the regeneration and renewal branch, with its focus on replacement and remodeling.

The early idea that aging is caused by the accumulation of mutations in the genome of somatic cells is clearly part of the simplistic branch and stems from the time when it was popular to explain aging as theories of a single cause[51]. Although such unitary theories have recently undergone some revival with the discovery of conserved pathways regulating longevity in multiple species, from invertebrates to mammals (discussed further below)[52], they lose a lot of appeal when confronted with the enormous complexity of the aging process as it takes place in different species[53]. Nevertheless, explaining aging in terms of one universal, driving force is not entirely unrealistic in view of the universality of the principles behind the living world on this planet. In spite of its bewildering variability, all life is based on combinations of the same, relatively few molecular species—sugars, fatty acids, amino acids, and nucleotides—and the single, universal, organizing principle of a genome that perpetuates itself as sequences of nucleic acid but functions by being expressed in the form of protein.

Is it possible that aging has its own inherent logic driven by a teleological process towards a specific goal? This is not to say that as a biological process aging could have some cosmic purpose. Such matter is beyond the scope of this book and beyond the scope of the biological sciences. However, whereas questions of what and how are often satisfactory in physics, biology is full of goal-directed processes guided by programs. Good examples are differentiation, development, and certain behavioral patterns. The question, then, of why we age—that is, the evolutionary causation of aging as distinguished from its proximate causes—is a legitimate one and should be actively pursued. Indeed, in the past it is this type of question that has led to the most important discoveries in biology. Whereas aging is different from most biological processes, in the sense that it resembles mostly random degeneration, which is unlikely to be programmed, it could still have its own logic, hidden in the depth of the history of life. If aging is inevitable, what then are the characteristics of a genome to let that happen?

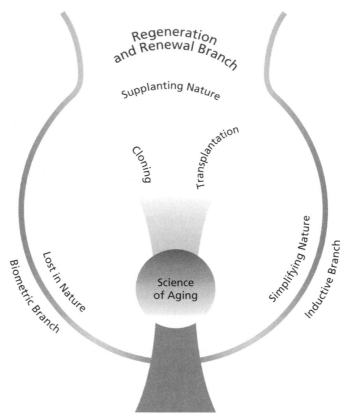

Fig. 2.1 The science of aging as a search for the Holy Grail: three different, parallel approaches, often difficult to reconcile.

2.1 **Aging genes**

One of the reasons why August Weismann's original idea of aging as an adaptive trait, which evolved to get rid of old, worn-out individuals (Chapter 1), makes little sense is the unlikely event of meeting such an organism in the wild. Whereas in nature animals do undergo aging, at least the early stages of the process, few if any of them ever reach the advanced state of decrepitude experienced in protected environments, such as by laboratory animals or humans in advanced societies[54]. Animals in the wild are likely to die early, from predation, starvation, or accidents, before the ravaging symptoms of aging can become manifest[55]. There are, therefore, very few animals to experience the advanced stages of aging in their natural habitat. In the past, the same used to be true for humans, only a selected few of which had the good fortune to reach old age. This lack of expression of the aging phenotype in nature makes it an unlikely target of natural selection. Hence, there are no genes that cause aging like there are genes that specify development and maturation.

That aging is nevertheless a real biological phenomenon common to most animal species can be derived from the observation that under protective conditions individuals of the same species, which in the wild never have the opportunity to get old, now reveal all the phenotypic characteristics of increased structural and functional degeneration, ultimately causing their death.

The aging phenotype and the similarities of many of its characteristics within and among species will be discussed in some detail in Chapters 5 and 7. Here I will only point out the main reason why we believe that aging is a genetically controlled process of intrinsic degeneration and not just a series of accidental changes over time. If aging as a biological process did not exist, one would expect the probability of death to stay the same with advancing age. However, in protective environments with the threat of external mortality effectively minimal, the probability of death increases exponentially with age once maturity is reached. This was first discovered by the British actuarian Benjamin Gompertz (1779–1865) in 1825, when the average lifespan of humans had just begun its dramatic rise. Gompertz showed that after about age 35 the probability of death doubles every 7 years. Of note, while this increase in mortality has been observed in all species that undergo aging, at very old age it no longer seems to hold. This has been most convincingly demonstrated for invertebrate species, large numbers of which can be studied relatively easily. Large-scale studies of the lifespans of tens of thousands of medflies, fruit flies, or nematode worms have shown a deceleration of mortality at old age, which may in fact also be true for humans[56].

The slopes of the Gompertz plots (Fig. 2.2a) reflect the age-specific death rates (effectively the rate of aging) and can be converted into survival curves (Fig. 2.2b). When comparing protected populations with non-aging populations in the wild, the Gompertz plots and the survival curves reveal dramatic differences. The mortality curve for wild animals no longer shows the exponential increase with age (because death is accidental and independent of age) and the survival curve is not rectangular but concave in shape. It would apply to animals that are subject to high rates of external mortality from predation and other environmental factors. For most animals in the wild the situation is probably somewhere between these extremes.

The survival curves reveal both the average lifespan of the population (the age at which 50% of the cohort has died) and the maximum lifespan, which is essentially the age of the last survivor or the average age of the longest-lived percentile or decile. The average lifespan or life expectancy is determined to a large extent by external factors. It can differ dramatically between the protected environment and the wild. The maximum lifespan is much less dependent on environmental conditions and is basically a function of the genetic make-up of the species. This explains why it does not change much when conditions improve. Careful comparisons of maximum lifespans for different animal species in captivity reveal significant differences, which indicate that lifespan is under genetic control (Fig. 2.3).

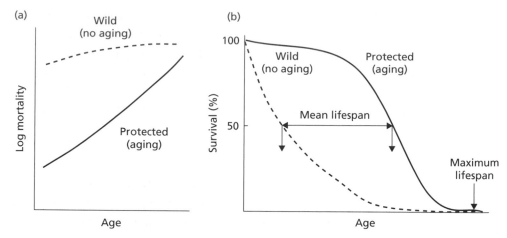

Fig. 2.2 (a) So-called Gompertz plot for animals in the wild (dashed line) and for animals under protective conditions (solid line). Both plots start in early adulthood. (b) The survival curves corresponding to mortality patterns in the wild (dashed line) and under protective conditions (solid line). Aging is revealed by the rectangularization of the survival curve.

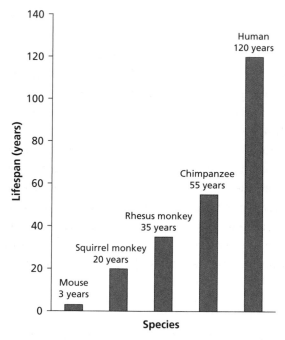

Fig. 2.3 Maximum lifespans for different animal species under protective conditions.

It should be kept in mind that lifespan values, especially maximum lifespans, are difficult to determine for animals in the wild because of difficulties in monitoring birth cohorts and the limited opportunities to survive beyond the age of reproduction. Mice, for example, may only live for a few months in the wild; they generally die early from predation or cold. For humans survival curves can be generated from census mortality tables. However, such data are only available for the last couple of hundred years and even then not always fully reliable. Hence, although it is often argued that maximum lifespan is fixed this is really not supported by strong data. In fact, there is evidence, for example from the observed deceleration of mortality at extreme old age mentioned above, that lifespan, even maximum lifespan, is more fluid that we tended to think (see Chapter 8 for a more detailed discussion of these issues).

At a population level, then, aging can be defined as a process that begins after maturity and results in an increasing probability of death. It is accompanied, at the individual level, by a gradual impairment of bodily functions, a host of structural changes and an increased incidence of certain diseases. In principle aging can include positive alterations and some have even defined it as all possible changes in an organism between conception and death. To indicate aging-related adverse changes exclusively some authors use the word senescence, which can be defined as those irreversible, deteriorative changes causing functional decline, disease, and death in aged organisms. This is somewhat confusing since, as we will see, senescence is also defined as the irreversible loss of proliferative activity of a population of cells, for example, yeast, protozoa, or mammalian cells. It is also used to simply indicate the aging of cells. In this book I will frequently use the word senescence in all these meanings. For more extensive discussions of aging in the animal world I refer to some outstanding works by some of my colleagues who are much better versed in the basic aspects of aging as a biological phenomenon. Examples are the monumental *Longevity, Senescence and the Genome* by Caleb Finch[57], *Why We Age: What Science is Discovering about the Body's Journey Through Life'* by Steve Austad[58], and *Time of Our Lives: the Science of Human Ageing'* by Tom Kirkwood[59].

The question has arisen as to how aging, which in evolutionary terms can be simply defined as the decline in fitness after the period of first reproduction, could ever emerge as a distinct, albeit complicated, phenotype. Indeed, direct natural selection should favor the suppression of senescence rather than its promotion. That is, the accidental inactivation of a gene causing a program of aging would provide an immediate selective advantage to its carrier, resulting in a rapid spread of immortality. In the next sections I will briefly explain why this is not the case and how aging may have evolved as a natural consequence of the logic of life.

2.1.1 AGING IS A BY-PRODUCT, NOT A GOAL OF NATURAL SELECTION

It is now generally accepted that the age-related decrease in fitness is non-adaptive; that is, it is not controlled by a purposeful genetic program similar to development. As we have seen in Chapter 1, August Weismann was the first to propose a non-adaptive explanation for aging. While, as just discussed, his name is generally associated with the idea of aging as a beneficial trait—evolved during evolution to cleanse the population of old, worn-out individuals, consuming resources without being reproductively active— Kirkwood and Cremer[7] pointed out that a little later Weismann also developed a non-adaptive explanation of aging. He recognized that if a specific function loses its usefulness, for example, due to a change in the organism's environment, natural selection will start ignoring it and the character will degenerate and disappear. Being ignored by natural selection is probably exactly what applies best to the supposedly useless period of life when reproductive duty is fulfilled and the theoretical chances of surviving much longer have become slim. Aging would therefore be the logical result of the declining force of natural selection after the period of first reproduction. Weismann, therefore, understood that aging by itself is unlikely to have an advantage, which contradicts his earlier idea of a beneficial cleansing mechanism.

The non-adaptive concept of how aging could originate in nature was first discussed systematically by Peter Medawar in the 1950s[60], going back to ideas of Fisher, Haldane, and Hamilton[10]. Medawar proposed that aging was the necessary result of constitutional mutations, accumulated in the germ line over evolutionary time, that reduce fitness late in life. As mentioned above, in the wild only a small fraction of a birth cohort will reach advanced age while continuing to reproduce. Therefore, the later the adverse effects of such mutations manifest after the period of first reproduction, the less likely that they will be weeded out by natural selection. The frequencies of such alleles, especially in small populations, will then drift randomly from generation to generation. In this way, natural selection acts to postpone adverse actions of genes until late age, when under normal conditions no animals can be expected to be alive anymore. Under such conditions natural selection does not 'see' the effect of these mutations, which only show up under more optimal conditions, such as in captivity.

In general, natural populations experiencing low mortality, from predation or disease, will postpone late adverse effects further than populations of the same species in a habitat with high extrinsic mortality. Indeed, under conditions of high extrinsic mortality, lifespan would already be short, pushing back the reproductive period and removing any evolutionary incentive to cleanse the genome from random mutations with adverse effects after this period. In captivity, under optimal conditions, such animals would display shorter lifespans. In other words, their intrinsic mortality would adapt to their extrinsic mortality.

Adaptation of intrinsic mortality to extrinsic mortality can also work the other way around; that is, lifespan of a population of organisms can be extended by artificially extending the period of reproduction. This has been demonstrated, for example, in outbred populations of *D. melanogaster*, the fruit fly. Clare and Luckinbill[61] restricted reproduction of this organism to late age, thereby increasing the intensity of selection during the later portion of the lifespan. They did this for 21–29 generations, at two different larval densities. Populations with high and uncontrolled numbers of competing larvae responded strongly to selection for late-life reproduction, with the length of adult life increasing by as much as 50%. Under such conditions selection produced true-breeding long-lived lines. When populations of developing larvae were held low, however, longevity fluctuated wildly during selection, showing little overall response. Interestingly, this experimental design allowed the simultaneous demonstration of the existence of natural pro-longevity gene variants and the strong influence of the environment on the selection of these variants. Later, the prediction that higher extrinsic mortality rates lead to the evolution of shorter lifespans in *Drosophila* was confirmed by directly comparing populations of wild flies under conditions of high and low imposed mortality[62].

Evidence that different levels of environmental hazards dictate longevity through evolutionary selection has also come from studies of vertebrate animals in the wild. Observations by Steve Austad (San Antonio, TX, USA) on two opossum populations, one on an island not subject to predation and one on the mainland where they were exposed to significant predation, confirmed the prediction that the island population had the highest longevity[63]. Environment is also not the only factor that appears to play a role. In general, attributes that reduce extrinsic mortality, such as wings (birds, bats), protective shells (turtle), or brains (humans) are generally associated with increased longevity[57,58]. However, the situation is not straightforward and other factors, such as ecology, play important roles. The aforementioned selection of long-lived flies through late-life reproduction only worked at high larval density. In a study of guppies derived from environments with high and low mortality rates, lifespan was found to be longest in streams where predators co-occur with guppies and shorter in individuals reared in the upper reaches of streams by waterfalls, from which predators are often excluded[64]. There are several possible explanations for this unexpected finding, including a reduction in population size of the guppies by the predators, which would result in an increase of the abundance of food and other resources leading to increased survival[65].

It should be noted that this (thus far) isolated finding does not falsify the evolutionary theory of aging. It merely illustrates that unlike the situation in the physical sciences theories in biology are often ambiguous and have no impenetrable mathematical basis. This is necessarily so when starting from the shared organizational features manifested by organisms. It is not possible in biology to reject theories on the basis of Popper's falsification principle. In biology exceptions are often the norm. This does not make biology unphysical. It simply means that in biology generalizations are often based on inadequate or incomplete theories.

The correlation between high extrinsic mortality and short lifespans may be a consequence of the trade-off most organisms face between investment in somatic maintenance and reproductive effort. This was first recognized by Tom Kirkwood and formulated in his disposable soma theory of aging[9], already mentioned in Chapter 1. This theory predicts that high extrinsic mortality would favor investment of scarce resources in early reproduction rather than somatic maintenance and repair, which would not be required for a population with a low risk of survival in the wild. This of course implies that age-related somatic degeneration and death is caused by the accumulation of unrepaired somatic damage, a reasonable explanation, which is now supported by a large body of evidence. For example, the extended longevity in late-reproducing fruit flies mentioned above appeared to be associated with a coordinated upregulation of genes specifying cellular defense mechanisms against free-radical attack. Consequently, these flies showed decreased levels of oxidative damage to proteins and lipids[66]. In addition, delayed senescence in the long-lived flies was accompanied by reduced fecundity, supporting the idea of a trade-off between reproductive effort and survival (see also below).

2.1.2 AGING OF UNICELLULAR ORGANISMS

From an evolutionary point of view, therefore, age-related decline and death is the effect of genetic variants (aging genes) that have escaped the force of natural selection by acting only post-reproductively. Ultimately, this is due to the separation of soma from germ, which creates a life cycle in which the soma is reconstructed in every generation from a germinal blueprint. As we have seen, reconstruction of the soma (or reproduction) is greatly favored by natural selection, especially under conditions of high external mortality. The fate of the old soma is not a priority. In essence, this means that the deleterious effects of mutations accumulated in the germ line are shed to the parents. However, this implies that unicellular organisms should not age because they do not have a distinction between the germ line, which is to be preserved, and the soma, which is expendable. Although for most unicellular species (as well as some metazoa without a clear distinction between germ line and somatic tissue, such as *Hydra*) there is indeed no evidence of aging[57], certain unicellular eukaryotes, such as budding yeast and protozoa, undergo a process of cellular senescence after a given number of population doublings. In such cases there usually is asymmetric division and it has been suggested that in parallel to the soma of multicellular organisms, organisms such as yeast segregate their senescent changes to the mother cell. Hence, like in most metazoa, the negative effects of aging are confined to the parent. A similar situation has been observed in asymmetrically dividing prokaryotes[67,68]. Even in a symmetrically dividing organism, like *E. coli*, which divides by binary fission into seemingly identical siblings, aging has recently been demonstrated. This was attributed to inheritance of the old pole; that is, the end of the cell that has not been newly created during division, but is pre-existing from a previous division. Old poles can exist for many divisions, and by following repeated cycles of reproduction by individual cells, through automated time-lapse microscopy, it was shown that the cell that inherits the old pole exhibits a diminished growth rate, decreased offspring production, and an increased incidence of death[68].

Protozoa have been studied extensively in the context of aging and reproduction. In his beautiful book, *Sex and Death in Protozoa*[69], Graham Bell (Montreal, Canada) has outlined in intricate detail many if not all key concepts of aging in unicellular organisms in comparison with multicellular ones. Protozoa reproduce asexually, by fission, and sometimes also sexually, which is termed conjugation. Although protozoa do not divide asymmetrically, they have a life cycle: immaturity, maturity, and senescence[69]. In a protozoan clone, senescence occurs in the absence of conjugation. Conjugation involves fusion of two paired individuals, meiosis of the diploid nucleus (termed the micronucleus as distinct from the macronucleus, which disintegrates) and exchange of one gamete nucleus. After fusion of gametes the diploid state is reconstituted and new individuals are generated. Conjugation prevents senescence, which only occurs in its absence. The zoologist Tracy Morton Sonneborn (1905–1981) discovered in the early 1950s that autogamy, the union of gametic pronuclei within the same cell, without conjugation, could substitute for conjugation in preventing senescence in protozoa[70].

Bell has interpreted senescence in protozoa as being caused by the irreversible accumulation of deleterious mutations in isolate lines and not as some adaptive response. However, it is possible that senescence offers a selective advantage to populations of protozoa. Indeed, Cui *et al.* used computer simulation to analyze two hypothetical species of protozoa, one with and one without senescence[71]. The two species were subject to the same rate of deleterious mutation and could undergo fission or conjugation. Conjugation was assumed to re-set the clock, whereas senescence would eliminate an individual reaching its age limit. The results of the simulation indicated that far fewer recessive deleterious mutations had accumulated in the species employing senescence. The authors speculated that senescence had emerged as an effective way of interrupting the pathway of mutation accumulation increasing the chance of sexual reproduction to produce high-fitness offspring. In this case, senescence has been explained as offering an advantage in combination with sexual reproduction. This would be in keeping with the more general idea that aging is the price we have to pay for sex, which is almost certainly false because, as outlined above, aging is not a major cause of mortality in natural animal populations[72]. However, senescence in protozoa is not necessarily identical to aging in mammals and it is certainly possible that the process has some adaptive value (see further below). It should be noted that Bell argued against senescence in protozoa as an adaptive response since the onset of senescent changes is highly variable between independent cultures in contrast to the onset of sexual maturation, which is invariant.

The basis of any discussion of mutation accumulation is the realization that in living organisms genetic information cannot be transmitted without error. In a finite asexual population deleterious mutations will accumulate over time because sooner or later those cells harboring the least mutations will be extinguished and cannot be restored. This phenomenon, called Muller's ratchet[73], is responsible for senescence in isolated cultures of protozoa and sets a limit to the longevity of germ lines. Sex would then act as a repair device for acquired mutations by re-shuffling genomes, effectively resulting in some genomes with low mutation loads. As explained by Bell, while this protection would be most effective during cross-fertilization, also autogamy will offer some protection, which explains the result obtained by Sonneborn, mentioned above.

It should be realized that population extinction by Muller's ratchet (germinal senescence) is essentially different from the concept of aging genes resulting from the accumulation of mutations in the germ line. As pointed out by Bell, the germ line ages because it accumulates mutations adversely affecting fertility. This can effectively terminate a population. The soma of multicellular organisms ages because selection will not cleanse their genome of genes with adverse effects on the soma at late age. Nevertheless, the ratchet would operate in all genomes over time and there is no reason to assume that it would not operate in somatic cells of metazoa, especially stem cells which would readily develop into clonal lines creating mosaics for the mutations which have arisen. This situation would be analogous to senescence of a protozoan clone, but significantly more complicated in view of the interdependence of different types of somatic cells in metazoan species.

In summary, Medawar's idea of aging as a result of the accumulation in the germ line of mutations creating gene variants with adverse effects on the soma late in life, after the age of first reproduction, is a logical explanation for the sheer universality of aging in the animal world. Such a random process of evolving aging genes suggests that aging is affected by many genes, each with a relatively small effect. This would make aging infinitely complex and different from species to species. While aging is indeed characterized by randomness, with variation within and between species the rule, this explanation is still unsatisfactory since it does not address the great similarities in the types of aging processes that are also a characteristic of aging in animals.

2.2 Pleiotropy in aging

The evolutionary logic of the Medawar theory begs the question of the type of aging genes we should be looking for when trying to identify the mechanisms that control aging and longevity. Before discussing this it is important to clarify what we mean by aging genes. First, Medawar's accidental mutations in the germ line are not necessarily in genes, but could be in DNA sequences that control the expression of genes. Second, as should be clear from the discussion above, there are no genes specifically selected to cause aging. Instead, most aging genes are variants of genes or DNA sequences that have adverse effects after the age of first reproduction. Finally, as discussed in Chapter 1, genes do not act alone, but in functional modules, which arise from interactions among components in the cell (proteins, DNA, RNA, and small molecules). This context should be kept in mind every time the term aging gene is used.

George Martin has distinguished private from public aging genes[74]. Private aging genes find their origin in Medawar's accidental germ-line mutations discussed above, which are neutral at early age, but start exerting their adverse effects later, after the period of first reproduction. In the absence of selection, the frequency of such mutations in the population would be determined by genetic drift. They will, therefore, be relatively rare and can also be expected to be specific for the population or species in which they arise. Hence, they were termed private aging genes. J.B.S. Haldane predicted such genes in the 1940s based on his observation that Huntington's disease, a genetic disorder, occurs late in life, after most people have already had their children.

Apart from Huntington's disease, in humans private aging genes could explain familial forms of age-related disease phenotypes, such as Alzheimer's disease. Martin systematically analyzed known loci involved in human genetic disease, which had been cataloged by Victor McKusick (Baltimore, MD, USA) as Mendelian Inheritance of Man (a phenotypic companion to the HGP and available online as OMIM; www.ncbi.nlm.nih.gov/entrez/query.fcgi?db=OMIM). Based on these phenotypic descriptions, Martin concluded that private mutations at a very large number of loci could play a causal role in the senescent phenotypes as they are observed among elderly[75].

In contrast to private aging genes, public aging genes are hypothesized to have arisen under the influence of natural selection, not because they cause aging but as a consequence of an early, beneficial effect. Their adverse by-products would only become manifest at later age. This involves the concept of antagonistic pleiotropy and was originally proposed by George Williams[76]. Since we know that genes usually have more than one effect (hence the term pleiotropy) it is more than likely that the same applies to the gene variants arising as accidental mutations. Since most mutations are bad, it would be unlikely that in the rare case of a beneficial effect of an accidental germ-line mutation early in life there would not be an adverse effect later. In this view, aging would be the result of the harmful side effects of genes selected for advantages they offered during youth. It is possible that there are many such trade-offs between early and late effects and, in contrast to Medawar's deleterious mutations, which are neutral at early age, pleiotropic mutations could involve conserved mechanisms of universal benefit. Indeed, mutational variants of genes affecting similar functions early in life could have arisen independently in different populations, and even in different, only distantly related species. For this reason one can refer to such evolutionarily conserved genes as public aging genes. A few theoretical examples underscore the existence of the trade-offs dictated by public aging genes.

To illustrate his pleiotropic gene theory of aging, Williams himself provided the first example, in the form of an arising gene variant with a favorable effect on bone calcification in the developmental period with the adverse side effect of depositions of calcium in the arterial walls at later ages[76]. As noted above, as an unfavorable character, senescence would always evoke direct action of selection to oppose it. Hence, additional gene variants would likely arise to suppress the adverse phenotype of calcium deposition in arterial walls (if environmental constraints permit). However, this mechanism would never succeed in suppressing the adverse trait completely due to diminishing selective pressure with age. Another example of a trade-off that may explain human aging-related disease is inflammation. Genetic pathways specifying a powerful response to infection have undoubtedly a high survival value, but may at later age contribute to disease-causing degeneration, such as atherosclerosis[77].

One could think of a large number of trade-offs in gene action, reflecting on age-related disease phenotypes. Cancer is one of them. Tumor-suppressor genes are obviously advantageous for young organisms and mutations inactivating such genes would not be passed on to many offspring. Apoptosis, or programmed cell death, and cellular senescence, irreversible mitotic arrest, are both critical processes for suppressing tumorigenesis in mammals[78]. Both responses are highly conserved in eukaryotic organisms also for reasons other than suppressing cancer. Apoptosis, for example, plays an essential role in embryonic development and also later in maintaining normal tissue homeostasis. As first realized by Judith Campisi (Berkeley, CA, USA), in mammals both apoptosis and cellular senescence are likely to be antagonistically pleiotropic, since they help to suppress cancer at early age, but possibly at the cost of promoting aging at later ages by exhausting progenitor- or stem-cell reservoirs[79]. This possible antagonism between cancer and non-cancer degenerative aging phenotypes will be discussed in more detail in Chapter 7.

Another example of antagonistic pleiotropy could be the production of ATP by oxidative phosphorylation, the main pathway of energy production. This effective method of harnessing energy arose early in evolution and has been conserved with relatively minor variation. However, cells produce reactive oxygen species (ROS) as by-products of oxidative phosphorylation and I already mentioned in Chapter 1 that it was Denham Harman who originally hypothesized that such free radicals are one of the major factors responsible for the aging of cells[23]. Since the accumulated damage from ROS usually only becomes manifest at late age, this could be another example of postponing adverse effects of a beneficial pathway. Like other processes of a conserved beneficial nature, genes controlling oxidative phosphorylation can be considered as public aging genes.

Probably the most ancient, hypothetical example of antagonistic pleiotropy is the complex of intertwined pathways to replicate, recombine, and repair DNA or, originally, RNA. While these systems guarantee life's continuity, there is a price in the form of the double-edged sword of mutations, the result of erroneous DNA transactions. Mutations are necessary as the ultimate source of genetic variation upon which evolution depends, yet most of them are harmful. We have already seen that mutations with late-life adverse effects but which are neutral at early age are likely to accumulate in the germ line. A high rate of germ-line mutation can lead to population extinction, which may have been the driving force behind the evolution of sex, mate choice, and diploid life cycles, all ways to limit the adverse effects of mutations. Since the adverse effects of mutations in somatic cells would only become manifest at late age, selection to improve the accuracy of DNA metabolism far beyond the age of first reproduction is absent.

Antagonistic pleiotropy can also act directly on reproductive success (sexual selection) rather than on the ability to deal with environmental factors (survival selection). Indeed, lower fecundity is often found associated with longevity, such as in the aforementioned insular opossum population experiencing the least extrinsic mortality[63]. The importance of sexual selection is further underscored by a series of experiments reported by Sgro and Partridge[80], the results of which elegantly demonstrated that antagonistic pleiotropy is likely to be more frequently responsible for aging phenotypes than Medawar's neutral mutations. In long- and short-lived lines of fruit flies, generated by making these flies reproduce late or very early (see above), Sgro and Partridge initially observed similar mortality. However, after 30 days, the short-lived lines, with eggs laid by very young adults, experienced a wave of higher mortality, which did not occur in the flies selected to live for longer[80]. Interestingly, this wave of higher mortality, causing the early reproducers to age quickly, could be abrogated by the ablation of egg laying, for example through irradiation. Hence, rather than late, adverse effects of mutations neutral at early age, these results suggest a damaging effect of early reproduction. Selection for life extension could be the result of a switch in resource allocation, from reproduction to somatic maintenance. Alternatively, early reproduction could have caused the damage directly with the effects accumulating over time, to result in the wave of mortality. In spite of this elegant experimental example of antagonistic pleiotropy in action, with sexual selection as the possible driver, an inverse relationship between fecundity and longevity is complex and not always found[81].

In summary, aging genes are not considered to have emerged on the basis of some selected value for fitness but are merely by-products of evolution. They are thought to dictate trade-offs between beneficial effects early in life and adverse effects later. In the next section we will discuss several examples of such genes. An adverse effect of any aging gene can be suppressed by another gene, which we can call longevity genes or longevity-assurance genes. It is conceivable that the emergence of aging genes and longevity genes go hand in hand, under the influence of environmental constraints. Although the existence of a large number of private genes, specifying aging mechanisms peculiar to certain species and to certain individuals within a species, cannot be ruled out, evidence points towards a major role of public mechanisms specifying highly conserved beneficial pathways, such as oxidative phosphorylation and apoptosis. This is in keeping with the many similarities of aging within and across species and, based on the ancestry of some universal pathways of life, opens up the possibility that aging has been with us from early times.

2.3 **Interrupting the pathways of aging**

One way to identify aging genes and their mechanisms of action would be to screen a population for individuals harboring inactivating or weakening mutations in a gene controlling a pathway of aging. This would work well for Medawar's gene variants that are neutral at young age and also for Williams' pleiotropic gene variants when depriving the individual of an early beneficial effect that is not lethal under the conditions the animal is studied. The latter may be difficult in mammals with their complicated genomes harboring numerous gene–gene interactions that can easily lead to adverse side effects. As we will see, most of the aging genes thus far identified have indeed been found in relatively simple invertebrates, although large-scale genetic screens are of course also much easier to perform in short-lived simple organisms than in mice or rats.

Whereas it was previously recognized that many single-gene mutants could shorten lifespan, a major increase in lifespan due to the effect of one gene was considered unlikely in view of the presumed multicausal nature of the aging process. This does not mean that mechanisms to delay aging and extend lifespan were not known. As described further below, the phenomenon of extending lifespan through calorie restriction has been known since the 1930s and the effect of temperature and reproductive status on lifespan of *Drosophila* was described as early as the 1920s by Raymond Pearl (1879–1940). Indeed, John Maynard Smith (1920–2004) was probably the first to describe an increase in lifespan due to the effect of a single-gene mutation in *Drosophila*. He found that the ovariless mutant (a mutant where the females entirely lack ovaries) as well as virgin or partially heat-sterilized females have extended survival compared to controls[82]. However, these results were not generalized in terms of the existence of many genes that could be manipulated to alter functional pathways controlling longevity. It is now well understood that single-gene mutants can affect a host of downstream processes and therefore significantly influence major prolongevity systems. This is the basis of the wealth of long-lived mutants that have now been discovered in nematodes, fruit flies, and, more recently, in mice[83].

2.3.1 THE NEMATODE WORM, *C. ELEGANS*

The mutant gene that transformed this field by its demonstrated pro-longevity effect in the nematode was first identified in the laboratory of Tom Johnson (Boulder, CO, USA)[50], who called it *age-1*. Later, this gene turned out to encode the *C. elegans* ortholog of the phosphoinositide 3-kinase p110 catalytic subunit, a central component of the *C. elegans* insulin-like signaling pathway, lying downstream of the DAF-2/insulin receptor (see below) and upstream of both the phosphoinositide-dependent protein kinase 1 (PDK-1) and thymoma viral proto-oncogenes 1 and 2 (AKT-1/AKT-2 kinases) and the DAF-16 forkhead-type transcription factor, whose negative regulation is the key output of the insulin signaling pathway[84] (see below).

Cynthia Kenyon (San Francisco, CA, USA) and co-workers subsequently demonstrated that mutations in *daf-2* also conferred longevity to the worm[85]. This gene was later identified in the laboratory of Gary Ruvkin (Boston, MA, USA) as a insulin/insulin-like growth factor 1 (IGF-1) receptor homolog, thus acting (upstream) in the same pathway as *age-1*[86]. Loss-of-function mutations in these genes and others acting in these pathways cause dauer formation, a state of diapause in response to food limitation and crowding. Weak mutations in such genes, however, allow these animals to become adults and live up to twice their normal lifespan. It appeared that the diapause-related genes confer their longevity-promoting effects through the action of a forkhead/winged helix transcription factor called DAF-16, identified independently by the Kenyon and Ruvkin labs[87,88]. DAF-16 relocates from the cytoplasm into the nuclei of different cell types in the nematode and affects the activities of genes involved in many processes, including metabolism, stress response, and antimicrobial action[89]. Indeed, it is possible that the mechanism underlying the increased longevity of these and other mutants in different organisms involves the upregulation of longevity-assurance mechanisms (see below).

The Kenyon lab also found that germ-line ablation increases the lifespan of nematodes[90], similar to what has been found in *Drosophila* (see above). While this effect is independent of the upstream *daf-2* gene, its signal does activate *daf-16*. DAF-16, therefore may be a master regulator integrating different longevity signals.

Among genes conferring longevity on the nematode upon mutational inactivation or weakening, those that impact on mitochondrial function are especially frequent. Examples are the *clk* genes, so-called because they regulate physiological, developmental, and behavioral clocks during the nematode life cycle. Four *clk* mutants have been identified that show a moderate increase in lifespan[91]. CLK-1 has been shown to localize in the mitochondria of all somatic cells of the worm and is required for the biosynthesis of coenzyme Q9[92]. Coenzyme Q plays a crucial role in the mitochondrial electron-transport chain, and *clk-1* mutants rely on the Q8 homolog of Q9 synthesized by the worm's diet of *E. coli*. Withdrawal of Q8 from the diet of wild-type nematodes extends adult lifespan by approximately 60%, probably by decreasing the release of free radicals from mitochondria or by increasing resistance to damage accumulation[93].

The first lifespan-extending mutations were discovered in nematodes in classical genetic screens, by mutagenizing worms to randomly disrupt gene function and subsequently recover long-lived mutants. However, with the emergence of the novel, high-throughput methods in functional genomics a more systematic approach became possible, which was quickly utilized by the main players in this by now highly competitive field. To identify in a systematic manner the different classes of genes that control *C. elegans* lifespan, both the Kenyon and Ruvkin labs began to apply the new RNAi approach (Chapter 1). Using so-called feeding RNAi libraries—that is, by feeding worms with bacteria expressing double-stranded RNA—over 17 000 nematode genes were screened by these investigators[94,95]. A number of known and unknown genes—as many as 89 in one of these studies—were found that extended nematode lifespan upon RNAi inactivation. Many of these genes turned out to encode components of the mitochondrial respiratory chain or were impacting mitochondrial function. Thus, the results obtained with this comprehensive RNAi screen confirm the idea that mutations that extend lifespan often involve genes participating in energy metabolism.

2.3.2 THE FRUIT FLY, *D. MELANOGASTER*

As expected from public aging pathways, when the interruption of a pathway leads to extended lifespan in one species, interruption of orthologous pathways in other species should give similar results. This has proved to be the case for the insulin/IGF-1 signaling pathway. As we have seen, reducing the activity of this pathway increases lifespan in the nematode. In *Drosophila*, complete inactivation of the gene that encodes the insulin receptor substrate (IRS) homolog (*chico*) extends lifespan by about 45%, but only in females. Heterozygous individuals of both sexes lived longer. There is also evidence that in *chico* mutants the longevity phenotype is associated with increased stress resistance. Similarly, hypomorph mutations in the insulin-like receptor (*InR*) gene extend the lifespan of females, but not of males[52]. In this case there is evidence that the longevity phenotype in the females is associated with infertility.

Another aging pathway identified in *Drosophila* on the basis of mutants of increased lifespan is the target of rapamycin (TOR) pathway, which has now emerged as a major regulator of growth and size, as well as longevity[96]. Inhibition of this nutrient-sensing pathway, which can modulate insulin signaling and growth, extends lifespan in the fly in a manner that may overlap with caloric restriction; that is, a reduction in nutrient intake increasing lifespan across different taxa.

Two partial loss-of-function mutations have been described in the fly, namely Methuselah (*mth*) and I'm not dead yet (*indy*), which increase lifespan by 30 and 100%, respectively. The effect is seen in both male and female flies without loss of fertility. However, the *indy* mutation only increases lifespan in the heterozygote flies. The *mth* gene encodes a G-protein-coupled transmembrane receptor[97]. In addition, mutations in the gene for its ligand, Stunted (*sun*), which encodes a subunit of ATP synthase, increase lifespan[98]. Whereas Stunted is found on the cell surface, the Methuselah–Stunted pathway may exert its role as a regulator of aging through the mitochondria. The *indy* gene encodes a dicarboxylate transporter, a membrane protein that transports Krebs-cycle intermediates in tissues participating in the uptake, utilization, and storage of nutrients. The life-extending effect of the Indy mutation may be due to an alteration in energy balance caused by a decrease in Indy transport function[99].

2.3.3 CALORIE RESTRICTION

Well before it was realized that single gene mutations could slow the rate of aging and increase lifespan in invertebrates, Clive McCay (1898–1967) reported in 1935 his unexpected discovery that rats that ate less lived nearly 30% longer[100]. Calorie or caloric restriction, as this is generally called, is underfeeding without malnutrition. Work from others, most notably a series of careful studies by Edward Masoro (San Antonio, TX, USA), subsequently demonstrated convincingly that reduced caloric intake rather than reduction of a specific nutrient was responsible for the beneficial effects, which also included a reduction or retardation of tumor formation and other aging-related phenotypes[101]. Importantly, a significant effect was still seen when calorie restriction in the mouse was initiated as late as 12 months of age[102]. Calorie restriction is unrelated to development since there was no effect when applied early, between 6 and 24 weeks in the rat[103]. Rodents on a calorie-restriction regimen have lower fasting levels of plasma glucose, insulin, and IGF-1. This suggests that decreased insulin signaling, similar to the situation in some of the long-lived nematode and fly mutants, could also contribute to the calorie-restriction longevity effect. Life extension through caloric restriction has been observed not only in rodents, but also in various species of worms, flies, and yeast. Studies to test whether similar effects can be found in primates are still under way[104].

Calorie restriction is often assumed to be caused by a lower metabolic rate (rate of energy utilization), often thought to promote longevity. The idea that aging is inversely related to metabolic rate goes back to a publication by Max Rubner (1854–1932) in 1908 showing for five mammalian species that in spite of large differences in lifespan total metabolic output (the amount of energy consumed during adult life) per unit body weight was similar[105]. The conclusion was that animals with a high metabolic rate would have a shorter lifespan than animals with lower rates. Raymond Pearl (1879–1940) subsequently presented his rate of living theory based on survival experiments with *Drosophila* and cantaloupe seedlings[106]. His conclusions were that an organism's duration of life is determined by its 'inherent vitality'. Inherent vitality was defined by Pearl as the total capacity of the organism to perform vital action in the complete absence of exogenous energy; under conditions of complete starvation. Inherent vitality is then lost at a rate that equals the rate of energy expenditure. Lifespan, therefore, is inversely related to the rate of living. Naturally, it is tempting to link metabolic rate to natural free-radical production and the accumulation of somatic damage as originally conceived by Harman[23]. Indeed, there is ample evidence that calorie restriction retards the accumulation of oxidatively damaged molecules in aging rodents[107].

Whereas metabolic alterations are likely to play a role in both caloric restriction and the longevity phenotype of most if not all the long-lived mutants in worms and flies, variation in lifespan appeared not to be attributable to reduced metabolic rate. For example, whereas in rodents immediately after the initiation of calorie restriction metabolic rate rapidly declines, it is subsequently restored and stays the same per lean body mass, as in control animals[108]. In flies and worms, calorie-restriction-induced life extension is not associated with a lower metabolic rate and the same is true for the longevity phenotypes of the insulin-signaling mutants of these invertebrate species[109–111]. This does not rule out the possibility of metabolic alterations facilitating a reduced formation of reactive oxygen molecules.

2.3.4 GENETIC MANIPULATION OF LIFESPAN IN MICE

Naturally, genetic screens for longevity mutants are difficult to carry out in mammals because of their long generation times and expensive husbandry. On the other hand, they offer the enormous advantages that they are evolutionarily very close to humans and that lifetime phenotypic information on mice and rats is extensive (albeit not as complete as for humans). Studies of some spontaneous and engineered mutants suggest that mechanisms similar to those acting in nematodes and fruit flies also control longevity in mammals. Mice homozygous for mutations in the *Prop-1* gene, termed Ames dwarf mice, are dwarfs and live 50–70% longer than wild-type mice[112]. These animals are deficient in growth hormone (GH), thyroid-stimulating hormone, and prolactin, and they show reduced levels of plasma insulin, IGF-1, and glucose. They also show a delayed occurrence of neoplastic lesions compared with their normal siblings[113].

Expression of the *Pit-1* gene, which controls development of the pituitary gland, depends on expression of the *Prop-1* gene. *Pit-1*-defective mice, termed Snell dwarf mice, are dwarfs and infertile like Ames mice, and they show a 40% increase in mean and maximum lifespan[114]. They also show a delay in collagen cross-linking with age and several age-sensitive indices of immune status. In Snell mice, evidence has been provided for reduced insulin/IGF-1 signaling in response to the GH deficiency[115], which could be a key factor in the lifespan extension of these mice, as it apparently is in longevity mutants of nematodes and flies and, possibly, in calorie restriction. Of note, both natural and engineered GH-defective mice show increased longevity and have also very low circulating IGF-1[114,116]. Interestingly, like the longevity mutants in nematodes and flies, both dwarf mice and calorie-restricted mice display increased resistance to stress, pointing towards an upregulation of longevity systems (see further below).

Reduced insulin signaling as a universal mechanism for delayed aging of mice is in keeping with the increased lifespan of female mice haploinsufficient for the IGF-1 receptor[117], although this finding has thus far not been reproduced independently. Interestingly, both male and female mice with a disruption of the insulin receptor gene only in adipose tissue show a decreased body fat mass and increased longevity[118]. These mice live longer with normal caloric intake and retain leanness and glucose tolerance with age. While it is tempting to suggest that based on this finding it is leanness rather than metabolic changes that causes the extension of life, it should be realized that as argued by Masoro[119] the elimination of the insulin receptor in fat would by itself lead to changes in metabolism. Indeed, reduced fat is not involved in the mechanism behind the phenomenon of calorie restriction. This can be derived from results obtained by David Harrison (Bar Harbor, ME, USA) and colleagues[120], who studied the effects of life-long calorie restriction in genetically obese (*ob*/*ob*) and normal mice of the same inbred strain. While the calorie-restricted *ob*/*ob* mice still had high levels of adiposity, their maximum lifespan exceeded that of the normal mice and was similar to the lifespan of calorie-restricted normal mice that were much leaner[120].

Interestingly, there have been attempts to mimic some of the lifespan-extending mutations in nematodes by targeted inactivation of orthologs of the same genes that conferred longevity in these invertebrate animals. Siegfried Hekimi (Montreal, Canada), who had previously discovered the *clk-1* mechanism of lifespan extension in nematodes (see above), inactivated the *clk-1* ortholog, *mClk1*, in the mouse[121]. It was found that homozygous inactivation of this gene (which is required for the synthesis of coenzyme Q9; see above) in mouse embryonic stem cells yielded cells that are protected from oxidative stress and contain lower levels of spontaneous DNA damage. Whereas the complete *mClk1* knockout is embryonically lethal, the heterozygous mice showed an increased lifespan. Interestingly, these investigators observed the loss of the remaining functioning allele of *mClk1* during aging in a subset of liver cells, as a consequence of so-called loss of heterozygosity; that is, through the loss of the chromosome or section of the chromosome (see Chapter 4).

2.3.5 INTERRUPTING AGING OR BACK TO NORMAL?

The current wealth of mutants of extended lifespan in various organisms appears to confirm that, as predicted by the antagonistic pleiotropy theory, there are genetic pathways that can be interrupted to attenuate aging. It would be logical to expect that such mutations would only lengthen life at the cost of some selective disadvantage, which may or may not be obvious under laboratory conditions. Typically, mutations conferring longevity result in decreased energy metabolism, growth, physical activity, and/or early-life fecundity. It is likely that such characteristics are not usually the preferred ones in nature with its high external mortality. For the dwarf mice, fitness costs are readily apparent in the form of infertility and hypothyroidism. Weak or sick nematodes or flies may not always be so easily recognizable, but as demonstrated by Gordon Lithgow (Novato, CA, USA) and co-workers, even under laboratory conditions, partial loss of function of DAF-2 results in dramatically reduced fitness as compared to wild-type worms[122]. However, a number of mutants have been described for which a price to be paid for longevity was not immediately obvious. For such mutants it is conceivable that in the wild there would be a strong selection in favor of the wild-type allele. It is now clear that this may indeed be the case. For example, the *age-1* mutation in nematodes, which affects the same pathway as *daf-2*, shows no obvious effects on fitness under standard laboratory conditions. However, the Lithgow lab exposed mixtures of *age-1* mutant and wild-type worms to more natural conditions: cycles of starvation, which are quite common in the wild. A large reduction in the frequency of the mutant allele suggested a substantial difference in relative fitness, which explains the selection of the wild-type allele[123].

The strong effect of laboratory conditions on the uncovering of aging genes naturally begs the question of whether such 'laboratory' genes reflect genuine pro-aging pathways, acting under natural conditions, or are merely artifacts. It is far from sure that loci with major effects on longevity in laboratory strains of various species will show segregating allelic variation in natural populations. Indeed, even if we adopt the premise that the basic mechanisms that control longevity and determine the rate of aging are common to all multicellular organisms, it is possible that the observed effects on longevity of candidate gene variants will prove to be exclusively associated with laboratory strains and not found in the same animals in the wild. Common laboratory situations generally select for high early fecundity and shorten the lifespan of the organisms under study. For example, Richard Miller (Ann Arbor, MI, USA) and Steve Austad (San Antonio, TX, USA) compared wild-derived mice with laboratory mice of a mixed genetic background and found significantly higher mean and maximum lifespans in the wild mice. Therefore, studies that use laboratory organisms to identify aging-related pathways might identify genes that simply restore the organism's original lifespan. Such results may not be fully relevant to wild populations[124,125]. In this respect, even the benefits of calorie restriction—still the only intervention that seems to increase longevity in a large number of species—may be exclusively associated with laboratory animals, which in the wild would forage continuously and endure cycles of starvation. It is conceivable that calorie restriction is the norm in wild populations of mice[126]. The only valid strategy to test the possibility that genetic variation at candidate loci contributes naturally to phenotypic variance for longevity and aging phenotypes is to demonstrate such associations in the wild (see also Chapter 7). Nevertheless, the discovery of pathways that appear to control aging rate through a synchronized retardation of multiple aging processes is obviously of extreme importance. It suggests that the underlying causes of aging may not be as diverse as often suspected.

2.3.6 HOW DO AGING GENES CAUSE AGING?

How could mutations that lead to mild inhibition of growth, reproduction and energy metabolism readily retard aging? We have already seen that many of these aging genes direct trade-offs, which greatly depend on the environmental conditions. Based on the available evidence, a logical explanation would be that the activities of such pro-aging pathways are associated with either an increased generation of somatic damage or a reduced investment in somatic repair and maintenance. In keeping with the theory of antagonistic pleiotropy, it is easy to see why gene variants promoting growth and reproduction are generally favored by evolution, even if their activity would lead to a more rapid rate of somatic damage and a shorter lifespan. Under optimal conditions, investing more in growth and reproduction is a sure bet, taking the ordinary environmental risks into consideration. However, there are situations where reducing growth and reproduction does offer some selective advantage. For example, under certain conditions, such as famine, it would help the organism to shut off reproduction and put all the resources in somatic maintenance and survival. This is exactly what happens when nematodes enter the so-called dauer stage, a growth-arrested, stress-resistant stage, analogous to spore formation in bacteria or protozoa or hibernation in vertebrates. Such shifts in allocation of resources from growth and reproduction to somatic maintenance and repair are predicted by the disposable soma theory (see above).

Are pathways related to growth and reproduction the only candidate aging pathways? Whereas it is certainly possible that all aging pathways may fall into these functional categories, at this stage it is impossible to rule out the existence of other, functionally unrelated pathways. Unfortunately, we will never know because reducing beneficial activities of pathways at early age might be lethal or cause major developmental defects. However, an alternative to genetic screening in finding aging genes and pathways has now emerged in the form of various computational approaches. As we have seen in Chapter 1, genes and proteins exert their physiological functions as networks of interaction rather than individually. In protein–protein interaction networks, or interactome networks, most proteins interact with few partners, whereas a small but significant proportion of proteins, the so-called hubs, interact with many partners. Such networks are called scale-free, which is the norm in cellular networks. Scale-free networks are particularly resistant to random node removal but are extremely sensitive to the targeted removal of hubs[127].

Daniel Promislow (Athens, GA, USA) compared patterns of connectivity for subsets of yeast proteins associated with senescence. He found that proteins associated with aging have significantly higher connectivity than expected by chance, even when controlling for other factors also associated with connectivity, such as localization of protein expression within the cell[128]. Similar observations were made by others[129] and are consistent with the antagonistic pleiotropy theory for the evolution of senescence. Importantly, they offer an *in silico* tool to discover new aging genes. However, it should be realized that such *in silico* approaches are still in their infancy. For example, also on the basis of protein–protein interactions in yeast, it has been suggested that the most highly connected proteins are essential genes[130]. However, recent results on such networks in mammals (assembled on the basis of interaction data derived from the literature) suggested no such correlation between highly connected genes and lethality of these genes when ablated in either mouse or yeast[131]. As more complete protein network data are now becoming available for multiple organisms these issues will be resolved and it is certainly conceivable that future *in silico* searches will either confirm that all aging genes are involved in some aspect of growth, reproduction, and energy metabolism or uncover other, unrelated pleiotropic genes.

2.4 **Longevity-assurance genes**

If aging is the result of the harmful side effects of genes selected for advantages they offer during youth, such effects can be suppressed by the action of other genes, which we can call longevity genes or longevity-assurance genes. As mentioned above, it is conceivable that the emergence of aging genes and longevity genes goes hand in hand, under the influence of environmental constraints. Longevity genes promote or ensure organismal survival without necessarily playing a direct role in development or maturation. They may encode components of stress-response systems that have apparently evolved to postpone adverse effects associated with endogenous or environmental sources of macromolecular damage. Longevity genes may be highly conserved since endogenous and environmental damage is ubiquitous in all living systems. In particular, DNA-repair systems are ancient, since from the first replicators all living organisms had to cope with agents damaging their nucleic acids (see below and Chapter 4).

As already mentioned, regulation of stress-response mechanisms is almost universally associated with increased lifespan. In the nematode, it has been demonstrated that the upregulation of DAF-16 in insulin/IGF-1 mutants affects expression of genes that increase resistance to environmental stress, including the ability to detoxify ROS, which could be the mechanism underlying the increased longevity of these mutants. Indeed, of the more than 40 single-gene mutants in *C. elegans* that display increased longevity, all increase the ability of the worm to respond to certain types of stress, for example heat, ultraviolet (UV) radiation, and ROS[132]. One longevity gene, which encodes the OLD-1 transmembrane tyrosine kinase, is stress-inducible and increases longevity by its overexpression. The expression of OLD-1 appears to be dependent on DAF-16 and is required for the lifespan extension of *age-1* and *daf-2* mutants[133].

Among various stress responses the heat-shock response has received special attention in view of the capacity of heat-shock proteins to protect cells against protein aggregation. Protein aggregation is caused by disruption of protein-folding homeostasis, which in turn has several possible causes, including oxidative damage, heat, and some forms of genome instability. Heat-shock genes encode chaperones to prevent this and other types of adverse effect (see Chapter 4 for a more detailed discussion of chaperones). Extra copies of the gene encoding the heat-shock protein HSP-16 conferred stress resistance and longevity in the nematode[134]. Also in this case, the DAF-16 transcription factor was essential for lifespan extension conferred by *hsp-16*.

Overexpression of heat-shock factor (HSF-1), a transcriptional regulator of heat-shock genes, increased longevity by about 20%[135]. Inactivation of *hsf-1*, but also inactivation of *daf-16*, accelerated the aggregation of polyglutamine expansion proteins[136]. As will be discussed in more detail in the next chapter, such aggregation is often caused by the amplification of small triplet repeats in the DNA, a form of genomic instability. Hence, these results suggest that DAF-16 and HSF-1 both act to prevent or attenuate the effects of such genome-instability events. Although *daf-16* was required for *hsf-1* overexpression to extend lifespan, HSF-1 can function independently of DAF-16[136]. What this tells us is that upregulation of longevity genes, such as *daf-16* or *hsf-1*, per se is sufficient to increase lifespan. This does not rule out the possibility that the targets of some of these genes are to be found in metabolism with the explicit purpose of reducing molecular damage at the source rather than fixing its consequences.

In addition *Drosophila* longevity mutants, such as Methuselah, were found to display enhanced resistance to heat, oxidants, and starvation[97]. In the fly, increased expression of genes involved in protein repair—those encoding protein carboxymethyltransferase and methionine sulfoxide reductase—increase lifespan[137,138]. Heightened activity of the antioxidant enzymes CuZn- and Mn-superoxide dismutase has also been demonstrated to increase longevity in this organism[139,140]. Hence, in both nematodes and flies, increased lifespan is caused by increased resistance to stress, either orchestrated in response to reduced growth and energy metabolism, or engineered by manipulating the longevity genes themselves.

In the mouse there is also convincing evidence that both the dwarf longevity mutants and mice subjected to calorie restriction display increased resistance to stress, including antioxidant defense, DNA repair, and heat shock[141,142]. Interestingly, results from Richard Miller and co-workers[143] indicate that the increased resistance to stress in the dwarf mice can be demonstrated at the level of skin fibroblasts in culture. Such cells are resistant to multiple forms of cellular stress, including UV light, heat, paraquat, H_2O_2, and cadmium[143,144]. Recent results from my own laboratory indicate that dwarf mice show a significantly lower rate of mutation accumulation with age. To detect spontaneous mutations in different organs and tissues of these mice, Ana Maria Garcia, a postdoctoral researcher in the laboratory, crossed a transgenic mouse harboring a *lacZ*-plasmid reporter construct in its germ line into the homozygous Ames dwarf background. By recovering the *lacZ* plasmids from genomic DNA of the hybrids and their wild-type littermate controls she was able to compare the mutation frequencies at the reporter locus (for a description of this system see Chapter 6). The results indicate a significantly lower mutation frequency in dwarf mice than in controls. She also subjected these mice to caloric restriction and demonstrated that this also reduced spontaneous mutation frequency, albeit to a lesser extent than dwarfism (Fig. 2.4). These results are in keeping with the delayed occurrence of total neoplastic lesions observed in Ames dwarf mice, which is a major contributing factor to the extended lifespan of these animals[113].

One reason why mutations conferring lifespan extension based on the inhibition of growth and reproduction show up so readily in a genetic screen could be the selective advantage of mechanisms to delay reproduction and increase survival, as originally proposed by Harrison and Archer[145] and Holliday[146] to explain the phenomenon of caloric restriction. While the possibility to temporarily halt or attenuate growth and reproduction would be advantageous, actual mutants with continuously downregulated insulin signaling would not survive for many generations, as we have seen from the results obtained by Lithgow (see above). What could be the mechanism underlying the upregulation of multiple longevity genes as a consequence of reducing growth and metabolism? To bridge the gap between metabolism and somatic maintenance, Leonard Guarente

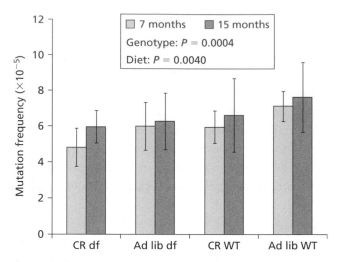

Fig. 2.4 Ames dwarf mice (df) and calorically restricted (CR) mice, both long-lived, display a lower spontaneous mutation frequency in different tissues compared to their littermate controls (WT, wild type). Here, only the data for the kidney are shown. The effect of dwarfism is stronger than that of caloric restriction. For details about the measurement of spontaneous mutation frequencies in these mice, see Chapter 6. (A. Garcia, submitted for publication.)

(Cambridge, MA, USA) proposed another master regulator, this time the silencing information regulator 2 (Sir2). Sir2 is the founding member of a phylogenetically conserved family of nicotinamide-adenine dinucleotide (NAD)-dependent histone deacetylases (HDACs), called sirtuins[147]. Sir2 is required for transcriptional silencing—transcriptional inactivation by altering chromatin structure through deacetylation of histones (Chapter 3)—at the ribosomal DNA (rDNA; the genes coding for rRNA), telomeres (the physical ends of linear eukaryotic chromosomes; Chapter 3), and mating-type loci (regions that differ in DNA sequence between cells of opposite mating-type) of yeast[148]. The specialized chromatin structure associated with transcriptional silencing is also repressive to recombination. Indeed, Sir2 also suppresses intrachromosomal recombination within the rRNA gene arrays, a process demonstrated by Sinclair and Guarente to lead to an accumulation of rDNA repeats in the form of extrachromosomal circles (ERCs)[149]. ERCs are self-replicating and are preferentially retained in the mother cells, eventually causing aging by limiting yeast replicative lifespan. Increased Sir2 dose extends yeast replicative lifespan by about 40%, whereas loss of function reduces natural longevity[150].

Other forms of increased genome instability during replicative aging of yeast, unrelated to rDNA loci or Sir2, have been observed. Most notably, evidence has been obtained for so-called loss-of-heterozygosity events by studying the loss of marker loci affecting colony color[151]. The steep increase of these loss-of-heterozygosity events was explained by an increasingly impaired ability to correctly detect and repair DNA double-strand breaks (DSBs), a major form of DNA damage in all organisms and one that can be induced by reactive oxygen (Chapter 4). As we will see in Chapter 6, such forms of genome instability are difficult to detect in mammals. On the other hand, ERCs should be easily detectable, but have not been observed in higher eukaryotes. However, extra-chromosomal circular DNA is not uncommon in mammalian tissues where it also is a sign of genomic instability[152].

Apart from yeast, Sir2 orthologs have been implicated in lifespan regulation of *C. elegans* and mammals. Tissenbaum and Guarente demonstrated that extra doses of Sir2.1 in the nematode extend adult lifespan by 50%[153]. This extension of life appeared to be dependent on DAF-16, linking Sir2 to the insulin-signaling pathway. Initially it was thought that Sir2 mediated the silencing of genes in the insulin-signaling pathway in the nematodes; that is, upstream of DAF-16. In mammals orthologs of DAF-16 are the four FOXO proteins, a subgroup of the Forkhead family of proteins, transcriptional regulators characterized by a conserved DNA-binding domain termed the forkhead box[154] (see Chapter 3). It turned out that FOXO itself is a direct, functional target for Sir2[155]. In mammalian cells, Sir2 (SIRT1 in humans) can deacetylate FOXO, increasing its ability to induce cell-cycle arrest and resistance to oxidative stress, but inhibiting FOXO's ability to induce apoptosis[156,157]. Similarly, SIRT1 can deacetylate p53 and attenuate its transcriptional activity[158]. Hence, Sir2 seems to specifically modulate activities that contribute to survival. The role of FOXO in increasing cellular defense is underscored by the observation that, in rat fibroblast cells, this transcription factor induces the repair of damaged DNA and upregulates the growth arrest and DNA damage response gene *Gadd45a*[159].

How can Sir2 regulate survival in different species? A possible answer can be found in the NAD hydrolysis, an integral step of the deacetylation reaction carried out by Sir2 and other sirtuins, which leads to the consumption of one NAD molecule for each deacety-lated lysine residue. It has been shown that the prolongevity effect of caloric restriction—accomplished in yeast by limiting glucose availability—is lost in the absence of Sir2[160]. This suggests that calorie restriction slows aging by activating Sir2. Guarente *et al.*[160], then, proposed that Sir2 may connect energy metabolism to lifespan through the absolute requirement of Sir2's activity for NAD. If cellular NAD levels are low, then Sir2's deacety-lase activity could be attenuated and vice versa. As demonstrated by the Guarente laboratory, calorie restriction decreases the levels of NADH, a competitive inhibitor of Sir2. Therefore, an increased NAD/NADH ratio in calorie restriction could underlie the increased Sir2 activity, which in turn would repress recombination at the rDNA locus, thereby slowing the formation of toxic rDNA circles and increase lifespan. This is exactly what has been shown[160]. However, it is not easy to determine the effective concentrations of NAD and NADH in living cells and the role of physiological variation in NADH in affecting Sir2 activity is controversial[161]. Also, the role of Sir2 itself as a key factor in caloric restriction has been disputed[162].

Interestingly, an increase in Sir2 activity by calorie restriction has been demonstrated in mammalian cells: expression of the ortholog of Sir2, SIRT1, is induced by calorie restriction in rats, as well as in human cells treated with serum from these animals[163]. It was also shown that SIRT1 suppresses apoptosis, suggesting that the induction of SIRT1 by calorie restriction represents a survival response, similar to its deacetylation of FOXO and p53 as described above.

Whether or not Sir2 is a master regulator, as some believe it is, there can be no doubt that intricate mechanisms have evolved to regulate the survival of an organism as part of its life-history strategy. As we have seen, these longevity mechanisms can be upregulated directly or by dampening pro-aging pathways through a reduction of growth and reproductive efforts. A more detailed understanding of the nature of the damage that limits our lifespan, and the evolutionary history of its interaction with the longevity systems that emerged to counter its effects, may give us insight into the causes of aging.

2.5 Somatic damage and the aging genome

Somatic damage accumulation appears to be a general characteristic of living organisms. In a broad sense such damage can vary from infectious agents to DNA damage and mutations. Exposure is dependent on a host of variables, including species-specific endogenous and ecological factors. Nevertheless, some universal principles basic to life or at least basic to most eukaryotes suggest similarities in exposure. For example all aerobic organisms are exposed to ROS. Many organisms are exposed to body heat and such environmental agents as UV and ionizing radiation. Such similarities in exposure to damaging agents is in keeping with the universality of pathways of life extension originally uncovered in nematodes and the resistance of virtually all these longevity mutants to ROS, heat, and other damaging agents.

The most ancient example of damage accumulation in the living world is damage to nucleic acids, first RNA then DNA. Genetic damage posed both a fundamental problem and an opportunity for living systems. A problem, because genetic damage essentially prevents the perpetuation of life, since it interferes with replication (and transcription); an opportunity, because it allows the generation of genetic variation through errors in replicating a damaged template. Similar to the trade-offs mentioned above between reproduction and somatic maintenance, the relative stability of a genome is optimized to the life history of the organism. The necessity of mutations from the first replicators onwards is a strong argument to consider DNA damage and genetic errors as the original instigators and major drivers of aging in the living world.

A second, logical argument to consider the DNA of the genome as the Achilles heel of an aging organism is the irreversibility of unwanted sequence changes in view of the lack of a back-up template. This is in contrast to proteins, which at least in principle can be easily replaced, with the corresponding gene as the template. Indeed, the maintenance of genomic DNA is of crucial importance to survival because its alteration by mutation is essentially irreversible and has the potential to affect all downstream processes. Third, as explained in Chapter 5, there is now overwhelming evidence that in both humans and mice heritable defects in genes involved in maintaining the integrity of the genome cause symptoms of premature aging.

The first replicating nucleic acids, almost certainly RNA, evolved almost 4 billion years ago in an environment with little molecular oxygen, but high fluxes of UV radiation due to the absence of an ozone layer. Based on this constant threat of DNA-damaging agents and the absolute need to increase replication fidelity from the primitive RNA-replicating systems to the much larger RNA and DNA genomes, it is now generally believed that recombinational repair was the first bona fide DNA-repair system that evolved[164,165] (see also Chapter 4). Although mutations are essential as the ultimate source of genetic variation upon which evolution depends, too many mutations will reduce fitness and can lead to population extinction. As described above, due to Muller's ratchet[73], asexual lineages will tend to lose mutation-free genomes (due to genetic drift) and inevitably suffer from loss of viability. This is especially relevant for small populations and relatively easy to demonstrate in simple unicellular organisms, such as *E. coli*. For example, Kibota and Lynch demonstrated that deleterious mutations of small effect escaped selection in *E. coli* lines with repeated population bottlenecks, resulting in decreased fitness[166]. Elena and Lensky, also in *E. coli*, demonstrated that randomly introduced mutations interact to negatively affect fitness; that is, the relationship between mutation number and decreased fitness was nearly log-linear[167].

Is it realistic to interpret loss of fitness caused by mutation accumulation in unicellular organisms as a parallel to aging of somatic cells of metazoa? As described by Graham Bell (see above), senescence-like phenomena have been observed in asexually propagated protozoa lineages, and were attributed to the accumulation of deleterious mutations[69]. Bell explained the recombination of genetic material between different lineages—sexual outcrossing—in terms of an exogenous repair mechanism functioning by creating variance on which selection can act effectively to reduce mutational load. However, Bell has argued against a parallel between mutation-induced senescence in protozoa and aging of somatic cells in metazoa. Although he did not rule out a role for mutation accumulation in aging of somatic cells of a mammal, he considered it unsound because senescence will evolve—as a by-product of selection for increased early reproduction—in the soma, but not the germ line of metazoa. In protozoa, germ cells and somatic cells are of course identical, but there is no *a priori* reason why Muller's ratchet would not apply to somatic cell lineages as well.

Mutations have been with us since the origin of life and were the primary condition for the evolution of the wide variety of species on our planet. Indeed, organisms with sufficient tolerance of their genome maintenance systems to allow for a certain level of mutations have the best chance to survive environmental challenges. Visible examples of this strategy are various infectious diseases in which the bacteria or viruses have the capacity to evade drugs and the immune system by rapidly altering their antibiotic resistance or cellular receptor repertoire. Indeed, pathogenic species, such as *Streptococcus pneumoniae*, harbor more plastic genomes with more potential for host adaptation than bacteria adapted to a non-threatened lifestyle, such as *Streptococcus thermophilus*, used for the manufacture of yogurt and cheese[168]. In this particular case the difference is due to the lack in the deadly pneumococcus of two enzymes involved in the repair of DNA DSBs (see Chapter 4). This situation is very similar to cancer. Tumor cells often gain the capability, like pathogenic microorganisms, to rapidly alter their genotypes, thereby creating new attributes to grow more efficiently, metastasize, and evade both the immune system and various drug treatments. While in microorganisms increased genetic variation is facilitated by stress-induced mutagenesis[169], in human cancers mutator phenotypes caused by the inactivation of genome-maintenance systems temporarily result in an increased opportunity to adapt to adverse conditions[170].

Whereas mutations in combination with recombination provide innovation at the population level, they can cause individual failure and decreased fitness. Organisms have several ways to protect themselves against mutations or their consequences. There is first of all a host of genome-maintenance systems acting to prevent mutations by efficiently, and most of all accurately, repairing chemical DNA damage. However, as we will see in Chapter 4, whereas these systems are able to repair most if not all of the various types of DNA damage continuously induced in living cells, they are imperfect. Indeed, apart from DNA-synthesis errors per se, virtually all mutations are due to so-called error-prone repair; that is, mistakes made during the repair of such chemical lesions. A less obvious defense against mutations and their consequences is the fault-tolerant way genetic information is encoded; that is, in a redundant manner with multiple copies of a gene or with different genes or pathways that can carry out similar functions. A third way to limit the impact of mutations is by avoiding extreme optimization of function, which implies tolerance for defects. Finally, organisms use so-called genetic capacitance, which involves proteins that buffer the effects of mutations, such as the heat-shock proteins or molecular chaperones that were demonstrated to increase nematode lifespan (see above).

Overall, therefore, organisms are robust: they are able to maintain function against various perturbations. Nevertheless, even robustness that is ubiquitous in biological systems has inherent trade-offs and failure patterns. These become especially apparent in multicellular organisms. While unicellular organisms can out-select mutation accumulation, in multicellular organisms this is only true for the germ line. Indeed, in animals germ-line cells are set aside very early in life, after which there is ample opportunity for selection against gametes and gamete combinations carrying deleterious mutations. (Plants do not set aside a germ-line early on in life, but instead rely on stringent selection during the haploid gametophytic phase and selection during somatic growth when cell lineages carrying deleterious mutations are impaired in growth and development and therefore less likely to contribute to gamete formation[171].) Even then, signs of system failure are evident even very early in life. In humans, for example, this translates in the relatively high percentage of spontaneous abortions and stillbirths, which are most likely caused by mutations[172] (see Chapter 6). It is estimated that every human newborn has acquired as many as 100 new mutations per genome, three of which might be deleterious[173].

In the somatic genome of multicellular organisms selection against deleterious mutations is more difficult than in the germ line. Obviously, cell loss or cellular loss of function due to deleterious mutations will contribute to functional decline of the organ and may cause disease, most notably cancer. In theory, optimization of genome-maintenance mechanisms for only the somatic cells could prevent any significant mutation accumulation. However, evolutionary theory would not predict a maximization of cellular maintenance and repair in view of the gradual decline of the force of natural selection after the age of first reproduction (see above). In most cases, our fault-tolerant genomes would allow mutation accumulation well into the reproductive period without appreciable loss of function. However, it is unlikely that even slight increases in mutation loads thereafter would have no consequences.

An observation that has often been used to discard the idea that mutation accumulation could adversely affect physiological functioning and cause aging and death is the identical lifespans of haploid and diploid wasps of the genus *Habrobracon*[174]. As in other hymenopterans, in *Habrobracon* unfertilized eggs become haploid males and fertilized eggs that are homozygous and heterozygous at the sex locus develop into diploid males and females, respectively. While X-irradiation of larvae, pupae, or adults reduced lifespan of the haploid variant more than that of its diploid counterpart, their normal aging rate was the same. Somewhat surprisingly, the results of this isolated observation, which has not been confirmed in other species for which both haploid and diploid variants exist, has been widely interpreted to discard the somatic mutation theory of aging.

As argued by Alec Morley (Adelaide, Australia) in a lucid paper not cited frequently enough, the *Habrobracon* results, even if reproducibly obtained in multiple species, do not exclude a causal role in aging of dominant or co-dominant mutations, which may be more important in terms of functional significance upon their random induction in somatic cells[175] (see also Chapter 6). Most importantly, although the authors assumed that mutations were the cause of the reduced lifespan in the wasps after X-irradiation, this is unlikely to be true. Indeed, adverse effects after an acute dose of X-rays are more likely to be caused by the toxic effects of DNA damage rather than mutations. As we shall see in Chapters 4 and 6, mutations are the result, mainly, of erroneous processing of DNA damage; their introduction requires time. Acute effects of DNA damage are likely to be mitigated by an additional copy of each gene as in the diploid situation. The absence of a

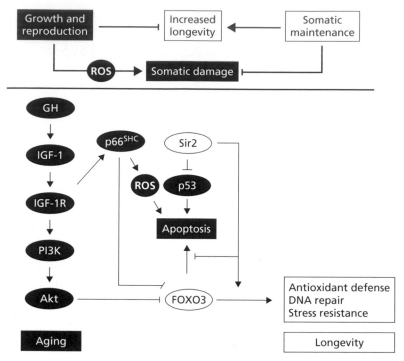

Fig. 2.5 Some of the factors involved in the genetic control of longevity and aging. Black represents pro-aging gene products for which reduced function or loss of function increases lifespan. In white are the pro-longevity gene products, increased expression of which extends lifespan (and vice versa). Arrows indicate agonists, and bars indicate antagonists. See text for further details. IGF-1R, IGF-1 receptor. Re-drawn with permission from ref. 443.

difference in lifespan between the unirradiated haploid and diploid animals is not inconsistent with a causal role of dominant mutations in aging. Indeed, as will be argued later (Chapter 7), mutations in the form of large rearrangements could easily lead to genome destabilization and an aberrant pattern of gene expression. This would be the case for both haploid and diploid genomes. An in-depth investigation of the haploid versus diploid aging wasps would be necessary to test this hypothesis.

In summary, from the earliest replicators onwards the logic of life with its need for evolvability has dictated the logic of aging. Species-specific lifespans reflect the balance between growth and reproduction and somatic maintenance and repair (Fig. 2.5). Somatic maintenance and repair initially only involved the need to preserve the genome, but later also other cellular and organismal structures became important. It is possible that even with the emergence of billions of different species of increasing complexity this basic principle is still valid and the genome may still be the most relevant target of the aging process. What has changed, however, is the complexity of genome structure and function, and with it the diversity of threats to its integrity. This will be discussed in the next chapter.

3 Genome structure and function

As described in Chapter 1, the science of aging has evolved in concert with an emerging insight into both the logic of life and its structure–function relationships. Insight into the logic of life gave us insight into the logic of aging, as outlined in Chapter 2. However, whereas we know why we age, the mechanistic basis of the process as it takes place in different species is still unclear. Full knowledge of how we age can only emerge hand in hand with the further uncovering of the basic principles underlying the nature of the living world. In the life sciences it is now slowly being realized that the unraveling of how life is ordered cannot be separated from enquiries into what causes its progressive disorder and ultimate demise. Genome structure and function are central to the organization of life. They determine our species-specific characteristics and provide the conditions for individual development and maturation. As such, the genome may bear the roots of its own destruction and with it the causes of age-related cellular degeneration and death.

The genome, a term coined in 1920 by Hans Winkler, a Professor of Botany at the University of Hamburg, to designate the haploid chromosome set, has more recently been considered to mean the whole complement of genes of an organism and sometimes also the sum total of coding and non-coding DNA sequences (see also Chapter 1). Here I will define a genome in somewhat broader terms as the organelle or physical entity that carries out all genetic transactions in a cell or organism. This necessitates a three-dimensional view of the genome as a coordinated ensemble of gene action in the context of a series of structure–function relationships, which ultimately represents the complete network of processes that defines a living organism. Such a genome interacts extensively with other molecules.

To understand how the genome exerts its function as the ultimate determinant of both order and progressive disorder it is necessary to know its structure and functional organization. There are recent textbooks and extensive review articles to which I refer for comprehensive information on various aspects of genome structure and function[176,177]. Here, I will focus predominantly on those components of genome structure that have the potential to drive its demise in somatic cells over time, based on our current knowledge of the sources of somatic damage in aging organisms and mechanisms of genomic instability. To this end, I will address genome structure and organization at three levels: first, DNA primary structure and sequence; second, higher-order DNA structure; and third, nuclear architecture. In all cases the focus will be on the impact of these structural levels on cellular function, especially patterns of regulation of gene transcription, which will be discussed in the last section of this chapter.

3.1 **DNA primary structure**

As discussed in Chapter 1, the landmark discovery of the double-helical structure of DNA by Watson and Crick in 1953 provided a logical basis for the role of the genome in both the perpetuation of life in evolutionary time and its functional organization (see Fig. 1.4). The DNA of the genome is a polymer consisting of four different types of monomer units, termed nucleotides. Each nucleotide consists of a five-carbon sugar (deoxyribose), a nitrogen-containing base attached to the sugar, and a phosphate group. The differences between the four nucleotides are determined by the different bases: adenine (A), guanine (G), cytosine (C), and thymine (T). Adenine and guanine are the purine bases and the larger of the two types, with cytosine and thymine the relatively small pyrimidine bases. RNA is different from DNA, since its nucleotides contain a ribose instead of a deoxyribose as the sugar, and uracil (U) replaces the thymine base. Apart from their major—amino and keto—tautomeric forms, the DNA bases can adopt alternate forms; the imino form for adenine and cytosine and the enol form for guanine and thymine. Although these structural isomers are rare, they do alter base-pairing properties, which makes them a source of mutations when accidentally present during DNA replicative or repair synthesis.

The monomer units in nucleic acids are connected through the phosphate residue attached to the hydroxyl group on the 5' carbon of one unit and the 3' hydroxyl on the next one (Fig. 3.1). This forms a phosphodiester link between two residues (shown as p; e.g. CpG), which can lead to very long nucleic acids—up to billions of units. The heteropolymeric character of nucleic acids is the basis for their role in information storage and transmission in the form of a base sequence code of nucleotides. As discovered by Watson and Crick[12], the DNA of the genome is a double-stranded macromolecule, with a double-helical structure of two polynucleotide chains, held together by specific pairing of AT and GC base pairs stacked on one another with their planes nearly perpendicular to the helix axis (Fig. 3.2). This configuration allows strong van-der-Waals interactions between the bases.

DNA has both a sense of direction and individuality. The phosphodiester linkage between monomer units is always between the 5' carbon of one monomer and the 3' carbon of the next. The two strands in the double-helical model run in opposite directions (Fig. 3.2). Individuality is, of course, determined by the sequence of its bases; that is, the nucleotide sequence. It is in the primary structure of a nucleic acid that genetic

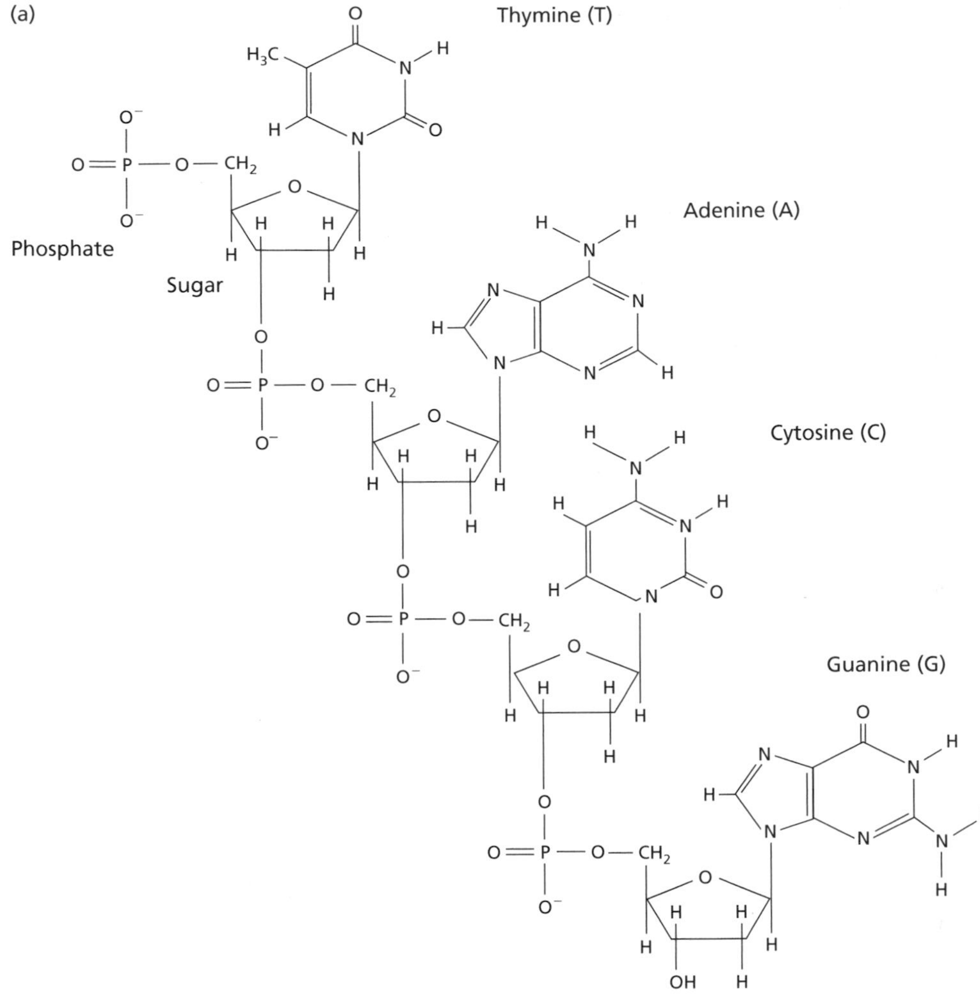

(a)

Thymine (T)

Adenine (A)

Cytosine (C)

Guanine (G)

Phosphate

Sugar

Fig. 3.1 Primary structure of DNA.

information is stored as a four-letter code, encoding genes and gene-regulatory or structural sequences.

Polynucleotides are thermodynamically unstable *in vivo* but their hydrolysis is exceedingly slow unless catalyzed. Similarly, whereas dehydration would readily allow adding nucleotide residues to a nucleic acid chain, the thermodynamics of this reaction are unfavorable and require high-energy nucleoside triphosphates (ATPs). Hence, both the breakdown and synthesis of nucleic acids require enzymatic processes; in their absence DNA is sufficiently stable to serve as a useful repository of genetic information. Nevertheless, DNA is not completely stable under physiological conditions. Apart from errors during its replication, DNA can be damaged through its interaction with a variety of reactive chemicals

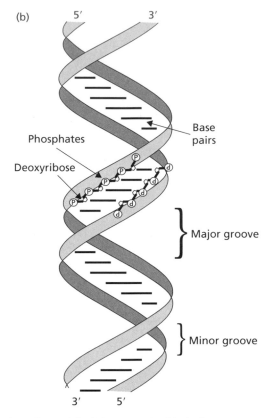

(b)

5′   3′

Base
pairs

Phosphates

Deoxyribose

} Major groove

} Minor groove

3′   5′

Fig. 3.2 The Watson and Crick model of the DNA double helix.

or radiation. Even in the absence of environmental challenges, the DNA in each cell of an organism suffers from a multitude of endogenous damage, resulting from spontaneous hydrolysis and oxidation (in aerobic organisms). Spontaneous DNA damage, DNA repair, and genome instability are discussed extensively in Chapters 4 and 6.

Nucleic acids can have several secondary structures, including the right-handed A and B helices, which differ mainly in the position of the bases with respect to the helix axis, and the left-handed Z helix, discovered by Alexander Rich (Cambridge, MA, USA) and co-workers in 1979[178]. Probably, most of the DNA in the aqueous milieu of cells *in vivo* is in the B form, whereas the A conformation is adopted by double-stranded RNA and DNA–RNA hybrids. The surface of the B helix contains two different grooves, called the major and minor groove (Fig. 3.2). Proteins that interact with DNA often make contact in these grooves, especially the major groove, which is larger and more suitable for sequence-specific recognition. Unfortunately, since these areas contain an abundance of reactive sites they are also often the target of DNA-damaging agents (see Chapter 6). Z-DNA occurs in regions of polynucleotides having alternating purine–pyrimidine sequences with a substantial fraction of the C residues methylated to form 5-methylcytosine.

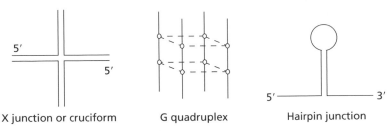

5'

5'

5' —————————— 3'

X junction or cruciform   G quadruplex   Hairpin junction

Fig. 3.3 Examples of multi-stranded DNA structures as they can occur *in vivo*. In the G quadruplex the four DNA strands can have different polarities, from all strands parallel to alternating antiparallel strands.

This modification of C occurs to a significant extent in natural DNA, where it plays a role in such transactions as the regulation of gene transcription (see below).

Apart from double-helical DNA, the genome *in vivo* may also contain multistranded DNA structures such as triplexes, quadruplexes, and junctions (Fig. 3.3). Some of these, for example Holliday junctions, can be intermediates of DNA transactions, such as homologous recombination (Chapter 4). A large variety of others may form spontaneously in an aqueous milieu and could have important biological implications. Examples of genomic DNA sequences with the potential to form such structures are the telomeric regions at the end of chromosomes and the so-called promoter regions that regulate gene action (see below). Most of these sequences have continuous stretches of guanine nucleotides, and the ability of the guanine base to form tetrads (DNA tetraplexes or G quadruplexes), with four guanines in a plane, lies at the crux of these complex structures. Among these, the structures formed by the telomere sequences have been the most widely investigated in relation to aging, especially in relation to the gene defects causing Werner syndrome, a segmental progeroid syndrome due to the inactivation of the *WRN* gene, a RecQ helicase thought to be involved in resolving such structures (Chapters 4 and 6).

3.1.1 GENES IN GENOMES

As we have seen in Chapter 1, the genes, hidden in the primary structure of the DNA of the genome, give rise to phenotypes through the generation of corresponding protein products by transcription and translation. However, the relationship between genes as the units of inheritance and their physical structure has become somewhat nebulous in view of the recognized need for accessory DNA sequences to regulate transcription. Indeed, genomes are complex and represent more than a bag of genes. This is illustrated by the results of the comparative analysis of the hundreds of prokaryotic and eukaryotic genomes that have now been decoded, including the complete sequences of five

Table 3.1 Genome features of several organisms

Organism	Approx. genome size (Mb)[a]	Diploid no. of chromosomes	Approx. no. of genes[b]	Human gene homologs (%)[c]
Homo sapiens	3400	46	30 000	–
Pan troglodytes	3690	48	30 000	87
Mus musculus	3450	40	30 000	79
Caenorhabditis elegans	100	12	19 100	31
Drosophila melanogaster	180	8	13 600	39
Saccharomyces cerevisiae	12	32	6300	11
Escherichia coli	4.6	1	3200	2

[a]*Source*: www.genomesize.com

[b]*Source*: www.ornl.gov/sci/techresources/Human_Genome/faq/compgen.shtml

[c]Percentage of human genes with homologs in the organism of interest. *Source*: http://eugenes.org/all/hgsummary.html

mammals: human, mouse, rat, dog, and cow. Genomes can vary in size from about 0.5 million bp (0.5 megabases or Mb) in *Nanoarchaeum equitans*, the smallest genome of a true organism yet found, with slightly more than 500 genes, to 3000 Mb in the human genome, with about 30 000 genes (see www.ncbi.nlm.nih.gov/genomes/ for all genome-related databases). In general, such comparative analysis has revealed increases in complexity from prokaryotes to multicellular eukaryotes.

Whereas genome size appears to be generally associated with organismal complexity this is much less obvious for the number of genes. For example, whereas humans have only about 10 000 more genes than *C. elegans*, the human genome is 30-fold larger than that of the nematode (see Table 3.1 for a comparison of genome features among selected species). Virtually all of the increase in genome size from prokaryotes to mammals is caused by the addition of non-genic DNA. Whereas *E. coli* has about 4000 protein-coding genes comprising almost 90% of its total sequence, in the human (or mouse) genome only slightly more than 1% is protein-coding sequence. Whereas there is an ongoing debate about the possible function of much of this non-genic DNA, one explanation that has been put forward for the origin of genome complexity in vertebrates is the enormous reduction in population size associated with the emergence of higher eukaryotes of larger size. This would permit an initially non-adaptive restructuring of eukaryotic genomes by genetic drift, providing novel substrates for the secondary evolution of phenotypic complexity[179]. However, it should be realized that genome size varies enormously among organisms without any obvious relationship to complexity. For example, *Amoeba dubia*, a small protozoan species of indisputably lower complexity than *Homo sapiens*, has a genome size of 670 billion bp, more than 100-fold the size of the human genome. Before drawing the conclusion that these figures demonstrate that large genomes and complexity are not associated it should be realized that these extraordinary large genome sizes are not based on complete sequence information but on biochemical measurements. In such cases genome size may be greatly exaggerated since they do not account for

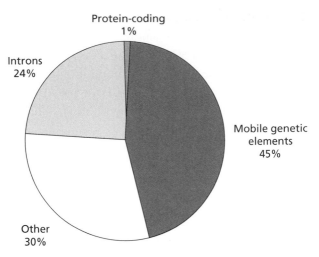

Fig. 3.4 Organization of the DNA sequence content of the genome. Other types of sequence, apart from mobile genetic elements and genes, are the various families of tandem repeats, non-coding RNAs, pseudogenes, and some as yet still unidentified DNA.

mitochondrial DNA (mtDNA), the amount of which can be substantial, and a sometimes high levels of polyploidy (see the Animal Genome Size Database, www.genomesize.com).

Non-genic DNA in the mammalian genome includes the relatively small 5' and 3' untranslated regions flanking the genes, spliceosomal introns, dispersed within genes and comprising as much as 95% of the genes, mobile and repetitive genetic elements, tandem repeats at centromeres and telomeres, and non-coding RNA genes (Fig. 3.4). Whereas in the past much of this type of DNA has been considered as junk DNA (because there seemed to be no obvious benefit to the host) or selfish DNA (because its only function seems to be to make more copies of itself), it is now clear that at least a sizable part of it must be functional[180]. This is suggested by the evolutionary conservation of many non-genic DNA sequences.

As explained in Chapter 1 once complete genome sequences were available it became possible to compare them in their entirety (e.g. human and mouse) for sequences of high similarity as a result of negative selection. Inter-species studies of homology revealed that a large part of the human or mouse genome—as much as about 20%—is subject to evolutionary constraint[180]. It should be noted that the situation for families of repeat elements (discussed below) has its own peculiarities. Indeed, such families, which are abundantly present in mammalian genomes, can be highly divergent between species. However, there is a high level of sequence conservation between repeats, which can include genes such as rRNA genes, within a species. This is called concerted evolution and based on what has been termed genetic drive[181]. At least some repeat families, certainly the satellites found at centromeres (see below) have a highly conserved biological function (in spite of their high divergence between species).

Interestingly, whereas high levels of conservation were found (as expected) in protein-coding parts of genes, 1–2% of the human genome represents equally highly conserved, non-genic sequences. The distribution of such sequences is negatively correlated with the distribution of genes[182]. This suggests that their function either involves long-distance regulation of gene expression or some other important role, such as in chromosomal transactions or in the interaction of the genome with other cellular and nuclear structures (see below).

Another characteristic feature of complex genomes, such as the human genome, is that while only about 1% may encode proteins, a much larger part (perhaps almost 50%[180]) is transcribed. It is possible that at least a part of these non-coding RNA molecules function in the regulation of gene transcription. For one category of non-coding RNAs (see below), the so-called miRNAs, this has now been demonstrated. What this picture of the genome suggests is that the structural and physiological complexity of organisms is dependent not so much on their genes, but on the way the activity of these genes is regulated. This can be very different from species to species even if most of their protein sequences and core regulatory regions are virtually identical. Indeed, the regulation of gene expression takes place at many levels, of which the relative position of genes and regulatory regions is of critical importance. As noticed by the late Alan Wilson (1934–1991), among placental mammals the variation in karyotype (chromosome number, large inversions or deletions) is much more dramatic than in the primary DNA sequence[183]. Indeed, based on early cytogenetic comparisons one would never believe that the mammalian genome is so highly conserved at the sequence level. During evolution of the different mammalian lineages, many genome rearrangements have scrambled the relative positions of genes and other homologous sequences beyond recognition. As noticed by several authors, including those active in the science of aging[184], it is unlikely that the considerable differences between human and chimpanzee, not least the approximately 2-fold difference in maximum lifespan, are due to differences in their protein sequences. Instead, there is ample reason to believe that much of the phenotypic difference between closely related species is explained by evolutionary changes in gene regulation. Apparently, a large number of genome rearrangements can occur over a relatively short evolutionary time span.

3.1.2 GENE DISPERSION AND MODULAR ORGANIZATION

In the mammalian genome, genes are clustered densely into small islands in desert-like expanses containing few or no genes. While their protein-coding sequences occupy only about 1% of the entire DNA sequence, their interruption by introns allows them to be spread over huge distances. Introns, which comprise about 24% of the genomic sequence, are non-coding stretches of DNA that are transcribed, but eventually excised from the primary transcript, thereby fusing the remaining parts, termed exons into the final, protein-coding mRNA. Introns almost never appear in prokaryotic cells and are rare in single-celled eukaryotes, but in multicellular animals and plants almost every gene has introns. Both terms—intron and exon—were coined by Walter Gilbert (Cambridge, MA, USA) in 1978, who postulated that exons were originally minigenes corresponding to current protein domains (structurally and functionally defined, semi-independent parts of a protein)[185]. At a later stage in evolution, the minigenes were assembled to make whole genes, with introns as the functionless pieces that held the exons together. Bacteria may have lost introns in later evolutionary stages. Now we know that exons do not always map to those domains. Indeed, another school of thought argues that introns are a relatively recent arrival in the eukaryotic lineage and necessary to help generate the diversity of regulatory mechanisms that are required to control gene expression in multicellular, highly differentiated organisms[186]. In this view, prokaryotes do not have introns because they never had them in the first place.

At first sight introns seem to impose a selective disadvantage on their host genes by increasing the chance of mutation to yield defective alleles. On the other hand, introns provide for alternative splicing (making different combinations of exons), which is advantageous to eukaryotes because it allows single genes to encode multiple protein iso-forms, a major source of genetic diversity. Introns may also have other functions, including a role in nucleosome formation and in anchoring chromatin loops to the nuclear matrix and to chromosome scaffolds (see below). The modular nature of genes in higher eukaryotes with so many introns has led to the substitution of the term transcription unit for the original gene or cistron, which remains valid in prokaryotes. The concept of a gene in mammalian genomes is blurred, not only by alternative splicing, which creates different proteins from one gene, but also by the discovery that different genes can physically overlap and that sometimes an exonic sequence in one gene is part of an intron in another.

3.1.3 MOBILE GENETIC ELEMENTS AND DISPERSED REPEATS

Originally discovered by Barbara McClintock (1902–1992) in maize[187], mobile genetic elements, such as Alu and L1 sequences, make up about 45% of the human genome. Mobile genetic elements are pieces of DNA that can move from one place in the genome to another. By insertion into a gene, they can be the cause of the mutations responsible for some cases of human genetic diseases, including hemophilia[188]. Like for introns there has been a debate as to whether such sequences have a function or should be merely considered as junk DNA. There is ample evidence that mobile genetic elements can contribute to genome evolution by providing regulatory elements to neighboring genes or by helping to create new combinations of exons, promoters, and enhancers[189]. It is also possible that these repeat elements have some structural role and help providing the necessary physical organization required for effective, integrated genome functioning.

There are four main families of mobile genetic elements in the human genome. First, the long interspersed nuclear elements (LINEs), of more than 6 kb; second, small interspersed nuclear elements (SINEs), of less than 500 bp, such as Alu; third, the retrovirus-like transposons; and fourth the DNA transposons. The first three proliferate via an RNA intermediate; only DNA transposons are real jumping genes, moving from one place to another by using a cut-and-paste mechanism. Here I will only briefly discuss the LINE1 element, also termed L1.

The L1 retrotransposon is the most abundant dispersed repeat sequence, comprising over one-third of the entire human genome[190]. This sequence generates a copy of itself, through transcription followed by reverse transcription, which then integrates elsewhere. About 100 L1 elements in the human genome are still active to retrotranspose and are therefore an ongoing source of mutations. Because L1 elements are so widespread and present in so many copies, they are also a target for erroneous homologous recombination (see Chapter 4 for a detailed description). In view of these adverse effects, one might wonder why L1 repeats, in spite of their evolutionary benefits, are tolerated in the genome and have not disappeared.

Apart from the fact that the efficacy of natural selection would be insufficient to prevent the spread of this evolutionary useful sequence—provided the activity of this aggressive mutator is dampened sufficiently long to allow its host to reach the reproductive age—L1 repeats are likely to have some benefits. L1 repeats may serve as attenuators of gene-transcriptional activity by spreading heterochromatin formation through cytosine methylation of its promoter[191] (see also below). Another suggested function of L1 elements involved their capacity to cause mutations. It has recently been demonstrated that an engineered human L1 element can retrotranspose in neuronal precursors derived from rat hippocampus neural stem cells[192]. The resulting retrotransposition events sometimes altered the expression of neuronal genes, affecting neuronal differentiation. The same investigators also showed that the L1 element was active in transgenic mice, resulting in neuronal somatic mosaicism. They suggested that L1 repeats could contribute in this way to neuronal somatic diversification in the developing brain, possibly explaining individual differences in brain organization and function. Whether or not this will prove to be correct (see also Chapter 6), these observations underscore the fact that somatic genomes are not static, but diverge during development, and possibly also during aging, because of *de novo* DNA sequence changes.

Totally different types of dispersed repeat are the simple sequence repeat families termed microsatellites and minisatellites. These elements, with a unit size of between about 2 and 20 nucleotides, are organized as loci of tandem repeats dispersed through the genome. They tend to be unstable, with some loci showing dramatically high mutation frequencies, often based on so-called slippage replication errors, leading to copy-number

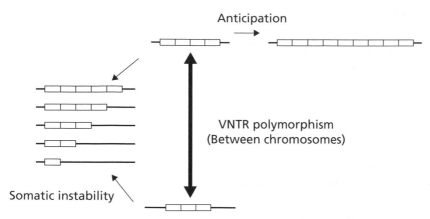

Fig. 3.5 Variable number of tandem repeat (VNTR) instability mainly resides in micro- and minisatellite loci. Extensive copy-number variation at these loci is present in the germ line, but also occurs spontaneously in somatic cells or can be induced by radiation. Anticipation is the increase in copy number of microsatellite repeats in the germ line, which can suddenly become manifest as a genetic disease. See the text for more details.

variation (Fig. 3.5). It explains why these loci are so polymorphic in the germ line, which makes them highly suitable for DNA-based identity testing (DNA fingerprinting) and as markers in genetic linkage studies aimed at discovering the location of disease genes[193].

Mini- and microsatellites are unstable in both the germ line (parent-to-offspring) and somatic cells. For microsatellite loci (including tracts of the same base) this may be due mostly to slippage replication errors, whereas minisatellite variation appears to involve mainly unequal exchange during recombination. Indeed, microsatellite instability is especially prominent in the presence of a defect in DNA-mismatch repair, the system that edits the newly synthesized DNA strand to correct mismatches, including those that result from slippage replication errors. Tumors arising in patients with a heritable form of cancer caused by a defect in DNA-mismatch repair often display this form of microsatellite instability since they are unable to repair the mismatches that are a consequence of slippage replication errors (see Chapter 4).

Microsatellite instability has also been associated with an intriguing set of heritable diseases, including Huntington's disease, fragile X syndrome, and myotonic dystrophy. These diseases are caused by an increase in the number of copies of the repeat beyond its normal range of about 6–55 repeats. This so-called genetic anticipation (Fig. 3.5), which occurs in the germ line, is responsible for the increasing severity of an inherited disease during intergenerational transmission. Studies of these diseases and the repeats involved have shown that this type of genomic instability is highly dynamic[194]. Disease-related instability of the repeats is not limited to the germ line but occurs also in somatic cells in a tissue-dependent manner. We have already seen that in *C. elegans* both DAF-16 and the regulator of the heat-shock response, HSF-1, protect against the adverse consequences of copy-number amplification of microsatellites as occurs in Huntington's disease; that is, polyglutamine aggregation[136] (Chapter 2). The possible contribution to aging of this form of genome instability will be discussed in Chapter 6.

3.1.4 TANDEM REPEATS AND CHROMOSOME STRUCTURE

Apart from the dispersed loci of short tandem repeats, most eukaryotes have many copies of tandemly repeated DNA sequences located around their centromeres and at the telomeres. Centromeres (the points at which spindle microtubules attach; see below) are responsible for chromosome segregation during mitosis and meiosis and contain mainly so-called α-satellite repeats. These highly homogeneous repeats, which are not transcribed, have a basic 170-bp unit and are required for centromere function in a way that is still unclear[195]. As mentioned above, these sequences are subject to concerted evolution, meaning that they show a high level of sequence conservation within a species but are divergent between species. However, while their sequences are divergent, the function of centromeres is conserved throughout eukaryotic biology and possibly based on a universal chromatin structure[196].

Telomeres are the nucleoprotein complexes that occur at the ends of eukaryotic linear chromosomes. In the mammalian genome they consist of several kilobase pairs of repetitive DNA sequences (TTAGGG) that attract a number of sequence- and structure-specific binding proteins. These chromosomal caps prevent nucleolytic degradation and provide a mechanism for cells to distinguish natural termini from DNA DSBs, which signal DNA damage and would result in cell-cycle arrest, senescence, or apoptosis[197]. The protective properties of telomeres were recognized by Barbara McClintock and Herman Muller, the latter of whom coined the term telomere. Telomeres terminate in 35–600 bases of single-stranded TTAGGG at the 3' end (the 3' overhang). This 3' overhang folds back into the duplex TTAGGG repeat array, forming a so-called T loop[198] (see also Fig. 4.7a). Telomeres are also thought to buffer the internal coding regions of the genome from the consequences of the end replication problem; that is, the inability to complete the 5' end by lagging-strand synthesis (Fig. 4.7b). While this may temporarily protect the genome against attrition, cells would inevitably lose terminal DNA with each cell division. However, telomere attrition can be countered by elongation mechanisms (see also Chapters 4 and 6).

3.1.5 NON-CODING RNA GENES

Non-coding RNA genes produce transcripts that function directly in regulatory, catalytic, or structural roles in the cell. They represent a major component of the transcriptomes of higher organisms and can be subdivided into housekeeping RNAs and regulatory RNAs[199]. Housekeeping RNAs include rRNAs, tRNAs, small nuclear RNAs, and small nucleolar RNAs, implicated in such functions as splice regulation and rRNA modification. Also the template RNA component of telomerase (TERC) should be considered as a housekeeping non-coding RNA. Regulatory non-coding RNAs include miRNAs, which can induce posttranscriptional gene-silencing activity, either by translational repression or by triggering mRNA degradation (Chapter 1).

In addition to the small non-coding RNAs, there are also larger regulatory non-coding RNAs. The best known example is the X-inactive specific transcript (*Xist*) gene. The 15–17-kb-long (in mice) *Xist* non-coding RNAs play a key role in transcriptional silencing of the X chromosome during early female embryogenesis, as part of the dosage compensation of X-linked gene products in mammals[200]. At the time of X-chromosome inactivation, *Xist* RNA becomes stable and accumulates. It spreads and eventually coats the whole chromosome. Whereas it is not exactly clear how this spreading takes place and how it brings about transcriptional silencing, evidence has been obtained that L1 elements, which are enriched on the X chromosome, serve to promote spreading. As mentioned above, methylation of the L1 repeats may play a role in stabilizing the inactive state, which originally may have been a silencing mechanism to defend the genome against parasites[191]. Modification of histone proteins may play a role as well (see below for a discussion of the role of methylation and histone modification in gene silencing).

3.1.6 MITOCHONDRIAL GENOMES

In a discussion of genome primary structure focused on mammals and other eukaryotes, it is important to mention the existence of a separate genome: the mitochondrial genome. Neither eubacteria nor archaea contain organelles, such as mitochondria, lysosomes, and endoplasmatic reticulum, which abound in eukaryotes. According to the endosymbiosis theory, the mitochondria of eukaryotes evolved from aerobic bacteria (probably related to the rickettsias) living within their host cell[201]. Mitochondria can arise only from pre-existing mitochondria and cannot be formed in a cell that lacks them because the nuclear genome encodes most but not all of the mitochondrial proteins. The remainder is encoded in the mitochondrion's own genome, which resembles that of prokaryotes, not that of the nuclear genome. The genome of human mitochondria contains 16 569 bp of DNA organized in a closed circle (Fig. 3.6). This encodes two rRNA molecules, 22 tRNA molecules and 13 polypeptides, which participate in building several protein complexes embedded in the inner mitochondrial membrane: seven subunits that make up the mitochondrial NADH dehydrogenase, three subunits of cytochrome *c* oxidase, two subunits of ATP synthase, and cytochrome *b*. Vertebrate transcription is initiated at two promoters, P_H and P_L for heavy and light strands respectively, located 150 nucleotides apart

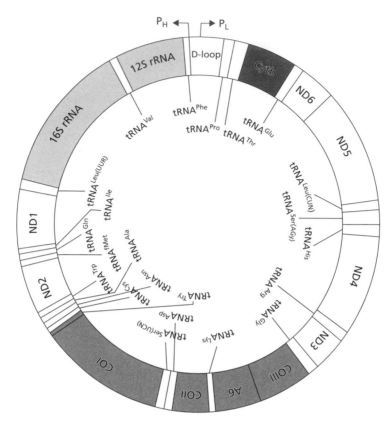

Fig. 3.6 Organization of the 16 569-bp human mitochondrial genome.

within the D-loop regulatory region. The H- and L-strand transcriptional units specify multiple genes, as in prokaryotes. However, like in eukaryotes, the mRNAs have poly(A) tails. The proteins encoded by nuclear genes (e.g. cytochrome *c* and the RNA and DNA polymerases used within the mitochondrion) are synthesized in the cytosol and then imported into the mitochondrion.

3.1.7 SUMMARY

In summary, the primary chemistry of the DNA of the genome dictates a highly stable structure harboring an abundance of functional elements of which protein-coding sequences are only a very small fraction. Most of the functional elements in a genome are probably involved in transcription regulation or in facilitating chromosome structure. How do such sequence elements exert their function? It is now clear that many of the functional DNA elements in the genome function through alterations in DNA higher-order structure, often through interactions with proteins. This is obvious for those sequences determining specific chromosomal structures relevant for processes such as cell division. However, the packaging of the DNA in the cell also controls accessibility of the sequences essential for basal or regulated gene expression. In Chapter 4 it will become clear that alterations in DNA packaging are important, not only for transcription regulation and basic structural maintenance, but also for the continuous repair of chemical damage inflicted upon the DNA from a multitude of endogenous and environmental sources. While attention thus far has been focused on alterations in DNA primary structure and sequence, changes in DNA higher-order structure, also termed epigenomic changes, are gaining increasing interest as possible causal factors in aging and disease (see also Chapter 6). Below, I will briefly discuss DNA higher-order structure and the mechanisms that facilitate genome reprogramming (or epigenetic control) during development, in tissue-specific gene expression, and in global gene silencing.

3.2 Higher-order DNA structure

One level up in the structural regulation of genome function from the DNA double helix and its primary sequence code is the folding of its DNA into chromatin. Chromatin is more than the genome and denotes the complex of DNA and proteins that form the chromosomes. As we shall see, by organizing itself as chromatin the genome maximizes its information content and simultaneously creates ways to make this information available whenever and wherever this is needed.

Most prokaryotic genomes are contained in a single, circular DNA molecule, tightly coiled in a compact structure called the nucleoid, with much less associated protein than eukaryotic chromosomes. The genome of eukaryotes is organized in the form of linear chromosomes, composed of euchromatin and heterochromatin, residing in a nucleus separate from the cytoplasm. Heterochromatin is compact, generally inaccessible to DNA-binding factors and transcriptionally silent. Euchromatic domains are the more accessible and transcriptionally active portions of the genome. What is the underlying structure that defines chromatin at the functional level? The basic unit of chromatin is the nucleosome. Nucleosomes consist of approximately 200 bp of DNA wrapped around an octamer of core histone proteins, consisting of two copies each of histones H2A, H2B, H3, and H4. This structure is the beads-on-a-string-conformation and can be condensed into the tightly packaged solenoid structure or 30-nm fiber. The molecular arrangement of the higher levels of chromatin organization is not well understood, but light- and electron-microscopic studies of interphase and metaphase chromosomes have revealed fibers ranging from 100 nm to the 700 nm structure seen in the metaphase chromosomes during cell division[202,203]. Figure 3.7 schematically depicts what is probably the current consensus with respect to the organization of the DNA in a mammalian cell nucleus.

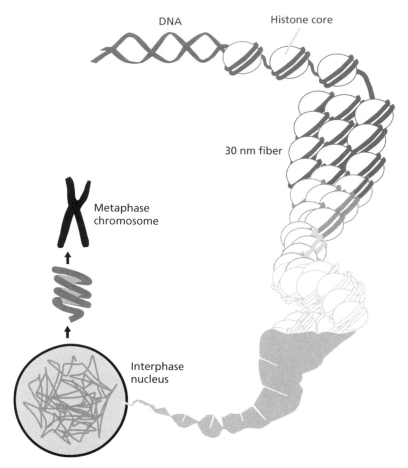

Fig. 3.7 Higher-order structure of nuclear DNA into chromatin.

Active or potentially active genes exist in the simple nucleosomal structure (euchromatin). In the regulatory regions of active genes nucleosomes are either removed or undergo structural alteration, which facilitates the binding of transcription factors (see below). These segments are sensitive to chemical and enzymatic attack because they are protected by histone proteins only poorly, if at all. They are called hypersensitive sites to indicate their extreme sensitivity to enzymatic digestion and typically appear and disappear in patterns that are coordinated with gene activity; that is, more hypersensitive sites appear as gene activity increases.

The highly compacted chromatin fibers represent heterochromatin domains, where genes are not transcriptionally active. Some heterochromatin is condensed chromatin that unfolds and becomes transcriptionally active during some portion of the cell cycle. Constitutive heterochromatin remains transcriptionally inert during the entire cell cycle. The bulk of constitutive heterochromatin is composed of repetitive DNA, such as at the centromeres and telomeres. Because DNA replication requires polymerases and regulatory proteins similar to those required for DNA transcription, condensed heterochromatin most likely unfolds into euchromatin before replication proceeds.

Using powerful new methods the genomic distribution of nucleosomes is now being determined. In the ChIP-on-chip method[204], protein–DNA complexes are cross-linked by the addition of formaldehyde to living cells. After lysis of the cells and mechanical shearing of the chromatin to yield fragments of 0.5–2 kb, the cross-linked protein–DNA complexes are immunoprecipitated using antibodies against invariant portions of histones. The DNA, recovered from the immunoprecipitated complexes, is then used as a probe to hybridize a DNA microarray harboring thousands of sequences covering the genome. The microarray signals indicate nucleosome density at that particular sequence. Application of such genome-wide methods is easiest in organisms with a small genome, such as yeast. Indeed, for this organism it could be concluded that upstream regions of highly active genes in yeast display a reduced nucleosome density compared to the upstream regions of inactive genes[205]. Hence, gene activation in yeast is associated with reduced nucleosome density. How is the situation in the genome of higher eukaryotes, which is so much more complex?

As mentioned above, human genes are not uniformly spread across the genome, but tend to be clustered together. To determine the chromatin architecture of the human genome, Gilbert *et al.* separated compact- and open-chromatin-fiber structures from a human lymphoblastoid cell line by sucrose sedimentation and analyzed their distributions by hybridization to metaphase chromosomes and genomic microarrays[206]. Their results indicate that gene-rich areas in the human genome are preferentially located in open chromatin fibers; that is, structures mostly devoid of fibers beyond 30 nm. However, this first study to globally map higher-order DNA structure in the human genome also revealed that there is no strict correlation between open chromatin and the activity of a gene. Indeed, not every gene in the open areas is being transcribed and active genes in regions of low gene density can be embedded in compact chromatin fibers. As will be outlined later, various protein factors are able to regulate transcriptional activity through chromatin remodeling at the level of an individual gene.

Various factors are involved in maintaining and altering higher-order DNA structure, which is exquisitely regulated and provides an additional layer of information to the primary sequence code of the DNA. This type of regulation involves epigenetic alterations, heritable changes in gene function that occur without a change in the DNA sequence[207]. The best-known example is genomic imprinting, which is a type of genomic modification that in mammals dictates the inactivation of either the paternal or maternal copies of a gene. Methylation is the epigenetic marker most likely responsible for repressing the activity of one of the alleles. Methylation of C residues at carbon 5 of the pyrimidine ring, primarily at CpG sequences, is widespread in mammalian genomic DNA and essentially an extension of the genetic code to a fifth base. Imprinted genes often contain parent-specific differences in DNA methylation within genomic regions known as differentially methylated domains (DMDs). Patterns of CpG methylation are maintained by the action of Dnmt1, the mammalian maintenance cytosine methyltransferase enzyme. Apart from imprinted genes, other regions of the genome with high levels of methylated CpG dinucleotides are the inactive X chromosome in female mammals and mobile genetic elements, all of which are associated with stable transcriptional repression. Gene silencing often occurs through hypermethylation of CpG-rich, promoter-associated regions, termed CpG islands (see below).

Yet another genome project, the Human Epigenome Project (HEP), aims to identify, catalogue and interpret genome-wide DNA methylation patterns of all human genes in all major tissues. The data, obtained by high-throughput methods, are deposited in a public database (www.epigenome.org). Since methylation patterns are known to undergo alterations during aging, such a systematic whole-genome study of DNA methylation at the sequence level could be a prelude to the systematic evaluation of aging-related patterns of DNA methylation levels of sites within the vicinity of the promoter and other relevant regions of a gene (see also Chapter 6).

Other factors that determine epigenetic regulation of gene expression involve histone proteins. DNA is bound to the histones through electrostatic forces between the negatively charged phosphate groups in the DNA backbone and positively charged amino acids (e.g. lysine and arginine) in the histone proteins. Histone proteins can be modified by the addition of acetyl, methyl, or phosphate groups, and this alters the strength of the bonding between the histones and DNA. Modifications such as these are usually associated with the regulation of DNA transactions such as replication, gene expression, chromatin assembly and condensation, and cell division. It has been suggested that, together, these modifications may form a complex, regulatory code: the so-called histone code. Attempts to decipher this code are currently underway, using the aforementioned ChIP-on-chip method with antibodies against specific histone modifications. The first results of this global analysis of histone modifications, in yeast and *Drosophila*, suggest very similar genomic distributions of virtually all tested histone modifications; for example, H3 and H4 acetylation and H3K4 di- and tri-methylation[208]. Moreover, most modifications appeared to be positively correlated with gene expression. This suggests that the different histone maps are linked, pointing towards the absence of a complex regulatory code of histone modifications. Multiple histone modifications employed in parallel to control transcription might be another example of redundancy, conferring robustness on transcriptional regulation. On the other hand, it is too early to rule out a role of histone modification patterns in the timing of specific gene activities.

How does the cell remember its epigenetic marks during cell division and how does it remodel the epigenetic modifications inherited from the transcriptionally inactive sperm and egg? Before a cell can divide, it must duplicate its DNA. In eukaryotes, this occurs during the S phase of the cell cycle. The process is schematically depicted in

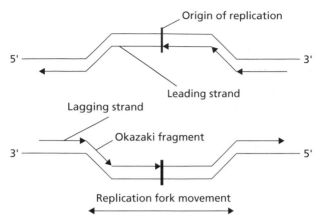

Fig. 3.8 DNA replication is bidirectional and semi-discontinuous, with the leading strand synthesized in the 5' → 3' direction as a single, continuous strand and the lagging strand synthesized discontinuously, but also in the 5' → 3' direction.

Fig. 3.8. After a portion of the DNA is unwound by a helicase, a molecule of DNA polymerase δ binds to one strand of the DNA, moving along in the 3' → 5' direction to assemble a leading strand of nucleotides and reforming a double helix. Then, a second DNA polymerase (polymerase ε in eukaryotes) binds to the other template strand as the double helix opens. It has to work the other way, because DNA synthesis can only occur 5' → 3'. This molecule must synthesize discontinuous segments of polynucleotides (called Okazaki fragments). Another enzyme, DNA ligase I, then joins these together into the lagging strand. Both leading and lagging strands need primers, generated by a primase. The two DNA polymerases carrying out leading- and lagging-strand synthesis are locked together in a replication machine with the DNA template for the lagging strand looping out from the twin polymerases.

This mode of replication, predicted by Watson and Crick and so elegantly confirmed experimentally by Meselson and Stahl (Chapter 1), is described as semi-conservative: one-half of each new molecule of DNA is old and one-half is new. Note that synthesis using the 5' → 3' strand as the template presents a special problem, which reveals yet another vulnerability of the genome for aging-related alterations. As the replication fork nears the end of the DNA, there is no longer enough template to continue forming Okazaki fragments. So the 5' end of each newly synthesized strand cannot be completed and each of the daughter chromosomes will have a shortened telomere. It is estimated that each mitosis event costs human telomeres about 100 bp of DNA. This so-called end-replication problem was recognized by Alexey Olovnikov in 1971, and independently by James Watson a year later[209].

An average 150-million-bp human chromosome can be copied within an hour because of the many places where it can begin. Indeed, whereas bacteria have one single specific origin of replication, eukaryotes can start replication at multiple origins on each chromosome. DNA replication proceeds bidirectionally from an origin of replication; that is, in opposite directions away from the origin (Fig. 3.7). At each cell cycle, during S phase, duplication of chromatin structure, which involves both redistribution of parental histones and histone neosynthesis, occurs in tight coordination with DNA replication[210]. There are several ways to maintain DNA chromatin organization. First, during DNA replication there are differences in replication timing. In general heterochromatin replicates later than euchromatin. It is conceivable that partitioning the genome into domains with specific replication times helps the cell to remember how to distribute histones between the daughter strands.

Second, specific proteins, such as the chromatin assembly factor-1 (CAF-1) complex, help in the assembly of nucleosomes onto newly replicated DNA. In this process, the proliferating cell nuclear antigen (PCNA) plays an important role as a sliding clamp, serving as a loading platform for many proteins involved in DNA replication (and DNA repair; see Chapter 4). Once DNA and chromatin replication is complete, nuclear division follows (see below).

Methylation patterns are perpetuated after replication by a maintenance methylase, DNA methyltransferase 1 (DNMT1[211]). Its importance is illustrated by the observation that inactivation of this gene in the mouse is lethal. The enzyme acts on the hemimethylated sites and converts them into fully methylated sites. Once DNA and chromatin replication are complete, nuclear division follows (see below).

In the cell divisions that give rise to the zygote, a crucial step is the decondensation and reorganization of chromatin of male and female gametes. This basically involves a largely genome-wide erasure of the germ-line-specific epigenetic modifications that had occurred during normal development in the embryonic primordial germ cell lineage[212]. The identification of epigenetic reprogramming mechanisms is a major current interest in view of the fact that successful cloning of animals from differentiated adult cells by somatic cell nuclear transfer also requires nuclear reprogramming. The latter is less successful than the natural reprogramming of gamete DNA, as testified by the widespread epigenetic defects in nuclear-transfer embryos[213]. A detailed discussion of this, as-yet insufficiently explored, topic is beyond the scope of this book.

Of note, while occurring with exceptionally high fidelity, the entire process of information transmission during cell division, including gametogenesis and embryonic development, is unlikely to be faultless, even under normal conditions in the absence of DNA chemical insults or toxicants. Errors are minimized by different and overlapping mechanisms (see Chapter 4), but cannot be entirely prevented. As mentioned in Chapter 2, in germ cells, the DNA mutations resulting from such errors are the driving force behind evolutionary change. Indeed, in times of stress, the propensity to mutate and to rapidly create variants that can escape selection pressures facilitates survival of a small fraction of the original population. However, in somatic cells they may lead to cellular degeneration and death (Chapter 6).

3.3 **Nuclear architecture**

After this brief discussion of the primary and higher-order structure of the DNA of the genome, it is important to review the physical space within which this complex set of structure–function relationships unfolds. Evidence is rapidly emerging for a critical role of nuclear architecture in determining gene activity. Nuclear architecture involves the chromosomal positioning in the nuclear space. We have already seen that in interphase cells the DNA of the genome is organized as linear chromosomes, which are really looped 30-nm solenoid fibers, alternating with beads-on-a-string structures. How is this organized in the nuclear three-dimensional space and what are the functional consequences of this organization?

The nucleus in eukaryotic cells is separated from the cytoplasm by a nuclear envelope (nuclear membrane). The nuclear envelope consists of inner and outer membranes separated by a perinuclear space (Fig. 3.9). The outer nuclear membrane is continuous with the endoplasmic reticulum and has ribosomes attached. So the space between the inner and outer membrane is directly connected to the lumen of the endoplasmic reticulum. In this way, ribosomal subunits and mRNA transcribed from genes in the DNA can leave the nucleus, enter the endoplasmic reticulum, and participate in protein synthesis. Nuclear pores are formed at sites where the inner and outer membranes of the nuclear envelope are joined. They allow free diffusion of small molecules as well as active transport through

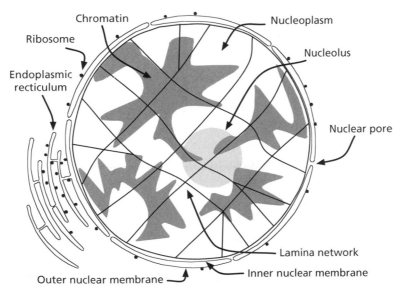

Fig. 3.9 Schematic depiction of the nucleus of a mammalian cell. Note that the lamins are not only restricted to the inner nuclear membrane but also form a network inside the nucleus, which may be involved in DNA transactions, such as replication and transcription. The dark gray areas represent the chromatin.

the interaction between import or export signals (nuclear localization signals, or NLSs, and nuclear export signals, or NESs, respectively) with receptors termed importins and exportins. The fluid within the nucleus is called nucleoplasm, which contains the linear chromosomes.

The inner nuclear membrane is linked—by integral proteins—to the lamina, a fibrous meshwork composed of intermediate filament proteins called lamins (lamins A/C and B). Lamins A and C arise from the same gene, *LMNA*, by alternative splicing. The lamins are coiled-coil structures that contain a small N-terminal head followed by a rod-like domain (α-helical coiled-coil) and a C-terminal globular tail. Via the coiled-coil regions, lamins can form parallel dimers, which in turn form polymers with other lamin dimers in an anti-parallel manner (head-to-tail). The nuclear lamina is dynamic and can depolymerize during mitosis and reform upon re-entry into interphase following rounds of phosphory-lation and dephosphorylation at residues flanking the lamin coiled-coil domains[214].

Lamins at the nuclear periphery are thought to maintain nuclear shape. This becomes apparent after treatment of cells with non-ionic detergents, a treatment that solubilizes the vast majority of cytoplasmic and nuclear proteins. In such cases, electron-microscopic images indicate that the nucleus of the treated cells retains its shape and the lamina remains at the nuclear periphery[215]. Lamins are also dispersed throughout the nucleoplasm, form-ing a thin fibrillar network proposed to be major structural elements of the internal nuclear matrix[216] (Fig. 3.9).

Heritable mutations in lamin A/C or lamin-binding proteins cause various diseases. For example, mutations in emerin (a lamin-binding protein) or A-type lamin cause Emery–Dreifuss muscular dystrophy. Mutations in *LMNA* can also cause Dunnigan-type partial lipodystrophy or Hutchinson–Gilford progeria syndrome (HGPS; see Chapter 5). The mutations that cause these diseases are non-overlapping and it is still unclear as to how different mutations in this gene can cause different diseases[217]. However, this is not unprecedented. For example, mutations in the RET proto-oncogene can cause multiple endocrine neoplasia and Hirschsprung disease[218] and we will see in Chapter 4 that differ-ent mutations in the *XPD* helicase gene can cause different diseases as well. Lamins may contribute to tissue-specific gene expression, through their role as key elements in nuclear architecture. Regions of chromatin appear to be anchored to the lamina and, at least *in vitro*, lamins have been demonstrated to bind directly to chromatin.

As discussed in more detail in Chapter 5, children with HGPS are born normal but start developing multiple symptoms of premature aging within 1 year. The gene product resulting from the dominant mutation that gives rise to this disorder disrupts nuclear architecture, which in a way that is still obscure is ultimately the cause of the disease. Interestingly, in both human skin fibroblast cell lines from old individuals[219] and non-neuronal tissues from aged nematodes[220] progressive deterioration of nuclear lamina and chromatin architecture has been observed. In the nematodes this degeneration was delayed in the long-lived *daf-2* mutants and accelerated in the short-lived *daf-16* mutants.

Whereas there is no complete understanding as to how transcription or other DNA-dependent transactions, such as replication and repair, are distributed within the nucleus, evidence is emerging that none of this is random. For example, whereas originally it was thought that the replication machinery was moving along the stationary DNA, it is now considered more likely that replication is organized at fixed positions in so-called replication factories[221]. Likewise, various models have been proposed in which RNA is transcribed on a solid substrate with an aggregate of all enzymes needed for transcription, processing, and transport in place. Such a transcription-factory model implies that different genes do not always assemble their own transcription sites *de novo* when they become active, but instead migrate to such a pre-assembled association. Compartmentalization of DNA-dependent transactions in the three-dimensional space of the nucleus would be in keeping with the organization of other cellular processes, such as the electron-transport chain in mitochondria, and is certainly a much more likely model than the assumption that all these processes happen at random in solution. A good candidate for such a solid support is the nuclear matrix, also termed nuclear scaffold, nuclear cage, or nuclear skeleton.

The nuclear matrix is a meshwork of proteins that connects to the cytoskeleton at the nuclear envelope. We have already seen that lamins could play a structural role in this meshwork. It has been suggested that in the interphase nucleus the DNA, organized in loops, is anchored to the nuclear matrix by means of non-coding sequences known as matrix-attachment regions. These matrix-attachment regions may constitute boundaries of independently controlled chromatin units within which the DNA packaging and function may be changed without affecting the neighboring regions (see also below). However, the existence of a nuclear matrix is controversial and an extensive meshwork of filaments in the interchromatin space could simply be an experimental artifact[222]. In contrast to the biochemically well characterized lamina, little is known about the protein composition of the internal nuclear matrix. Others, however, argue that a nuclear matrix is a readily observed cellular structure and that the concept of a matrix to provide architectural support for higher-order chromatin packaging is sound[223].

It is conceivable that chromatin is itself responsible for nuclear structure. In this respect, emerging evidence suggests that the distribution of chromosomes in mammalian interphase cell nuclei is non-random. Whereas the physically distinct nature of chromosomes is clearly visible during mitosis—when chromosomes condense and appear as separate entities—the technique of chromosome painting demonstrated that also during interphase each chromosome occupies a well-defined nuclear sub-volume, called chromosome territory[224]. In chromosome painting, probe sets for specific chromosomes, labeled with different fluorescent dyes, are hybridized to metaphase plates or interphase cells so that each of the chromosomes shows a different color when viewed with a fluorescence microscope. Chromosome territories are non-randomly arranged within the nuclear space and occupy preferential positions relative to the center of the nucleus and relative to each other. Work by Tom Misteli and co-workers, who carried out a systematic analysis of the spatial positioning of a subset of mouse chromosomes in several tissues, suggests a pattern that is tissue-specific[225].

The genes within chromosome territories appear to be non-randomly positioned relative to each other, or to nuclear landmarks, such as the nuclear envelope. An attractive model has been presented in which the clustering of genes or other functional sequences into contiguous regions of the three-dimensional space of the nucleus, termed neighborhoods, is explained by similar requirements for optimal function, such as coordinated and efficient expression[226]. As yet, it is unclear whether this model is generally applicable, but there are some intriguing examples of such potential compartmentalization. For example, ribosomal genes cluster together in physical space due to the congregation of the various chromosomes harboring the tandem arrays in which these genes occur, to form the so-called nucleoli (Fig. 3.8). Each nucleolus contains genetic material from multiple chromosomes. An area of DNA called the nucleolar organizer directs the synthesis of rRNA, which subsequently combines with ribosomal proteins to form immature ribosomal subunits that mature in the cytoplasm after they leave the nucleus by way of the pores in the nuclear envelope. Apart from nucleoli, other such compartments include Cajal bodies, a congregation of small nuclear RNA and histone gene clusters, nuclear speckles or spliceosomes, the so-called promyelocytic leukemia (PML) nuclear bodies and, possibly, repairosomes.

Interestingly, the gene-neighborhood model would indicate an additional concept of gene position effects. That is, the actions of a gene are not only influenced by its position in the one-dimensional DNA sequence, relative to regulatory elements, but also by its particular location in the nucleus. Position effects here may include not only transcription, but also replication, repair, and recombination. For example, when DNA undergoes more than one DSB the spatial proximity of the broken ends is positively correlated with the probability of illegitimate joining[227].

Similar to the transmission of the DNA sequence and its methylation patterns and histone code, nuclear compartmentalization must also be accurately perpetuated during cell division to prevent cell death or phenotypic change. After generating two identical chromosomes, termed sister chromatids, which remain attached at the centromere, nuclear division begins with the prophase. In this stage the nucleolus disappears and the nuclear envelope fragments. The latter happens, as mentioned already, by disassembling nuclear lamins, which is regulated by their phosphorylation. The mitotic spindle begins to assemble as the two centrosomes (duplicated just before mitosis) migrate away from each other until they are on opposite sides of the nucleus. Each centrosome contains a pair of centrioles, consisting of nine microtubule triplets surrounding a hollow cylinder. The centrioles are replicated at the same time as the DNA is replicated, also in a semi-conservative manner; that is, a daughter centriole grows out of the side of each parent centriole. The centrioles may function to orientate the spindle and anchor the microtubules radiating out to the chromosomes with some of them extending to the other pole. The point of connection on each chromosome is the kinetochore, a protein complex in the constricted regions defined by the centromere. The centromeric satellite repeat sequences discussed above bind to the kinetochore through specific proteins[228].

During pro-metaphase the chromosomes are attached to the spindle and move to align at the metaphase plate or equator of the spindle. During metaphase, nucleus and chromatin have essentially become spindle and sister-chromatid pairs, with the DNA at its highest level of compaction. Mitosis has now reached the essential point, to correctly distribute identical chromosomes over the future daughter cells by pulling the sister chromatids of each pair toward opposite poles. This process is guided by a quality-control mechanism called the mitotic spindle checkpoint. This checkpoint allows every chromosome to send a stop signal, arresting cell growth until all the chromosomes are appropriately distributed. Defects in this checkpoint provoke chromosome mis-segregation and aneuploidy (gain or loss of chromosomes), which can have adverse functional consequences that include cell death and cancer. For example, defects in different components of the mitotic spindle checkpoint have consistently been observed in cancer cells, characterized by chromosomal instability. As described in detail in Chapter 5, partial inactivation of genes encoding proteins of the checkpoint machinery cause an increase in aneuploidy and are associated with increased cancer and symptoms of accelerated aging.

In the anaphase, then, the two sister chromatids separate at the centromere and move to opposite poles, driven by the depolymerization of microtubules at the kinetochores. Finally, in the telophase the spindle disappears, the chromosomes decondense and the nucleoli reappear. Also the nuclear envelope reforms and at the same time the lamins are dephosphorylated and reassembled. It is unclear as yet what the exact role of lamins is in assembly of the nuclear envelope.

3.4 **Transcription regulation**

The elaborate three-dimensional structure of our genome equipped with sophisticated tools for the transfer of the information it contains, as described here, has developed over the last 3–4 billion years since the emergence of our ancestors, the first replicators. Its purpose has never been any other than to make its information available for providing the functions that ultimately serve its own perpetuation. This has resulted—through the inevitable errors in the transmission of genetic information—in the spectacular level of evolutionary diversity that can be witnessed all around us. The information encoded in the genome is not solely the digital information specifying the amino acid sequences that underlie the complicated protein machines that run our cells. More importantly, the code dictates the time and place of protein expression as well as the interactions that result in the regulatory networks that control the function of the molecular machines in the first place. Whereas protein expression can be regulated at various levels and through different means, control at the gene-transcriptional level is by far the most important. It is at this level of regulation where decisions are made as to why some proteins are expressed in one tissue and not in others, why particular stimuli activate expression of a certain protein and have no effect on others, and how the complex protein machines come together, often acting in concert with many others, to execute their functions. It is also at this level where we can expect to find the most significant adverse effects of the aging process, for example, as a consequence of genomic alterations due to DNA damage or errors during information transfer.

Our knowledge of transcription in the relatively simple prokaryotes is much deeper than of that in eukaryotes, which have to operate in a less gene-dense genome. Nevertheless, it is clear that in both types of species transcription regulation is fundamentally similar, with more complexity and intricacy in eukaryotes. In prokaryotes transcription and its regulation is dominated by operons, groups of adjacent, co-expressed, and co-regulated genes that encode functionally interacting proteins (discovered by François Jacob and Jacques Monod in the early 1960s). Operon transcripts always code for more than one protein, and prokaryotes can start translation of the mRNA into amino acids separately at the beginning of each protein-coding section. By contrast, in eukaryotes, mRNA is not directly used in protein synthesis, but must undergo processing, including the removal of introns, and the addition of a 7-methylguanylate cap structure at the 5' end and a poly(A) tail at its 3' end, prior to export from the nucleus to the cytoplasm, where it is attached to the ribosomes for translation.

Whereas functionally related genes in eukaryotes are often clustered, eukaryotic transcription machinery generally cannot handle polycistronic transcripts. Operons are rare in eukaryotes, but the nematode *C. elegans* occasionally does make polycistronic transcripts and can process them[229]. The myriad of processes that regulate transcription, in both prokaryotes and eukaryotes, all converge on the promoter, a term introduced in 1964 by Jacob, Ullman, and Monod[230] for a site on DNA that is upstream (5') to coding sequences to which RNA polymerase will bind and initiate transcription. In contrast to prokaryotes, which require only one kind of RNA polymerase, eukaryotes require three RNA polymerases: RNA polymerase I synthesizes rRNA (90% of the RNA in a cell); (2) RNA polymerase II synthesizes pre-messenger RNAs; and (3) RNA polymerase III is responsible for the synthesis of tRNA, 5 S RNA, and small nuclear RNA. RNA polymerase II is the most delicately regulated eukaryotic RNA polymerase with an extensive need for accessory proteins.

In prokaryotes the promoters of genes are approximately 200 bp long and consist of two conserved regions, called -35 and -10, because they are approximately 35 and 10 nucleotides upstream from the transcription initiation site, which is called $+1$ (base number 1). The -10 region contains the so-called Pribnow box with a sequence similar

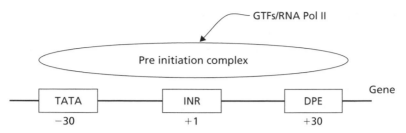

Fig. 3.10 Schematic representation of a core promoter of a mammalian gene. A preinitiation complex is assembled at the core promoter through the binding and interaction of the general transcription factors (GTFs) and RNA polymerase II (RNA Pol II). DPE, downstream promoter element; INR, initiator element; TATA, TATA box. See text for details.

to 5'-TATAAT-3', which is one of the signals to initiate transcription. The expression level of metazoan genes is regulated by the core promoter, *cis*-acting sequences, and *trans*-acting factors. A prototypical metazoan promoter is assembled from a modular array of relatively short sequence motifs (about 7–20 bp), each of which independently represents a binding site for specific transcription regulatory proteins. The core promoter contains DNA sequence elements that are recognized by the general transcription machinery, which help to direct and orient the preinitiation complex at the promoter and play a critical role in the regulation of transcription (Fig. 3.10). The best characterized element is the TATA element, an AT-rich sequence located about 25–35 bp upstream of the transcription-initiation site (equivalent to the Pribnow box in bacteria), which is recognized by the TATA-binding protein subunit of the transcription factor IID TFIID, initiating preinitiation-complex formation. Other elements include the pyrimidine-rich initiator element (INR), spanning the transcription start site, and the downstream promoter element (DPE), which is recognized by components of TFIID other than TATA-binding protein. Not all promoters contain all these elements. For example, the DPE is often present in promoters that do not contain a TATA element. The strongest core promoters contain both TATA and INR elements. Weaker core promoters, such as promoters of housekeeping genes, expressed in all tissues at a low level, generally lack a TATA element, an INR element, or both. Such differences in core promoter do not affect the assembly of the RNA polymerase II preinitiation complex.

Bacterial RNA polymerase can initiate transcription *in vitro* from a core promoter. This is in contrast to eukaryotic RNA polymerases, which need initiation factors, termed general transcription factors, that are required to assemble the stable transcriptional complex needed for all three eukaryotic RNA polymerases. Of these three polymerases, RNA polymerase II is associated with a wide variety of regulatory events affecting the genes transcribed by this enzyme. Also the general transcriptional complex for RNA polymerase II contains far more components than the two other RNA polymerases. The first step in the formation of this complex is the binding of TFIID to the TATA box or equivalent region, which is facilitated by TFIIA. TFIIB then joins the complex by binding to TFIID, allowing the recruitment of RNA polymerase to the complex, in association with TFIIF. Following polymerase binding, two other general transcription factors, TFIIE and TFIIH, associate with the complex. TFIIH, which has a helicase activity, then unwinds the double-stranded DNA, thereby permitting its being copied into RNA. TFIIH is also important in DNA repair and genetic defects in the helicase component of this and other proteins in mammals have been demonstrated to cause defects in DNA-repair processes and are often associated with the premature appearance of symptoms of aging (see Chapter 5).

In metazoa, recruitment of the general transcription factors to the core promoter completes the formation of the preinitiation complex. This allows low levels of accurate transcription *in vitro*, but not *in vivo*. *In vivo* transcription utilizes binding sites for transcription factors[231], which can function as activators or repressors. Whereas the core promoter with its preinitiation complex of RNA polymerase II provides the basic machinery of transcription, it is the ensemble of transcription factors that orchestrates the patterns of gene expression in an organism. Within eukaryotes there is a great variety in the number of transcription factors, reflecting the differences in complexity. In yeast, there are about 300 transcription factors, including subunits of general transcription complexes such as TFIID. *C. elegans* and *Drosophila* have at least 1000 transcription factors, with as many as 3000 in humans. Transcription factors are structurally organized as combinations of a regulatory domain (for activation or repression) with a DNA-binding domain. Examples of DNA-binding domains are the helix-turn-helix domain, the homeodomain, the zinc-finger domain, the winged helix domain, and the leucine zipper domain.

Many transcription-factor-binding sites are within about 200 bp upstream of the core promoter, an area referred to as the proximal promoter. There are binding sites here for at least three types of transcription activation. First, constitutive elements, such as Sp1, bind transcription factors for ubiquitously expressed genes, such as housekeeping genes. Second, inducible elements bind transcription factors that are involved in responses to intra- or extracellular signals, such as the heat-shock-response element and steroid-response elements. Third, tissue-specific elements bind transcription factors that regulate the expression of tissue-specific genes. An example is the CArG element, a motif repeated four times within the proximal promoter of the human α cardiac actin gene and responsible for heart- and muscle-specific transcription.

Other *cis*-acting sequences are often much further away from the core promoter. Examples are enhancers, which are sequence elements located upstream, downstream, or within a transcription unit that can influence the level of gene expression by increasing the activity of a promoter. A special group of enhancer elements is a so-called locus-control region, which regulates the expression of functional gene clusters, such as the

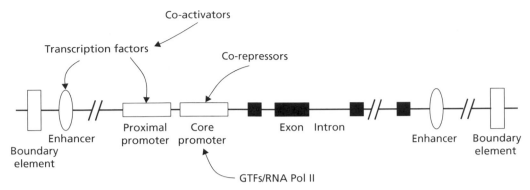

Fig. 3.11 Schematic representation of the transcription unit of a typical mammalian gene. Indicated are the introns and exons of the gene, its core promoter and various *cis*-elements that contribute to the regulation of gene expression. GTF, general transcription factor; RNA Pol II, RNA polymerase II.

β-globin gene cluster in the red-blood-cell lineage. Finally, there are silencer elements, which act to inhibit gene transcription, and insulator or boundary sequences, which prevent enhancers associated with one gene from inappropriately regulating neighboring genes. All these *cis*-regulatory sequences can be scattered over distances as great as 100 kb in mammals (Fig. 3.11).

How do transcription factors activate or repress transcription from the core promoter in the context of all these regulatory sequences? This comes down to the question of how transcription-factor binding to a recognition site that is often some distance away from the core promoter conveys its activation or repression signal to the general-transcription-factor-based transcription-initiation complex at the site of the core promoter. Whereas there are some examples of direct interaction between transcription factors and general transcription factors, indirect interaction through co-activators and co-repressors seems to be the norm. Co-activators or co-repressors provide a connection between the transcription factors and the preinitiation complex. There are different classes of transcriptional co-regulator, some intrinsic to components of the core machinery, others facilitating enhancer–promoter contact, sometimes over large distances. This requires looping between a gene and a distal enhancer, which is determined by higher-order DNA structure. This indicates the importance of cofactors that can alter this higher-order structure. Many transcriptional co-activators and co-repressors have now been identified as factors that either covalently modify the N-termini of core histones through phosphorylation, acetylation, or methylation, or actively effect nucleosome positioning. For example, gene-specific transcriptional activation often involves the targeting of two types of chromatin-remodeling enzyme to the core promoter: histone acetyl transferase (HAT) and the ATP-dependent SWI/SNF-like complex[232]. HATs have been demonstrated to

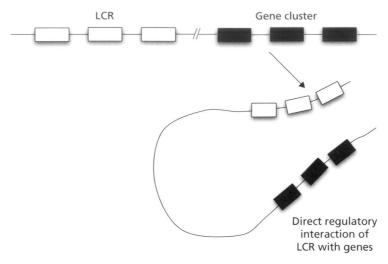

Fig. 3.13 Long-distance transcriptional regulation. DNA can form loops that directly juxtapose enhancer-bound regulatory proteins with promoter-bound transcription complexes. LCR, locus-control region.

halting the spread of transcriptionally repressive condensed chromatin structures. Long-range gene-regulatory interactions and the organization of the genome in active chromatin domains immediately explain the large variability in transgene expression levels normally observed after integration of a gene-expression construct at diverse chromosomal positions. Disruption of long-range regulatory action, for example, as a consequence of deletions or other types of genome rearrangement, can result in gene-expression alterations. Indeed, position-independence of gene expression is lost when one essential *cis*-regulatory region is deleted[235]. The possibility that this type of event contributes to increased aging-related stochasticity of gene expression will be discussed in Chapter 7.

In summary, gene-transcription regulation depends on the three-dimensional interaction of a large variety of DNA sequences—which are not limited to the immediate vicinity of a gene—interacting with a multitude of protein and RNA cofactors. Transcription activation and repression through these cofactors appear to depend on higher-order DNA structure, with most of the genome in a permanently repressed, heterochromatin state. This view of gene-transcription regulation significantly differs from earlier ideas and has immediate relevance for potential mechanisms of aging-related cellular dysfunction. Indeed, the enormous advantage of having so many factors impacting on the transcriptional activity of a gene, often in parallel, is the opportunity to provide redundancy as well as fine regulation of cellular function. However, it also creates a myriad of access points to bring about slight, but cumulative, defects that may ultimately adversely affect function.

3.5 **Conclusions**

Almost five decades after it was first postulated that we age because our genomes age, what can we learn from this brief overview of our most recent insights into the genome's structure and functional organization? For decades, attention has been focused on a genome consisting primarily of protein-coding genes without seriously considering a function for the vast majority of DNA sequences in our genome, which do not encode proteins, but are often transcribed. It is only very recently that this focus has shifted to non-coding DNA sequences, RNA-coding DNA, their transcripts, and their possible role in the systems engineering process that operates our genomes as the information organelles that control development and functional maintenance in adulthood.

The first lesson clearly is the realization that in this vast amount of functional primary sequence, with the extra dimensions of higher-order DNA and nuclear architecture, errors—some significant, many subtle—are inevitable. This is especially true during the perpetuation of the entire structure during cell division, including those divisions that lead to the zygote and its development into an adult. We can see the consequences of such errors already at a very early stage. In humans, chromosomal aberrations and possibly other types of mutation are responsible for the vast majority of spontaneous abortions and still births. This is in spite of the selection opportunity to eliminate genetically aberrant germ cells. The fact that most of the time adverse effects of genomic errors are not immediately obvious, at least not during development and early adulthood, does not mean there are no genomic errors or that their effects are unimportant. The difficulties in detecting small differences in organismal fitness limit the identification of subtle effects, which may be especially important in somatic cells on the long term. As we can deduce from Chapter 2, heritable deficiencies in maintaining the genome code causing harmful effects only at old age are not eliminated by natural selection.

A second lesson that can be learnt from our current insights is the highly sophisticated manner with which the genome regulates its transactions, resulting in unprecedented levels of precision. Yet one cannot escape the notion that this entire symphony of cellular function is in some way inefficient, with multiple, often overlapping and haphazard, levels of control. This serves as another reminder that evolution does not work as a designer that pursues a long-term plan for the most straightforward operational system. The way we now understand Darwinian evolution gives us reason to expect complex, sometimes messy, interactions between all possible levels of genome functioning, including transcription regulation, maintenance and replication. The system, in spite of its many safeguards, including active repair, redundancy, and mechanisms for cell elimination, certainly has not been designed to last the ages.

Finally, from what we have seen, the genome's Achilles heel seems to reside in its need to provide function through transcription. Maintenance of both the DNA primary sequence and the higher-order regulatory punctuation of the genome are critically important for its long-term performance in delivering the right protein at the right time in the right place. As we shall see, in spite of the sophistication of genome-control systems, it is not going to meet this requirement for very much longer than strictly necessary to perpetuate itself. Indeed, threats to its integrity come from both outside and from within, with chemical DNA damage the most prominent source of genomic errors. This becomes apparent when genetic defects, from subtle to severe, predispose the organism to making mistakes. In the next chapter I will discuss the many overlapping systems that are available to correct chemical damage to the DNA of the genome.

4 Genome maintenance

Genomes exist by virtue of their capacity to self-replicate as the key attributes for organismal perpetuation. Whereas the possibility of genetic errors has been recognized since the discovery of the DNA double helix, the magnitude of lapses in information transfer was not immediately clear. Indeed, a major hindrance in gaining a full understanding of the potential impact of genome instability on a living organism was the initial lack of insight into the mechanisms by which genomes may disintegrate over time and the factors influencing such process. Only in the early 1960s, after the discovery of excision repair in bacteria by Richard Setlow (Upton, NY, USA)[236], later confirmed by Boyce and Howard-Flanders[237], did suspicion arise that spontaneous DNA alterations would be much more frequent, were it not for the continuous monitoring of the genome for changes in DNA chemical structure. Once the critical importance of maintaining genome integrity was realized, the concept was borne that the genes specifying genome maintenance may belong to a category of genes called longevity-assurance genes[238,239]. However, at the time, the enormous complexity of the various interconnected systems for detecting and repairing damage to the genome was still unknown[240].

Genome maintenance is a general term and includes systems for sensing and signaling the presence of DNA damage, the repair of such damage or its tolerance, as well as the proper reconstruction of DNA higher-order structure after repair is completed. It also includes systems to continuously monitor the key processes of genome information transfer, from DNA replication to mitosis, to prevent or correct errors and in a broad sense also those molecules and enzymes that act to prevent damage to the genome or buffer its adverse effects. The core of the genome is of course DNA and a major source of genome instability is the continuous introduction of lesions in this molecule by a variety of exogenous and endogenous physical, chemical, and biological agents. The removal of such lesions occurs through DNA repair and is considered so important that DNA repair and genome maintenance are often used as synonyms. I will do that too, occasionally.

As we have seen in Chapter 3, DNA is a very stable molecule. However, under physiological conditions, DNA readily reacts with a variety of agents, most notably water and oxygen, to gradually disintegrate. Damage to the genome, resulting in errors in its information content, is not something that has emerged over time in modern organisms, but has from the beginning been intricately associated with life as the driver of evolutionary change (see Chapter 2). The first replicating nucleic acids, probably RNAs, evolved over 3.5 billion years ago in an environment with little molecular oxygen. The absence of an ozone layer around the primitive earth would have subjected any exposed early replicators to high fluxes of damaging ultraviolet radiation. Once initiated, life based on nucleic acids had to cope with the problem of maintaining a balance between a sufficient amount of genetic variation and its integrity as an individual entity; it had to balance evolvability against longevity. Probably the most ancient longevity genes selected must have been those involved in catalyzing the chemical reactions useful to maintaining the genomes of the individual protocells sufficiently long for them to multiply and compete with other individual protocells. Hence, some form of DNA repair was essential from the very beginning, which is reflected by the wide diversity of genome-maintenance pathways in all current species.

Since all organisms are derived from a common ancestor, with new species arising by a splitting of one population into two or more populations that do not cross-breed, the evolutionary relationship of systems such as DNA repair can be studied by creating evolutionary trees of gene sequence data. As discussed in Chapter 1, such phylogenetic analysis is greatly facilitated by the availability of complete sequence information on the genomes of a great many species in all three kingdoms of life: Archaea, Bacteria, and Eukarya. The evolutionary history of DNA-repair enzymes is complex and characterized by extensive gene duplication, gene loss, and horizontal (also termed lateral) gene transfer between species. In horizontal transfer genetic information is passed on, not just to progeny but to other individuals in the same population or even to members of other species. When this occurs, the introduced gene brings into the cell its own history, which does not reflect that of the host cell. An extreme case of horizontal transfer took place about 1 billion years ago when Eubacteria entered into a symbiosis with Archaea-type host cells to become the mitochondria of eukaryotes[201]. This explains why some eukaryotic DNA-repair proteins can be traced to bacterial and archaeal roots.

Another characteristic of the evolutionary history of DNA repair is the relative scarcity of DNA-repair proteins that are truly universal across the three kingdoms. It is possible that truly universal DNA-repair proteins—present in the last common ancestral genome—are limited to a RecA-like recombinase, a few helicases, nucleases, and ATPases[241]. There are profound differences in core replicative enzymes (but not the Y-family polymerases, which are bona fide homologs in bacteria and eukaryotes; see below), which is in striking contrast with the universal conservation of the translation machinery. It is possible that such differences in DNA-repair enzymes between the three kingdoms reflect early changes in both internal and external environment, requiring new DNA-repair functions.

Genome maintenance may have been the first system that became subject to natural selection because of its critical importance to life as we know it[242]. Originally the most ancient DNA-repair systems may have functioned in protecting an RNA-based ancestral cell. The inherent instability of RNA—because of the presence of the 2'-hydroxyl group of ribose, making its phosphodiester bonds very susceptible to hydrolysis—could have been the primary reason for the transition to a DNA/protein world, which eventually permitted the evolution to complex, multicellular organisms. Genome maintenance, now to a large extent driven by DNA-damage signaling and repair processes, remained of the utmost importance as a cellular defense mechanism. This is evident from the large investment that cells make in DNA-repair enzymes, with several percent of the coding capacity of organisms devoted solely to DNA-repair functions. In some cases, one entire protein molecule is sacrificed for the removal of a single lesion.

Whereas random changes occur in all biological macromolecules, either as errors in their synthesis or from the reaction with environmental or endogenous damaging agents, the consequences for the DNA of the genome are especially dire. For most biopolymers, including RNA, proteins, and lipids, the effects of damage or synthetic errors are minimized by turnover and replacement of altered molecules. DNA is distinctive in that its information content must be transmitted virtually intact from one cell to another during cell division or reproduction of an organism. Moreover, even in non-dividing cells, major portions of the genome are continuously needed as regulated templates to carry out specific functions. It is not surprising, therefore, that defects in genome-maintenance systems are incompatible with normal life. Such defects are often embryonically lethal and defective or diminished DNA-repair capacity can cause disease, including cancer, and accelerate normal symptoms of aging (see below and Chapter 5).

The specific need of DNA for stability is satisfied by the various genome-maintenance pathways, which will be reviewed in this chapter with a focus on eukaryotes and mammalian cells. Similar to Chapter 3, this overview is not exhaustive and mainly serves in providing the context for the main purpose of this book: to sketch the rationale for genome alterations as a possible molecular basis of aging. For more detailed information the reader is referred to the excellent textbook, *DNA Repair and Mutagenesis*, by Friedberg, Walker, Siede, Wood, Schultz and Ellenberger[240] and several recent reviews referred to in the text.

4.1 **Why genome maintenance?**

The unique position of the DNA of the genome among biological macromolecules, that is, its lack of opportunities to easily replace the genetic information primarily embedded in its sequences through natural turnover like proteins and lipids, does not automatically imply the need for advanced maintenance systems. Indeed, genome integrity can be protected through redundancy, when the same function can be performed by other, identical elements, and degeneracy, when elements that are structurally different are able to perform the same function. It is also possible that large sections of the genome are dispensable, which is suggested, but not proven, by the generation of viable mice homozygous for large deletions of the non-coding DNA referred to as gene deserts[243].

However, structural redundancy and degeneracy of a genome are apparently not enough to survive natural levels of DNA damage. This is illustrated by the bacterium *Deinococcus radiodurans*. This organism is extremely resistant to ionizing radiation and many other agents that damage DNA. This is likely to be an adaptation to extreme desiccation followed by rehydration, which causes very high levels of DNA damage. This bacterium is highly polyploid (4–10 genome equivalents), which would provide it with additional gene copies to complement inactivated ones. However, this is likely not the main function of the polyploidy. Instead, this form of redundancy serves to increase the efficiency of homologous recombination (HR) repair, and not primarily to provide back-up copies of genes. In fact, apart from the polyploidy, there is also a redundancy in DNA-repair functions[244]. Indeed, almost all possible repair mechanisms that have been identified in prokaryotes, many of them overlapping, are encoded in the genome of *D. radiodurans*. The genetic cost of maintaining genome function solely by structural redundancy (creating back-up copies of all possible genetically encoded functions) is apparently too high, illustrating the magnitude of the problem. Structural redundancy can also be dangerous, for example, because it could cause undesirably high levels of gene expression or, in vertebrates, promote the activation of oncogenes when more copies of such genes are available.

DNA repair is desperately needed because damage to the genome is a very frequent event, with all organisms continuously exposed to a large variety of genotoxic agents. UV radiation has already been mentioned as a major exogenous DNA-damaging agent, especially relevant during the early stages of life in the absence of an ozone layer. Modern humans are also exposed to UV radiation in the form of sunlight, but to other agents as well, including ionizing radiation and a range of genotoxic chemicals in our environment, including dietary components and air-borne pollutants. However, the main reason for all organisms to invest so heavily in DNA repair involves an enemy from within rather than from outside. All organisms dependent on water and oxygen for their existence are continuously threatened by endogenous DNA damage. It has been estimated that many thousands of spontaneous lesions are induced in the DNA of the genome of a cell each day, solely as a consequence of hydrolysis. Another major endogenous source of spontaneous DNA damage in aerobic organisms is oxidation. Oxidation, as we have seen, was first recognized by Denham Harman as a logical explanation for the various forms of cell and tissue damage observed to occur during aging[23,74,245]. As by-products of oxidative phosphorylation and other biological and physiological processes, oxygen radicals can inflict a variety of damages on cellular DNA and other biological macromolecules like proteins and lipids. The lesions induced in DNA by free radicals are diverse and include a variety of adducts[246] as well as abasic sites, cross-links, DNA single-strand breaks (SSBs) and DNA double-strand breaks (DSBs). Figure 4.1 schematically depicts the different types of DNA chemical damage that can be induced by exogenous and endogenous agents: inter- or intra-chromosomal cross-links, DNA SSBs or DSBs, and a variety of base adducts subdivided into bulky and small adducts.

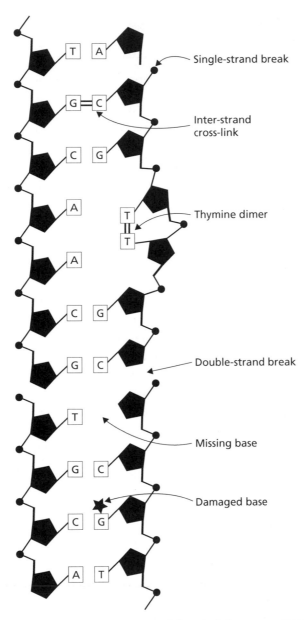

Fig. 4.1 A simplified depiction of the spectrum of chemical damage to DNA as can be found in the living cell.

Many DNA lesions, including the ubiquitous oxidative base alterations, have been identified and measured in mammals, both upon their induction in the DNA and as adducts, often as excretion products in the urine of rodents or humans. The excretion of adducts is a result of their excision by DNA-repair systems[247]. Hence, spontaneous DNA damage reflects a steady-state situation and is generally not permanent.

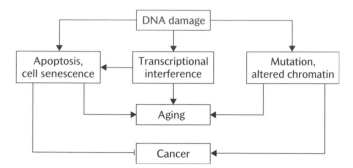

Fig. 4.2 Adverse effects as a result of DNA damage that can contribute to aging. Note that cancer, itself an important component of aging, can be suppressed by the same cellular end points that promote degenerative senescent changes.

In principle, DNA damage can result in three main types of adverse effect (Fig. 4.2). First, DNA damage can interfere with transcription. Second, DNA damage can initiate the cellular signals that lead to programmed cell death (apoptosis) or permanent cessation of mitotic activity: replicative senescence. Third, DNA damage can have long-term effects in the form of mutations: irreversible alterations in DNA sequence information. These consequences often depend on various factors, such as the type of cell, the stage of differentiation or the stage of the cell cycle, the amount and type of DNA damage, and the genomic location of the damage, for example, transcribed or non-transcribed DNA sequences, repeat elements or unique sequences. The events that may lead from these lesions to the aging phenotype are discussed in detail in the following chapters. What is clear, however, is that without efficient systems to rapidly address genome structural deficiencies and restore the original situation, spontaneous DNA damage would soon overwhelm the cell and result in a total collapse of its functional integrity.

Genome maintenance can be roughly subdivided into two components: DNA-damage signaling and DNA repair per se, including tolerance (Fig. 4.3). Whereas DNA-damage signaling is primarily involved in orchestrating a large variety of cellular responses to DNA damage, DNA repair is a straightforward damage-removal system. Both branches of genome maintenance require proteins to recognize or sense the damage, which then in turn recruit other proteins to carry out the task at hand, being the signaling of a cellular response or a given repair activity. As described in Chapter 3, many of these proteins are kept at storage sites, which may include nucleoli, telomeres, and PML bodies. Upon damage infliction these proteins undergo intranuclear relocalization and translocate to nuclear foci thought to represent sites of DNA damage and repair; such sites are sometimes called repairosomes.

As yet, little is known about the exact mechanisms of damage recognition and the nature of the sensors for various types of damage. Indeed, considering the fact that a diploid genome comprises about 2 m of DNA, how is the cell alerted to the presence of

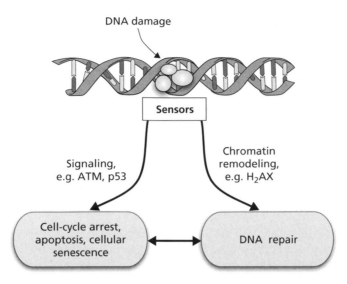

Fig. 4.3 The two major branches of genome maintenance. DNA repair *sensu stricto* aims to restore the original situation by removing the lesion. The complex of DNA-damage signaling pathways assists in these repair activities or activates cellular responses that kill or terminate mitotic activity of a cell when it is beyond repair. The elements of the two pathways depicted on the figure are discussed later in this chapter. ATM, ataxia telangiectasia-mutated.

damage? As we will see, there are indirect ways of 'seeing' DNA damage, but the problem of finding a DNA lesion by a repair or signaling protein can to some extent be generalized by comparing it with the recognition of specific DNA sequences by transcription factors or restriction endonucleases. In this context a model has been developed, termed facilitated diffusion, in which the protein first binds DNA non-specifically and then slides towards its target along the DNA molecule. Already in 1965 Hanawalt and Haynes postulated that DNA-damage recognition was based on proteins that they compared to 'close-fitting sleeves', through which the DNA was threaded to detect deviations from the formal Watson and Crick structure[248]. However, it is unlikely that such models reflect reality and it is now considered more likely that damage recognition involves an iterative process of transient binding and dissociation. It is possible that the repair enzymes initially recognize discontinuity of the DNA double helix due to base unstacking, kinking, or nucleotide extrusion[249]. For lesions such as DNA breaks the recognition process may be different and based on chromatin distortion. It is as yet unclear whether the recognition process of DNA damage is similar for DNA repair and DNA damage signaling. Whereas in DNA-repair damage is recognized through specific protein–DNA interactions, the DNA-damage sensors activating cellular responses to the damage are more general since they must respond to a wide variety of damage as well as problems during replication. Overlap between these two branches of genome maintenance in terms of damage recognition is nevertheless conceivable.

4.2 **DNA-damage signaling and cellular responses**

Whereas the details of the initial stages of DNA-damage signaling are not clear, it is assumed that the presence of most lesions is detected by sensor proteins that convey the damage signals to transducers, transmitting them to numerous downstream effectors. The signaling mechanism, which depends on the type of damage and whether or not the DNA sequence is transcribed, has thus far been approached almost exclusively from the perspective of the actively proliferating cell with a strong focus on the concept of cell-cycle checkpoints, introduced by Hartwell and Weinert[250]. Originally, a cell-cycle checkpoint was defined as a regulatory pathway that controls the order and timing of cell-cycle transitions by checking at the beginning of each new step whether the previous one is completed. For our present discussion this basically involves an examination of genome integrity at specific points in the cell cycle. Upon sensing the damage, mechanisms are activated that arrest cell-cycle progression at checkpoints in the G_1/S, intra-S, or G_2/M phases to allow time for repair. This somewhat narrow concept of DNA-damage checkpoints has now been extended to include the whole range of cellular responses to DNA damage, which comprise: (1) activation of checkpoints to allow time for repair; (2) recruitment of and/or participation in various damage-repair systems; (3) activation of systems for damage tolerance; (4) activation of programmed cell death or replicative senescence to eliminate the affected cell or prevent it from further replication; and (5) the activation of a host of stress-response genes, the function of which is still far from clear.

Since cultured, actively proliferating cells, often in the form of bacteria, yeast, or mammalian cell lines, have become our major model systems in cell biology, a bias towards studying DNA damage and repair in cycling cells is not surprising. However, most tissues *in vivo* contain a very small number of cells in the S phase or no proliferating cells at all, and even stem cells, the presence of which in some normal adult tissues has now been amply confirmed, exist mainly in a quiescent state. Recently, evidence emerged for cell-cycle activation and apoptosis in postmitotic neurons upon treatment with genotoxic agents[251]. While intriguing, the generality of such mechanisms remains to be demonstrated.

Below, I will briefly summarize our current knowledge as to the nature of the various DNA-damage-sensing and -signaling mechanisms and how they interact with the multiprotein complexes specialized in different DNA-repair pathways.

4.2.1 ATM/ATR-DEPENDENT CHECKPOINTS

An important role in the damage signaling process is played by the DNA-damage-inducible phosphoinositide 3-kinase-related kinase (PIKK) family. For example, the presence of the highly toxic DNA DSBs in mammalian cells is often signaled by the ataxia telangiectasia-mutated (ATM) kinase, which phosphorylates proteins that initiate cell-cycle arrest, apoptosis, and DNA repair. Humans lacking functional ATM suffer from a syndrome called ataxia telangiectasia, characterized by cerebellar neurodegeneration, immunodeficiency, extreme sensitivity to ionizing radiation, and increased susceptibility to cancer. Ataxia telangiectasia is sometimes considered as a segmental progeroid syndrome, a disease characterized by the premature appearance of multiple symptoms of aging. As we will see in Chapter 5, most of these syndromes are caused by heritable mutations in genes involved in genome maintenance, by itself evidence that genome instability is a causal factor in aging.

Since DSBs are so toxic, ATM is ever present near the DNA, albeit in an inactive dimeric state or as a higher-order multimer. Inactive ATM dimers are converted into catalytically active monomers, through autophosphorylation, which leads to dimer dissociation and the initiation of cellular ATM kinase activity within minutes of exposure to an agent that induces DSBs, such as ionizing radiation. According to one model, a DSB causes alterations in chromatin structure that can affect very large regions of genomic DNA, thereby activating hundreds of ATM molecules[252]. However, at least *in vitro*, ATM activation appears to be dependent on the Mre11–Rad50–Nbs1 (MRN) complex, which may sense the DSBs, recruit ATM, and dissociate the ATM dimer[253]. How MRN does this is unclear but it may also involve a conformational alteration. Of note, MRN unwinds the DNA ends of a break to generate single-stranded DNA, which is essential for ATM stimulation. This is of interest since the activation of ATM- and Rad3-related (ATR) kinase also requires single-stranded DNA (see below), which is an evolutionarily conserved signal for DNA damage. Whereas the exact mechanism of ATM activation is still debated there is general consensus that, once activated, ATM activates downstream cellular targets, such as p53 and Chk2, through phosphorylation.

Another PIKK, the ATR kinase, has a broader specificity and participates in responses to a wide range of DNA damage, probably exclusively during the cell cycle in response to replication blockage. Unlike ATM, ATR is essential for both embryonic development and somatic cell growth. ATR exists as a complex with ATR-interacting protein (ATRIP), which is recruited to DNA single-stranded regions coated by replication protein A (RPA)[254]. Most DNA-repair pathways process DNA damage through RPA–single-stranded DNA intermediates and stalled replication forks expose extended regions of RPA–single-stranded DNA. Once recruited, ATR activates other checkpoint proteins through phosphorylation, including proteins of the 9–1–1 and Rad17–replication factor C (RFC) complexes (see below) and claspin. The latter, in complex with the breast cancer-susceptibility protein BRCA1, is required for the ATR/ATRIP-dependent phosphorylation of Chk1. All these checkpoint proteins must be recruited independently. Eventually, ATR-dependent phosphorylation of Chk1 and p53 helps to stabilize stalled replication forks by establishing arrest at critical cell-cycle checkpoints, including intra-S- and G_2/M-phase checkpoints, and possibly signals apoptosis.

By comparing different sources of DNA damage—oxidative stress, ionizing radiation, and UV radiation—it has been demonstrated that the ATM and ATR pathways overlap. However, regardless of the genotoxic agent, ATM or ATR always phosphorylated the checkpoint proteins p53, Chk1, and Chk2. These proteins are important transducers of the damage-response signal to effector proteins, including p21 and Cdc25A, that finally result in arrest at critical cell-cycle checkpoints. Thus, for all three sources of damage, the downstream targets for phosphorylation are similar; rather, it is the requirement for ATM or ATR that is different. Taken together, these findings suggest that the ATM and ATR kinases each evolved to respond to unique forms of genotoxic stress[255].

While primarily regulating checkpoint activation in response to DNA DSBs, there is evidence that ATM is also directly involved in the repair of such breaks by activating component proteins of the process of non-homologous end-joining (NHEJ; see below)[256]. In addition, ATM may function in the maintenance of the telomeres, described in the previous chapter and further below, that cap the ends of chromosomes. It may do this through direct binding to short telomeres, phosphorylation of telomere-binding proteins, or other types of regulation of telomere proteins[257] (see also Chapter 5).

4.2.2 DNA-PK

Another member of the PIKK family of kinases involved in sensing and signaling of DNA damage in response to genotoxic insult is the DNA-dependent protein kinase DNA-PK, best known for its role in NHEJ. DNA-PK is composed of a large catalytic subunit, DNA-PK$_{CS}$, and the Ku70–Ku80 dimer with high affinity for DNA ends, which guides the kinase subunit to the site of the damage, usually a DSB. DNA-PK$_{CS}$ homologs are present in all vertebrates, but not in invertebrates, such as yeast, nematodes, and fruit flies, in contrast to other PIKK family members, such as ATM and ATR. Apart from its role in NHEJ, DNA-PK$_{CS}$ acts as a genuine DNA-damage sensor, transmitting signals to p53, to regulate apoptosis and cell-cycle arrest, or mediate the recruitment of DNA-repair complexes[258]. There is evidence that DNA-PK activity in rodent cells is much lower than in human cells[259], which could point towards some role of the protein in determining lifespan differences between species (see further below). Interestingly, a fourth member of the PIKK family is the target of rapamycin (TOR), a kinase that, as we have seen in Chapter 2, plays a role in nutrient sensing. Its downregulation extends lifespan in the fly and the worm, but it seems to play no role in DNA-damage sensing or signaling.

4.2.3 THE 9–1–1 COMPLEX

Another possible DNA-damage sensor is the heterotrimeric complex of Rad9, Rad1, and Hus1, known as the 9–1–1 complex. The complex is targeted to the nucleus and to sites of damaged DNA following genotoxic stress. The complex is structurally similar to proliferating cell nuclear antigen (PCNA), the sliding DNA clamp encountered in Chapter 3 in relation to DNA replication, and it is loaded onto DNA via the RFC complex. In this case Rad17 replaces the largest subunit (subunit 1) in the RFC complex to form Rad17–RFC$_{2-5}$. Rad17 is essential during mammalian development since its inactivation in the mouse is embryonically lethal[260]. However, it is possible that this is not because of its participation in sensing DNA damage, but a consequence of its role in the repair of damage by HR. At this stage the action of the 9–1–1 complex is still speculative, but it may work alongside ATM and ATR, which are independently recruited to the sites of DNA damage. ATM and ATR have been implicated in the efficient loading of the 9–1–1 complex through phosphorylation of the Rad17–RFC clamp loader. RPA-coated single-stranded DNA enhances Rad17 binding, resulting in the recruitment of 9–1–1 to these sites, similar to the ATRIP, which is also targeted to single-stranded DNA in an RPA-dependent manner. Like ATM, there is evidence that 9–1–1 is directly involved in DNA repair[261]. In this respect, it is possible that the 9–1–1 complex serves as a general recruiting platform for DNA-damage checkpoint proteins as well as for DNA-repair enzymes.

4.2.4 POLY(ADP-RIBOSE) POLYMERASE-1

Poly(ADP-ribose) polymerase-1 (PARP-1) is an abundant nuclear enzyme that can be activated by ionizing radiation, alkylating agents, or oxidative stress. Like DNA-PK$_{CS}$, it recognizes distortions in the DNA helical backbone; first of all DNA strand breaks, but also other structures, such as single-strand regions, four-way (Holliday) junctions, and DNA hairpins. Such binding, through its double zinc-finger DNA-binding domain, leads to PARP's catalytic activation and the display of the function first described by Pierre Chambon (Strasbourg, France) and co-workers in 1963; that is, the posttranslational modification called poly(ADP-ribose) polymerization[262]. This is the most extensive post-translational modification known. The process consumes NAD, not unlike the (NAD)-dependent HDAC Sir2, discussed in Chapter 2. The enzyme targets glutamate or aspartate residues in a number of proteins, including PARP-1 itself, histones, and topoi-somerases. It is possible that PARP's action facilitates access of DNA-repair enzymes to the site of damage, for example, by the covalent modification of histones or interaction of histones with auto-modified PARP. Whereas PARP-1 accounts for most of the total cellular poly(ADP-ribose) formation, other PARPs exist, including PARP-2, which can also be activated by DNA strand breaks. Interestingly, whereas mice with either PARP-1 or PARP-2 deleted from their genome are viable and fertile, the combined ablation of these genes is embryonically lethal.

Upon activation, PARP-1 can signal cell death, probably after severe genotoxic stress, through the depletion of cellular NAD pools. Normally, at low levels of DNA damage, poly(ADP-ribose) polymerization is a transient modification, which is rapidly reversed by the action of poly(ADP-ribose) glycohydrolase (PARG). In this context, PARP-1 is a survival factor that has been implicated in multiple DNA-repair pathways, including base-excision repair (BER; see below), and also DSB repair. PARP-1, PARP-2, and two other PARPs, TANK1 and TANK2 (called tankyrases), also play a role in telomere maintenance (see below). As yet the mechanistic details of PARP-1's involvement in DNA-repair pathways are unclear, but modification of chromatin structure to provide accessibility to DNA-repair proteins could be important. Results thus far indicate that PARP-1 can facilitate both the compaction and decondensation of chromatin through poly(ADP-ribose) polymerization with histone and non-histone chromosomal proteins or itself as the targets, depending on the signals available.

PARP-1 is one of several genome-maintenance factors that have been linked to aging. The activity of PARP was tested in white blood cells of different mammalian species stimulated with either γ radiation or double-stranded oligonucleotides and was found to correlate with the lifespan of the species[263,264]. Although correlations can be misleading and nothing is known about PARP activity in different organs and tissues, this may point to a genuine relationship between genome-maintenance capacity and species lifespan.

4.2.5 REPLICATION AND TRANSCRIPTION AS DNA-DAMAGE SENSORS

As we have seen, the sensing of DNA damage has been studied most extensively in proliferating cells, in which it is strongly associated with DNA-damage checkpoints and DNA repair. Replication is of course an ideal way of sensing DNA damage, since its machinery removes the histone protein cover from the DNA and cannot proceed immediately when it encounters DNA damage. Outside the S-phase, other ways of DNA-damage sensing must come into play. Another form of DNA processing that lends itself well to monitoring DNA integrity and to activating DNA-damage signaling, also in non-replicating cells, is transcription[265]. Of the three RNA polymerases, RNA polymerase II covers the largest and possibly the most relevant part of the genome; that is, transcribed protein-coding genes. DNA lesions blocking the elongation of RNA polymerase II can signal apoptosis through p53 or independent of p53[266]. As further described below, the process of transcription-coupled repair can prevent this situation by quickly removing the lesions blocking the polymerase.

4.2.6 DNA-MISMATCH REPAIR AS A DAMAGE SENSOR

DNA-mismatch repair (MMR) is principally an enzymatic pathway for editing newly replicated DNA (see below). However, apart from correcting nucleotide mismatches as a result of replication errors, MMR also recognizes DNA lesions, most notably

O^6-methylguanine, the major lesion produced by methylating agents. Cells with a defect in the process that normally repairs such lesions undergo programmed cell death (apoptosis), but only in the presence of an active MMR system. When such cells are also defective for MMR they are highly resistant to killing by methylating agents. This has been explained by MMR signaling of apoptosis. In addition to apoptosis, treatment of mammalian cells with methylating agents induces cell-cycle arrest in the G_2/M phase, which also is dependent on a functional MMR system. This is accompanied by the activation of ATM and ATR[267]. The idea is that the MMR proteins MSH2 and MSH6 would recognize the damage and directly transmit the signal to the checkpoint machinery.

At this point it is not clear if the MMR-dependent apoptosis and cell-cycle arrest represent genuine DNA-damage checkpoint responses or are merely by-products of MMR's attempt to process adducts resembling true mismatches. If the repair pathways that normally repair DNA methylation damage are inactive, the MMR proteins MSH2 and MSH6 may recognize this damage and confuse it with a mismatch. Since MMR can only excise bases from the newly synthesized strand, the lesion will remain. This will result in repeated cycles of futile repair, which will keep the methylated, template strand single-stranded for much of the time and may lead to a DSB. This may then provide the ultimate signal for apoptosis[268]. However, alternative models in which the MMR proteins act as direct sensors capable of signaling to the apoptotic machinery have been proposed[269]. Hence, it is as yet unclear whether MMR is a DNA-damage sensor or if the observed signaling merely reflects its structural limitations in damage processing.

4.2.7 ROLE OF CHROMATIN STRUCTURE IN SENSING DNA DAMAGE

An efficient cellular response to DNA damage requires changes in higher-order genome structure—through nucleosome rearrangements or chromatin remodeling, as it is also called. The signals eliciting responses to DNA damage are not necessarily a detection of the primary damage but could be a response to modified chromatin structure. Similar to other nuclear processes, such as the establishment of heterochromatin or transcription, chromatin remodeling in the DNA-damage response is accomplished through modifications of histone proteins. These marks include phosphate, methyl, or acetyl groups and even small proteins, such as ubiquitin or SUMO, a small, ubiquitin-related modifier. One of the best-characterized chromatin modification events in DNA-damage responses is phosphorylation of the SQ motif found in histone H2A or the H2AX histone variant in higher eukaryotes[270]. This modification is an early response to the induction of DNA damage, and occurs in a wide range of eukaryotic organisms, suggesting an important conserved function. This phosphorylation can be carried out by ATM, ATR, or DNA-PK$_{CS}$ in a redundant, overlapping manner.

H2AX activation is important for the repair of DSBs, or of SSBs converted into DSBs by replication. The key role of H2AX in genome maintenance is underscored by increased tumor susceptibility, genomic instability, and sensitivity to ionizing radiation in H2AX-haploinsufficient mice[271]. Foci of phosphorylated H2AX (designated γH2AX) are readily formed at the site of DNA DSBs. Indeed, the formation and loss of γH2AX has proved a sensitive measure for DSB formation and repair. Interestingly, γH2AX foci have been found to accumulate in senescing human cell cultures and in aging mice[272], contributing to the now widely held belief that cellular senescence is a DNA damage response. γH2AX may exert its function as a genome-maintenance factor through chromatin remodeling. H2AX phosphorylation in mammalian cells spreads into genome regions of millions of base pairs flanking a break. This influences higher-order chromatin structure and may make the DNA more accessible to DNA-repair enzymes.

4.2.8 SUMMARY

In summary, there are different overlapping signaling mechanisms for DNA damage. Their utilization depends on the type of damage, the phase of the cell cycle and, to some extent, the species. DNA-damage signaling appears to be well integrated in the actual repair processes, with some sensors or transducers being actual components of a repair pathway. Sensors may work on the basis of alterations in DNA higher-order structure, which are also part of the response to DNA damage and tightly coordinated with DNA repair to facilitate access to the lesion and subsequent restoration of chromatin structure after repair is complete. The proteins that sense or detect DNA damage may be shared between pathways that signal damage to enact cell-cycle arrest, apoptosis, or replicative senescence and those that repair the damage. There may be a branch point, with the number of lesions as the main determinant of the option that is chosen. However, it is also possible that there is a distinction between DNA-damage sensors that signal repair and those that signal to enact a cellular DNA-damage response. Also in this case, the level of DNA damage could determine the choice between these two branches of the DNA-damage response. At low levels of DNA damage, signaling proteins may not be quick enough to reach the damage before it is repaired. At high levels of damage the repair systems may be saturated and signaling proteins, such as ATR, may then be able to reach the damage and signal cell-cycle arrest, apoptosis, or cellular senescence[273]. These options are not mutually exclusive and DNA-damage checkpoint proteins, as we have seen, can get involved in DNA repair. It is possible that cells rarely suffer from such high levels of DNA damage that an apoptotic or senescence response is necessary and that most of the time DNA-repair systems are capable of handling the situation. This may be especially true for non-dividing cells, which probably rarely need—if ever—to take action as abruptly as cells in the middle of DNA replication.

4.3 **DNA-repair mechanisms**

Before discussing the different mechanisms for repairing damage inflicted on the genome from various endogenous and exogenous sources, it is first necessary to contemplate the options of not repairing damage at all, which is in a sense the simplest solution, especially when there is reason to assume that the damage is likely not to have major adverse consequences. Such cases exist and some of them are, somewhat paradoxically, a part of the normal DNA-damage response. They will be discussed here under the heading of DNA-repair mechanisms. Then it is important to consider the next easiest form of DNA repair, which is known as direct reversal of DNA damage. The fact that there are only very few types of damage that can be repaired through such mechanisms already suggests that the problem is very complex and that such direct solutions may work for some major lesions, but do not reflect in any way the more general strategies in genome maintenance that have evolved. Indeed, most DNA damage is repaired through excision-repair pathways. The double-helical structure of DNA lends itself well to such mechanisms because of its two complementary strands. When one strand is damaged the other one can serve as the template for restoring the original situation. This offers yet another advantage of using double-stranded DNA as the carrier of the genetic information rather than single-stranded RNA or DNA and may explain why virtually all species use the DNA double helix for this purpose. Also, the DNA bases are ideally suited for repair. They cannot be easily converted, for example, by spontaneous alkylation or deamination into other natural bases. This may explain why DNA uses thymine rather than uracil as in RNA. Indeed, the repair system would be unable to distinguish a deaminated cytosine from a uracil. Such uracils are now effectively removed by uracil DNA glycosylase. One exception involves methylated CpG sites. 5-Methylcytosine can spontaneously deaminate into thymine, resulting in a mismatch with guanine. This explains why so many spontaneous mutations are found at such sites[274]. They also accumulate with age in different organs and tissues of the mouse[275] (see also Chapter 6).

The presence of two complementary strands is especially useful for excision-repair systems, which work by cutting part of the damaged strand containing the lesion out of the DNA and subsequently pasting a replacement fragment using the non-damaged strand as the template. However, there are multiple overlapping DNA-repair systems, some of which are also capable of repairing lesions opposite each other at one site on each strand, for example, DSBs or DNA interstrand cross-links. Such lesions may be less frequent, but they are also highly toxic and present a considerable problem to the cell's arsenal of repair tools. As discussed above, possibly all these systems interact with DNA-damage checkpoints and chromatin-remodeling systems. Figure 4.4 schematically depicts the major DNA-repair pathways in mammalian cells.

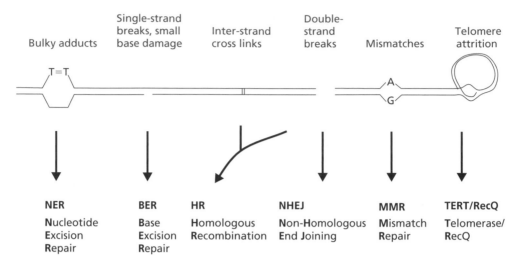

Fig. 4.4 Schematic depiction of the main DNA-repair pathways in mammalian cells subdivided on the basis of the specific forms of DNA damage that prompt their action. TERT, telomerase reverse transcriptase.

4.3.1 TOLERANCE OF DNA DAMAGE

Upon sensing DNA damage, the easiest response for the cell is to do nothing and either tolerate the damage or, when in the process of transcription or replication, activate systems allowing it to bypass the lesion. The first option may be selected by terminally differentiated cells in mammalian organisms, such as neurons. Nouspikel and Hanawalt observed that neurons do not efficiently remove DNA damage from their genome[276]. In general, DNA repair appears to be less active in terminally differentiated cells than in proliferating, undifferentiated cells. For example, in my own laboratory, we showed in 1986 that senescence of rat skin fibroblasts in culture (reminiscent of terminal differentiation) is associated with an approximately 50% reduction in excision-repair-associated DNA synthesis[277].

Interestingly, as subsequently demonstrated in Hanawalt's laboratory for UV-irradiated human neurons, whereas the bulk of their genome is not repaired, lesions from active genes were removed swiftly. In these cells they also observed proficient repair of lesions caused by UV irradiation in the non-transcribed, opposite strand, which is unusual (see below). They called this type of repair DAR, or differentiation-associated repair, and hypothesized that in this way neurons maintain a clean template for repairing their active genes while ignoring the remaining part of their genome. While still speculative, this strategy may be yet another example of a sound allocation of resources by the organism. Indeed, genome alterations in these cells may only start having adverse effects well after the reproductive period, in which case it is only prudent not to invest any major resources in repairing non-essential parts of the genome in such cells.

Ignoring lesions also occurs in proliferating cells, both of prokaryotic and eukaryotic origin, during replication. This is called translesion DNA synthesis, a process to avoid cell death when encountering unrepaired DNA damage. For example, the major mutagenic oxidative DNA lesion, 7,8-dihydro-8-oxoguanine (8-oxoG; a modification of the C8 position of guanine with its tautomeric form called 8-hydroxyguanine), is bypassed efficiently by high-fidelity DNA polymerases. However, this frequently results in the incorporation of an adenine rather than a cytosine opposite the damaged G[278]. This A–8-oxoG mismatch is often not recognized by the mismatch-repair systems that normally proofread DNA, resulting in G $\rightarrow$ T transversion mutations, frequently observed in human cancers.

Lesion bypass, however, can also be regulated by utilizing specific enzymes dedicated to operate with high fidelity on DNA harboring specific types of DNA damage. For example, the Rev1 protein incorporates cytosines opposite abasic sites, allowing lesion bypass in both yeast and mammalian cells. In humans, polή, encoded by the *XPV* gene, is able to insert the correct adenine base across a cyclobutane thymidine dimer, the main UV-induced lesion. When encountering replication-blocking lesions, the use of such enzymes, often referred to as Y-family polymerases[279], increases the chance of survival while maintaining a reasonable level of genome integrity. This is illustrated by the increased mutagenesis and reduced survival of cells from patients with human xeroderma pigmentosum (XP) harboring a genetic defect in the *XPV* gene[280].

The mechanism of translesion synthesis is not as straightforward as one might expect. Replication of the lagging strand is discontinuous and a lesion will not compromise replication-fork progression. However, if a replication block is located in the leading-strand template, the polymerases can become uncoupled and continued template unwinding allows nascent lagging-strand synthesis to proceed past the blocked nascent leading strand (see Chapter 3). A possible scenario for translesion synthesis in bacteria, derived in part from reconstitution experiments *in vitro*[281], is as follows (R. Fuchs, personal communication). The lesion is bypassed to reinitiate leading-strand synthesis downstream of the lesion, which leaves a gap. Such a gap, which can be 1000 nucleotides long, can then be filled by the specialized Y polymerases in an error-prone manner. However, an alternative way of gap filling was already described by Rupp and Howard-Flanders in UV-irradiated bacteria lacking nucleotide-excision repair (NER)[282]. Such so-called postreplication repair involves template switching—using the newly synthesized complementary strand as a template rather than the damaged one—and recombinational gap filling[283]. This latter, recombinational strategy is a genuine mode of DNA repair, most notably in the repair of DSBs, and will be discussed further below. In yeast, the lesion-bypass pathway is part of the RAD6 pathway, which contributes to this organism's resistance to DNA damage. This class of mechanisms for tolerance to DNA damage should not be confused with another form of postreplication repair, MMR, which will be discussed below.

The story of the Y-family polymerases, the so-called mutator polymerases, really begins with the discovery of SOS repair in bacteria. In the SOS response about 30 genes are coordinately induced as a consequence of the proteolytic inactivation of LexA, which is greatly enhanced by the RecA protein bound to single-stranded DNA. The latter explains how the SOS response is induced by DNA-damaging agents. However, LexA can also be inactivated or its cellular concentration decreased under stressful conditions, such as starvation and bacterial senescence (elicited by starvation-induced arrest of proliferation). SOS induction and mutagenesis have been observed in aging *E. coli* colonies, in the absence of exogenous sources of DNA damage[284]. The rationale for such a survival strategy of unicellular organisms under adverse conditions has been discussed in Chapter 2: a transient increase in mutation rate provides some members of the population with a new genotype that allows them to survive and proliferate. Such adaptive mutagenesis can even take the appearance of being directed to specific genes. This phenomenon was first noted by John Cairns and Patricia Foster (then both in Boston, MA, USA) as a high frequency of mutations reversing a mutationally inactivated *lacZ* gene in stationary cultures of *E. coli*[285]. LacZ encodes the enzyme β-galactosidase, permitting the use of lactose as a nutrient. The mutations in these stationary cultures appeared to be directed by the selection process, since they only occur after these cells had been exposed to lactose. Rather than indicating some hidden causality, however, these mutants are singled out since only they, and no others, allow the cell to resume growth.

It is important to realize that a similar strategy in multicellular organisms is not very useful. Whereas there may be evolutionary advantages for all species in increasing the amount of genetic variation in the germ line (temporarily, because the fitness of such mutators rapidly declines as a consequence of Muller's ratchet; see Chapter 2), random mutations in somatic cells offer no selective advantage, but contribute to disease and aging, the topic of this book. An exception is the highly regulated process of somatic hypermutation of immunoglobulin genes, which is critical for unfolding the full power of the immune response. In this form of adaptive mutagenesis mutations are targeted to the variable regions of rearranged immunoglobulin (*Ig*) genes to generate a wide variety of memory cells with high affinity for a specific antigen. The enzyme activation-induced cytosine deaminase (AID) plays a major role in this process by deaminating cytosines in *Ig* genes, leading to the formation of uracil. It is likely that the subsequent mutations are then caused by Y-family DNA polymerases that bypass these lesions[286]. There is evidence that the same enzyme also deaminates mRNA in a process called editing. RNA editing is the co- or posttranscriptional modification of the primary sequence of RNA through nucleotide deletion, insertion, or base-modification mechanisms. In mammals editing seems to exclusively use base modification. RNA-editing processes are known to create diversity in proteins involved in various pathways like lipid transport and metabolism[287]. For example, deamination of cytosine generates a shorter form of the apolipoprotein-B mRNA in mammalian intestine and liver.

The genes encoding the Y-family polymerases induced by the SOS response in bacteria—pol IV and pol V—led to the discovery, by searching genome databases, of multiple orthologs and paralogs in eukaryotes, including mammals. All these DNA polymerases display increased error rates on undamaged DNA (as compared to replicative polymerases) and are capable of translesion synthesis of damaged DNA. In contrast to the situation in bacteria, these enzymes are constitutively present in mammalian cells and tissues and highly regulated to avoid their access to undamaged DNA[280].

The mechanisms to ignore DNA damage or bypass lesions at the risk of mutation are highly relevant for late-life genomic integrity. Ignoring DNA damage in postmitotic cells leads to an increased DNA damage load at old age, which may contribute to cellular degeneration and death. Likewise, mutations as a consequence of lesion bypass will accumulate with age and can easily lead to increased risk of cancer and other adverse effects. This will be further discussed in Chapter 6. However, DNA-damage tolerance can also have immediate consequences for the cell at the level of gene transcription. Indeed, translesion synthesis not only occurs during replication, but also during transcription, which has been observed in both bacteria and mammalian cells, and shown to lead to mutant transcripts (transcriptional mutagenesis)[288]. Since quiescent, non-replicating cells are the norm in tissues of multicellular organisms, this was an important discovery, the consequences of which are still not entirely clear. Indeed, whereas replication errors in certain genes can conceivably lead to cancer, a clonal disease, the effect of transcriptional errors may be transient and have a limited impact on an organism. Nevertheless, it is easy to see that whereas the effects of infrequent events, such as occasional misincorporation of amino acids or ribonucleotides, are likely to be limited, consistent bypassing of a persistent lesion in a gene could alter the entire transcriptional output of that gene in a given cell.

Most studies on transcriptional bypass have been done in bacteria. The results indicate that one of the most frequent base changes—deaminated cytosine or uracil—is readily bypassed by *E. coli* RNA polymerase, resulting in mutated transcript. Fortunately, uracil is normally rapidly removed by BER, which is also the case in mammalian cells (see further below). In mammals, altered proteins as a result of transcriptional mutagenesis have been demonstrated in the brains of patients with Alzheimer's disease and aged control individuals without dementia, but not in the brains of young control subjects[289]. This increase in altered protein with age may remind the reader of Leslie Orgel's error catastrophe hypothesis (Chapter 1), which was discarded years ago, mainly because of a lack of evidence for a high frequency of altered proteins with age. To implicate transcriptional mutagenesis as a causal factor in aging it would be important to study transcriptional infidelity in the synthesis of components of the transcriptional machinery itself, which would create the positive-feedback loop originally suggested by Orgel[290].

4.3.2 DIRECT REVERSAL

Apart from ignoring it altogether, the next simplest way of dealing with DNA damage is its direct reversal. The main reason why this is not utilized more generally is that it involves the need for too many specialized enzymes, which is genetically costly when dealing with hundreds and perhaps thousands of different types of chemical damage in DNA. Nevertheless, three major systems of direct reversal exist: enzymatic photoreactivation of UV-induced pyrimidine dimers, reversal of alkylation damage, and ligation of DNA SSBs.

In photoreactivation, the nature of which was first described by Jane and Dick Setlow[291], a single enzyme, photolyase, is able to repair UV-induced DNA lesions—cyclobutane pyrimidine dimers (CPDs) or pyrimidine 6–4 pyrimidone photoproducts (6–4PPs)—by effectively reversing their formation using blue light[292]. Based on what has been mentioned at the beginning of this chapter regarding UV radiation as perhaps the main source of DNA damage early in the history of life, it is not surprising that such a highly specific mechanism as enzymatic photoreactivation for the speedy direct repair of the main UV-induced lesions has evolved. The system also occurs in yeast, but most likely not in placental mammals, which may have lost it during evolution. It is possible that the need for such a specialized mechanism has become much less in multicellular organisms, in which only a fraction of the cells—and then not even in all organisms—is regularly exposed to UV. In mammalian cells UV damage is generally repaired through the NER pathway (see below), which in humans is especially important for protecting the skin. Interestingly, transgenic mice have been generated expressing marsupial CPD-photolyase, which dramatically reduced acute UV effects like erythema (sunburn), hyperplasia, and apoptosis[293].

The direct reversal of alkylation damage in human cells can be carried out by three proteins. The O^6-alkylguanine alkyl transferase (encoded by the *MGMT* gene in humans) catalyses the transfer of alkyl groups, varying from methyl to benzyl, from the O^6 of guanine to one of the repair enzyme's own cysteine groups. This protein is itself expended in the reaction. Less widespread direct repair enzymes in mammals are the ABH2 and ABH3 proteins. Orthologs of *E. coli* AlkB, these demethylases catalyze the oxidation and release of the methyl group from 1-methyladenine and 3-methylcytosine. Interestingly, ABH3 can also repair RNA[294], which is in fact the main target for these lesions.

O^6-Alkylguanines are lesions of considerable biological importance in mammals. They are highly mutagenic and may occur spontaneously at a high rate, possibly through intracellular methyl donors, such as *S*-adenosyl-L-methionine. Mice lacking MGMT are highly susceptible to tumorigenesis by alkylating agents[295]. Interestingly, transgenic C3HeB male mice overexpressing the human *MGMT* gene in liver and brain were found to have significantly lower levels of hepatocellular carcinoma, a frequent spontaneous tumor in this mouse strain[296]. This indicates that O^6-alkylguanines occur spontaneously and contribute significantly to the risk of liver cancer in animals of this strain.

E. coli and other prokaryotes are able to rapidly increase the amount of their O^6-methylguanine-DNA methyltransferase (O^6-MGT I) activity upon chronic exposure to alkylation stress[297]. Discovered by Leona Samson (Cambridge, MA, USA) in the laboratory of John Cairns, then in Boston, MA, USA, this response is called the adaptive response, which can lead from the 1–2 molecules of the protein that are normally present to several thousands of such molecules in adapted cells. There is no evidence for such a response by the mammalian *MGMT* gene, which in this respect resembles the *E. coli* O^6-*MGT II* gene, which is also not inducible (at least, not in an O^6-*MGT I* mutant).

DNA SSBs can be repaired through direct reversal, simply by rejoining the ends using ligase I. However, SSBs, for example, as inflicted by ionizing radiation or ROS, possess abnormal 3' and 5' termini, which need to be restored to proper 3'-hydroxyl and 5'-phosphate moieties. Therefore, most SSBs are repaired through the BER process (see below).

4.3.3 DNA-EXCISION REPAIR

There are three excision-repair pathways: BER, NER, and MMR. They all employ the double-helical nature of the DNA to remove lesions from one strand and restore the original situation using the second strand as a template (Fig. 4.5).

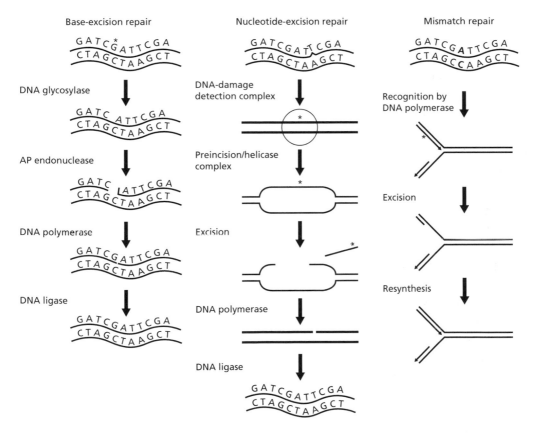

Fig. 4.5 Simplified version of the three excision-repair pathways: base-excision repair (BER), nucleotide-excision repair (NER), and DNA mismatch repair (MMR). They all repair DNA damage through excision and resynthesis using the opposite strand as a template. AP, abasic (see text for an explanation).

4.3.3.1 **Base excision repair (BER)**

BER is specifically devoted to the repair of small base damage, SSBs, and abasic sites, as a consequence of such processes as oxidation and non-enzymatic hydrolysis of base–sugar bonds. These lesions occur at high frequency; there are tens of thousands of lesions per day in a typical cell. Spontaneous base damage can occur, for example, as a consequence of hydrolytic deamination of cytosine, resulting in uracil. Reactive oxygen species (ROS) can cause a large number of different types of small base damage, including the 8-oxoG lesion mentioned above and thymine glycol[298]. To repair the large variety of small base alterations, BER relies on a battery of enzymes, termed glycosylases, that can recognize a specific type of damaged base and cleave the N-glycosylic bond between the base and the sugar. There are glycosylases specific for deaminated bases, alkylated bases, oxidized bases, and base mismatches. Examples are uracil-DNA glycosylase (UNG), recognizing deaminated cytosine (uracil), and TDG and MBD4 DNA glycosylases, excising thymine when it is mispaired with guanine as a consequence of deamination of 5-methylcytosine. There are many others, partially overlapping in their activities and also including a number of enzymes that recognize oxidative lesions, such as OGG1 DNA glycosylase excising 8-oxoG from DNA (its ortholog in bacteria is Fpg), and NTH, removing oxidized pyrimidines. When occurring in transcribed sequences, oxidative base damage can also be repaired by transcription-coupled repair (see below). The repair of endogenous DNA base damage was reviewed extensively by Barnes and Lindahl in 2004[298].

Removal of damaged bases by hydrolytic glycosylases leaves abasic sites, or AP sites, which can also be a direct product of damage. Once initiated, BER can proceed along two mechanistically different lines: short- or long-patch BER. The distinction involves the size of the repair patch, the number of nucleotides replaced. For short-patch BER, which is the major pathway, this is only a single nucleotide. Long-patch BER involves between two and 10 nucleotides. The enzyme that recognizes and cleaves the 5' phosphodiester bond at AP sites is AP endonuclease (APE1), the key component of BER after the glycosylase has done its work. There is another enzyme that can cleave at the AP site, this time at its 3' site. This enzyme, designated AP lyase, is associated with some glycosylases as well as with the main polymerase involved in BER, polβ. The exact mechanistic details as to how the cell decides which activity and which branch of the BER pathway to use for which lesion, is not yet entirely clear and a discussion is beyond the scope of this book. The reader is referred to several excellent recent reviews[283,298,299].

Briefly, in the short-patch pathway polβ is recruited to fill the one-nucleotide gap. Ligation occurs through the Lig3–XRCC1 complex. In the less utilized long-patch sub-pathway, the replicative DNA polymerases polδ/ε or polβ, in a complex with RFC/PCNA, displace the strand, generating a flap of between two and 10 nucleotides, which is subsequently cut off by FEN1 endonuclease. The nick is closed by ligase I. As already mentioned, SSBs, such as those induced by ionizing radiation or ROS, are also repaired through the BER pathway. For this purpose, the enzyme polynucleotide kinase (PNK) or

APE1, in a complex with XRCC1, converts the damaged termini into 5'-phosphate and 3'-hydroxyl moieties[300].

Apart from the core players, accessory factors likely play a role in BER. As mentioned above, the ring-like PCNA DNA clamp in combination with the RFC clamp loader organizes various proteins involved in DNA replication, DNA repair, DNA modification, and chromatin remodeling. As such, it stimulates FEN1 cleavage of the flap in long-patch BER. A similar role can be played by the DNA-damage sensor complex 9–1–1, which is structurally and functionally similar to PCNA (see above). As already mentioned, members of the PARP family may be involved in BER by binding to SSBs, either those arising during BER or breaks that are induced by ROS or ionizing radiation. PARP is then automodified, as described above. Since BER proceeds efficiently in extracts of Parp-1-null cells, such a role is not essential. In this respect it has been postulated that PARP is merely facilitating access to the DNA through local chromatin relaxation.

As it turned out, BER is not only important in the maintenance of the nuclear genome, but also highly proficient in the repair of small base damage in mitochondrial DNA (mtDNA)[301]. This is not surprising since mitochondria are the main sites of ROS production, and ROS in turn cause the lesions that are the main substrate for BER, as we have seen. Only the short-patch variant of BER appears to be active in repairing lesions in mtDNA. As discussed further below there is some evidence that this activity declines with age in rodents.

BER must be a critically important genome-maintenance system, since complete absence of one of its core components is incompatible with life. For example, the homozygous deletion of APE1, XRCC1, ligase I, or polβ in the mouse germ line is embryonically lethal. Indeed, in contrast to other repair systems, there are no known human diseases caused by a complete defect in BER, although it remains possible that slight deficiencies are associated with disease or accelerated aging. On the other hand, elimination of one of its glycosylase functions has only a limited impact. For example, mice with an engineered defect in uracil-DNA glycosylase have an increased level of uracil in their genome, but do not exhibit a greatly increased frequency of spontaneous mutation. They also do not display any signs of premature aging. This lack of a major phenotype may be due to the existence of a compensatory uracil-DNA glycosylase activity in the form of SMUG uracil-DNA glycosylase[298].

Targeted inactivation of the OGG1 DNA glycosylase in the mouse led to the accumulation of 8-oxoG and a modest increase in spontaneous mutation frequency in non-proliferative tissues, but no increased spontaneous tumors or other marked pathological changes[302]. The lifespan of these animals was not significantly shorter and there were no signs of premature aging. The nature of a possible backup activity for this enzyme is not known, but transcription-coupled repair or NER may contribute.

The results with these and other mouse glycosylase mutants suggest that there is significant overlap between glycosylases in recognizing different lesions as well as overlap with other repair pathways. This overlap may to some extent compensate for the inherent inefficiency of BER in needing so many different glycosylases to accommodate the wide range of DNA lesions in a cell. A battery of different enzymes may be unavoidable because it may simply not be possible to design a general recognition system for all possible types of small base damage, which is in striking contrast to the next system to be discussed, NER.

4.3.3.2 Nucleotide excision repair (NER)

NER, a general repair system that removes damaged DNA bases by dual incision of the affected strand followed by resynthesis of 24–32 bp (in eukaryotes), is inextricably linked to the heritable human disease xeroderma pigmentosum (XP). XP patients develop fatal skin cancers when exposed to sunlight and in 1967 it occurred to James Cleaver (San Francisco, CA, USA), then a postdoctoral fellow at the laboratory of Bob Painter at the University of California at San Francisco, that the high susceptibility to cancer in these patients could be related to failure of DNA repair of sunlight-induced DNA lesions[303]. At the time, DNA repair had just been discovered, in bacteria (see the beginning of this chapter), and the first methods to measure the repair of UV-induced damage had been developed. Pettijohn and Hanawalt were the first to describe the patching step in the process by showing after irradiation that bromouracil incorporation occurred in small patches and did not give the density shifts expected from semiconservative DNA replication (see the description of the Meselson and Stahl experiment in Chapter 1). They called this repair replication[304]. Painter himself had developed the techniques of autoradiographic and equilibrium density-gradient detection of repair (unscheduled DNA synthesis and repair replication)[305]. Unscheduled DNA synthesis is a dramatically visual method of measuring NER, which has a far larger patch size than BER (Fig. 4.6). Using these methods Cleaver found that XP cells were indeed repair-deficient[306].

The next key player in this field was Dirk Bootsma of the Erasmus University in Rotterdam, the Netherlands, who fused cells from different XP patients, to see if they were able to complement each other by restoring UV-induced unscheduled DNA synthesis in the heterodikaryons. If so, then the genetic defect in DNA repair in each of the two complementing cells was in different genetic complementation groups required for excision repair[307]. Soon, this kind of somatic-cell genetic analysis using cells of XP patients from different families indicated multiple genetic complementation groups for XP, which were designated *XPA–XPG*. These observations, in turn, suggested that multiple genes are required for the particular mode of excision repair demonstrated in normal human cells and that mutational inactivation of any one of these genes results in XP.

NER is essentially different from BER in the sense that it has no battery of enzymes to recognize specific chemical groups that make up the lesion, but instead recognizes major backbone distortions of the DNA. Such bulky lesions include the aforementioned UV-induced lesions, especially 6–4PPs, and damage created by electrophilic metabolites of carcinogenic polycyclic hydrocarbons, formed during inefficient combustion of fossil

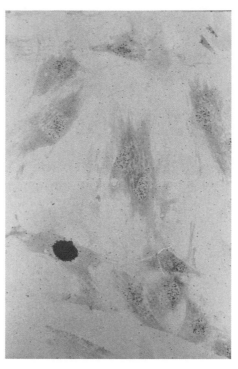

Fig. 4.6 UV-induced unscheduled DNA synthesis in rat fibroblasts. After UV irradiation the incorporation of [³H]thymidine is visualized by autoradiography. The occasional S-phase cell is indicated by a completely black nucleus. The significant, but much lower levels of [³H]thymidine incorporation, revealed by the black grains above the nuclei, indicate excision repair of the UV-induced lesions.

fuels and found in cigarette smoke, car exhaust fumes, and barbecued meat. The best known example is benzo(*a*)pyrene, which has been studied extensively as a genotoxic agent (Chapter 6).

NER has two sub-pathways, differing in damage recognition but sharing the same downstream repair machinery: global genome NER (GG-NER) for the removal of distorting lesions anywhere in the genome and transcription-coupled NER (TC-NER) for the elimination of distorting DNA damage blocking transcription. As already briefly discussed above, in TC-NER lesions are sensed by RNA polymerase II itself and the resulting signals can activate either repair or apoptosis. As the name suggests, TC-NER involves the removal of lesions from the transcribed strand in active genes, which for some lesions, such as CPDs, is much faster than GG-NER. The pathway was discovered by Vilhelm Bohr (Baltimore, MD, USA), then in the laboratory of Philip Hanawalt (Stanford, CA, USA)[308]. Interestingly, it is now clear that TC-NER also functions in removing non-NER lesions. Whereas GG-NER can also remove some types of oxidative lesion, TC-NER acts on those oxidative lesions thought to be the exclusive domain of BER[309]. Whereas this is controversial in view of the recent retraction of several major publications[310], it remains a likely possibility. In this sense, TC-NER is now to some extent recognized as a repair pathway in its own right and often just referred to as transcription-coupled repair.

In TC-NER, DNA-damage sensing occurs by RNA polymerase II, thereby arresting transcription. Transcription arrest is followed by the recruitment of the products of the *CSA* and *CSB* genes, with the *CSB* gene product possibly acting in a complex with the basal transcription factor IIH TFIIH, and perhaps the XPG protein. TFIIH, also needed for transcription initiation (see Chapter 3), has then already dissociated and needs to be recruited again. Its XPB and XPD helicase protein subunits are again needed to open up the helix, in this case a 24–32-nucleotide stretch containing the damaged region. The details of the early steps in TC-NER are not yet known, but the CSA and CSB proteins are thought to be involved in displacing RNA polymerase stalled by a DNA lesion and recruiting the NER (and perhaps BER) machinery to the site of the lesion. The CSB protein is a member of the SWI/SNF family of ATP-dependent chromatin-remodeling factors (see Chapter 3) and changing DNA higher-order structure may be an important component of its role in TC-NER. In addition, various subunits of TFIIH and also the XPG nuclease may assist in the removal of the RNA polymerase and RNA. At this point TC-NER merges with GG-NER.

Of note, heritable mutations in *CSA* or *CSB* cause Cockayne syndrome, an autosomal recessive disorder characterized by progressive postnatal growth failure, neurological dysfunction, and a short lifespan of about 12 years on average. This disease, which shows no signs of increased cancer, is often considered as a segmental progeroid syndrome and will be discussed in more detail in Chapter 5. It is only one of several human disorders associated with NER genetic defects, underscoring the importance of the pathway for human health, but also pointing towards a certain tolerance for its absence, which is in striking contrast with BER.

In GG-NER the situation with respect to DNA-damage sensing is far more complex than for TC-NER. As we have seen, TC-NER is able to quickly recognize lesions, including those that are normally a substrate for BER, on the basis of transcription arrest. In this respect, TC-NER functions, to some extent, as a backup system for BER, by acting on non-bulky lesions, such as oxidized or methylated bases. However, certain helix-distorting oxidative lesions, such as cyclopurines, fall within the realm of GG-NER. It has been speculated that accumulation of these lesions in the brain may explain the observed progressive atrophy of the cerebral cortex of XP patients[311].

Because its recognition system is based on distortions of the DNA double helix, GG-NER may also occasionally attack natural configurations of undamaged DNA, evidence for which has indeed been obtained using cell-free extracts acting on synthetic DNA fragments[312]. These findings suggest that even in non-dividing cells there can be considerable DNA turnover due to this kind of gratuitous repair, which may then contribute to mutation accumulation during aging, due to errors during resynthesis. However, this is merely speculation because GG-NER is likely to abort when no lesion is present, based on the damage-verification role of XPA (see below).

A key role in lesion recognition in GG-NER in the context of chromatin organization is probably played by the damaged-DNA-binding protein (DDB), at least in the case of UV-induced pyrimidine dimers. This protein may activate the DNA-damage checkpoint protein ATR and also initiates GG-NER, by bending the DNA to the lesion, so as to allow XPC to bind to it. The important role of DDB in NER in a situation *in vivo* is indicated by the fact that a heritable mutation in the gene *XPE*, encoding its small, p48 subunit (DDB2), can cause XP. The gene encoding p48 is regulated by p53[313] (see also below). Somewhat similar to the role of PARP in BER, DDB is not required for human NER *in vitro*, as became evident in a reconstitution experiment. Similar to PARP, it has therefore been speculated that DDB is especially important in facilitating repair in a chromatin context. For example, a role of DDB in nucleosome unfolding may be the key in providing access to XPC and XPA and other components of the NER machinery to damaged DNA.

It is now generally assumed that GG-NER begins with the binding at the site of the lesion of a heterodimer consisting of the XPC and HR23B proteins, facilitated as we have seen by DDB (at least, for pyrimidine dimers). Binding of XPC/HR23B and, possibly, its action in increasing the DNA backbone distortion, permits the binding of TFIIH, which plays a very similar role as in TC-NER; opening up the helix through the helicase activity of two of its 10 subunits, XPB and XPD. Here again we should pause to mention the existence of human heritable diseases caused by mutations in some of these genes. Mutations in XPC or XPD can result in XP, whereas mutations in XPB cause XPCS, a combination of XP and Cockayne syndrome. Interestingly, different mutations in the same gene can lead to different diseases. An extreme example is XPD, mutations in which can not only lead to XP, but also to Cockayne syndrome or trichothiodystrophy. This strong, complicated disease link of NER explains much of its enormous appeal as a subject of study, to both basic scientists and clinicians. A detailed discussion of NER-associated diseases is beyond the scope of this book and the reader is referred to recent reviews[314,315]. However, the subject will come back repeatedly.

Opening up the DNA helix by the action of TFIIH is followed by the recruitment of XPA, RPA, and XPG; XPC and HR23B are released. RPA binds to both strands and keeps them apart. XPA probably acts in damage verification and in guiding subsequent cleavage of the damaged strand. At this stage GG-NER has come together with TC-NER. Hence, the difference between the two sub-pathways is really in the damage-recognition steps. Therefore, defects at this and the next stages affect NER as a whole. For example, whereas defects in XPC affect only GG-NER and defects in CSA or CSB only TC-NER, XPA inactivation silences the entire NER pathway. However, CSB (and possibly CSA) may also be involved in general transcription[316] and could be directly linked to BER for the repair of oxidative damage[317], which may explain why Cockayne syndrome is so different from XP.

The downstream part of NER involves the incision at both sides of the lesion, at the 3' site by XPG and at the 5' site by a dimer composed of XPF and ERCC1. The distance between the incisions is about 30 bases, but can vary with each repair event. The ERCC1–XPF endonuclease is also essential for interstrand cross-link repair (see further below). It should be noted that ERCC1 represents the formal nomenclature of the NER genes; that is, excision repair cross complementing. The reason that it has no XP-derived name is the absence of a patient with a mutation in this gene, which is in contrast to its sister subunit XPF and the other endonuclease XPG. After dual incision, the gap is filled in by DNA synthesis, using the opposite, normal DNA strand as a template, by DNA polymerases δ and ε, followed by ligation.

NER in *E coli* is very similar to NER in mammals, but simpler. However, it should be noted that with respect to UV damage, bacteria, bacteriophages, and some eukaryotic viruses contain up to three distinct mechanisms to initiate the repair of UV-induced dipyrimidine adducts: NER, BER, and photoreversal. Photoreversal has been discussed. To initiate BER, glycosylases, such as T4 pyrimidine dimer glycosylase or the *Micrococcus luteus* UV endonuclease, are necessary[240].

Similar to the situation in BER, genetically modified mice are now playing a major role in unraveling the mechanisms of NER and the pathophysiological consequences of deficiencies in its performance. One of the first of these mouse models for NER was a knockout of the gene *XPA*, made independently in the laboratories of Harry van Steeg (Bilthoven, the Netherlands)[318] and Kiyoji Tanaka (Osaka, Japan)[319]. Similar to their human counterparts, these mice were highly sensitive to genotoxic agents inducing bulky adducts, such as UV or benzo(*a*)pyrene. By now, almost all gene defects involved in NER-related diseases have been modeled in the mouse, most of them in the laboratory of Jan Hoeijmakers (Rotterdam, the Netherlands). Hoeijmakers and collaborators were able to model, in the mouse, many of the exact same mutations in NER genes that caused these different diseases in humans, thereby mimicking their human phenotypes to an often remarkable extent. Interestingly, it was noted in the course of this work that several of these mouse mutants displayed multiple symptoms of premature aging, a phenomenon extensively discussed in Chapter 5.

Taken together, there can be no doubt about the importance of NER as a major repair pathway, especially in humans. However, the fact that elimination of this pathway is not embryonically lethal (except when the gene is also involved in some other basic function, such as general transcription) indicates that NER is not as important as BER. This may be especially true in the mouse, since in the above-mentioned mutant mouse models symptoms were often considerably less severe than in their human counterparts. There are two possible explanations for this. First, it is conceivable that NER in humans is mainly important as a defense against UV. Since rodents are nocturnal animals and have a fur, they normally do not encounter any UV damage. This is illustrated by the aforementioned XPA-knockout mouse, which only showed symptoms when its (shaved) skin is challenged with UV or when it is fed compounds that induce bulky adducts, such as benzo(*a*)pyrene.

Second, it is possible that NER gains in importance at older ages, for example, in a possible backup role for BER. Rodents generally have a short lifespan and may need NER to a much lesser extent than humans. Indeed, there are now a number of cases in which the consequences of a DNA-repair defect appear to be much less severe in mice than in humans. This will be further discussed in Chapter 5.

The argument that NER is less important for rodents than humans can be illustrated by the so-called rodent repairadox, which has direct relevance for the relationship between DNA repair and aging[320]. Shortly after the clinical importance of NER was realized—with its absence in cells from XP patients so dramatically illustrated using the new autoradiographic assay—the wide diversity of DNA-repair processes was not immediately known. Indeed, UV-induced unscheduled DNA synthesis in cultured skin cells was often considered as a measure of the total DNA-repair capacity of the organism. Subsequent studies by Ronald Hart in the laboratory of Dick Setlow revealed a roughly linear correlation between UV-induced unscheduled DNA synthesis in skin fibroblasts cultured from different species—shrew, mouse, rat, hamster, cow, elephant, and human—and the lifespan of the species (plotted on a logarithmic scale)[238]. Most striking of these results was the dramatic difference between human and rodent cells. Indeed, around this time data from a number of laboratories indicated the almost complete lack of UV-induced CPD removal from rodent cells, an activity clearly present in human cells under the same conditions[321,322]. Although also in human cells the repair of CPDs is slow (about 80% in 24 h in cultured fibroblasts), the almost complete absence of this activity in rodent cells was considered strange in view of the fact that the survival rate of human and rodent cells upon UV irradiation is very similar[322].

The difference in CPD repair between rodent and human cells is likely due to a deficiency in rodent cells of the stimulation of NER by p53. The p53 tumor suppressor, an important mediator of cellular responses to DNA damage, affects the efficiency of NER by transcriptionally regulating the expression of the *DDB2* gene (encoding the p48 protein) and the *XPC* gene, two important components of the NER pathway involved in DNA damage recognition. Whereas heritable mutations in the human *DDB2* gene generate the E subgroup of XP (XPE), in rodent cells this gene is not induced by p53. The absence of a fully functional DDB causes a deficiency in global genome repair of CPDs (as well as 6–4PPs and possibly other types of lesion). By contrast, TC-NER in rodent cells is fully functional and it is likely that this proficiency in repairing UV-induced pyrimidine dimers in transcribed DNA strands explains why the survival of rodent cells after UV irradiation is not different from that of human cells. Indeed, arrested RNA polymerase II at a lesion is likely to induce apoptosis unless repair is swift[266].

Based on the above, one would expect that defects in TC-NER alone would lead to increased levels of apoptosis, thereby eliminating cells with a severely damaged genome and reducing cancer risk. On the other hand, defects in GG-NER alone may lead to increased genome instability but not to elevated levels of apoptotic cells. This is in keeping with results obtained after comparing mice with defects in XPC (no GG-NER), CSB (no TC-NER), and XPA (no NER) for signs of increased genome instability or apoptosis. XPC-deficient mice were shown to rapidly accumulate somatic mutations at the Hprt locus[323] in T lymphocytes as well as in liver, spleen, kidney and lung at a *lacZ* transgene locus (S. Wijnhoven, personal communication). These results are in keeping with the observation that XPC-deficient mice develop multiple spontaneous lung tumors[324]. CSB-null mice displayed no such increased genomic instability at a *lacZ* transgene locus[325], but instead increased levels of apoptotic cells (Y. Suh, personal communication). In XPA-deficient mice both an increased level of genomic instability in liver and kidney[325] and increased numbers of apoptotic cells (Y. Suh, personal communication) were found. Whereas in the XPA mutant the absence of TC-NER would cause apoptosis, as in the CSB-deficient mouse, this may not be enough to entirely eliminate cells with accumulated damage due to the defect in GG-NER, which is also present in this mouse model.

At the time the Hart and Setlow paper appeared, the mechanistic details discussed above were not known. It is possible that the low level of UV-induced GG-NER in rodents simply reflects an adaptation to nocturnal life. Interestingly, Kato *et al.*[326], who greatly expanded the Hart and Setlow study, observed very low UV-induced repair synthesis in cells from bats, nocturnal animals that are long-lived. Of note, unscheduled DNA synthesis as a measure for NER can be influenced by numerous confounding factors, such as cell geometry, nucleoside pool sizes, and [³H]thymidine uptake. Based on the results thus far it is not possible to conclude that species-specific lifespan correlates with DNA-repair capacity. The lesson here is probably that we still lack comprehensive insight into the various ways an organism maintains its genome and how the various characteristics of the pathways to preserve the integrity of our genes impact on the process of aging (see below). In this respect, as we shall see, NER remains a very interesting system.

4.3.3.3 Mismatch repair (MMR)

The third excision-repair pathway, MMR, is primarily devoted to the excision of mispaired nucleotides or short loops generated by insertions or deletions of nucleotides. These are typically errors made by the DNA-replicative machinery and MMR serves the purpose of correcting such errors. This makes MMR a postreplication repair system which complements an error-correction mechanism that is already part of many DNA polymerases. This built-in correction system is called proofreading and it is the primary line of defense against mistakes in newly synthesized DNA. When an incorrect base pair is recognized, DNA polymerase reverses its direction by 1 bp of DNA. The 3' → 5' exonuclease activity of the enzyme allows the incorrect base pair to be excised. Following base excision, the polymerase can re-insert the correct base and replication can continue. MMR, therefore, is an additional safeguard against errors during information transfer. The process is distinct from the two other excision-repair pathways in the sense that it relies on a way of distinguishing the newly replicated from the parental DNA strand. The mechanism by which MMR in eukaryotes is directed to the newly synthesized strand is as yet unknown. In *E. coli*, which has a very similar system to repair replication errors, the strand harboring the mispaired base is distinguished from the parental strand by adenine methylation on the latter. However, methylation in eukaryotes is different from prokaryotes and it is possible that part of the increased complexity of MMR in eukaryotes is due to the need for alternative mechanisms of strand discrimination.

MMR in eukaryotes involves homologs of the *E. coli* MutS and MutL proteins, but is much more complex. A dimer of the MutS homologous MSH2 and MSH6 proteins (MutSα) recognizes mismatches and single-base loops, whereas the MSH2–MSH3 dimer (MutSβ) recognizes insertion and deletion loops. Then, the MutL-like protein complexes MLH1–PMS2 and MLH1–PMS1 bind to the MSH2–MSH6 and MSH2–MSH3 MutS homologs, respectively, after which the complexes can migrate in either direction. It is not only strand discrimination that remains unclear in eukaryotes; the excision step does as well. Evidence has been obtained, using a cell-free system, that the PCNA/RFC clamp and clamp-loader pair interact with MSH and MLH proteins to direct excision of the mismatch on the nascent strand through exonuclease[327]. The details of this process are still unknown in mammalian cells *in vivo*.

Errors in DNA replication are not the only substrate for MMR. MMR plays a role in preventing strand exchange between divergent DNA sequences during HR, and in this sense functions as a monitor of meiotic and mitotic recombination. MMR acts as a barrier against the exchange of DNA molecules from unrelated organisms by aborting mismatched heteroduplexes. MMR is also capable of recognizing DNA damage other than mismatches. An example is its action in stabilizing hairpin structures that arise during BER of SSBs in repetitive stretches of microsatellite triplet repeats[328]. Such action may allow repeat expansion in postmitotic cells, such as in the brain of patients with Huntington's disease (see Chapter 3). Lack of MSH2 interrupts the progressive repeat expansion that has been demonstrated in the mouse model for this disease[329]. As already briefly discussed in the section on DNA-damage signaling, MMR appears to be active also in DNA-damage signaling, including the signaling of an apoptotic response. Finally, Donald MacPhee (Melbourne, Australia) has suggested that MMR also acts in non-dividing cells in the repair of mismatches generated during error-prone repair or as a consequence of mitotic recombination[330]. Repair of such mismatches by MMR in a non-dividing cell would result in mutations some 50% of the time because there is no way to distinguish the correct from the wrong base.

Whereas NER deficiencies will always be associated with XP, MMR genetic defects are responsible for hereditary non-polyposis colorectal cancer (HNPCC), a human autosomal dominant disorder characterized by a strong predisposition to develop tumors, most notably tumors of the colon[331]. Most HNPCC cases are caused by mutations in MLH1 or MSH2. Cells from tumors of these patients display instability of the small repeat-element clusters called microsatellites (see Chapter 3). This is due to the tendency of the replication machinery to slip when encountering stretches of short repeat elements, resulting in the loss or addition of one or more repeat units. Such slippage replication errors result in the short insertion or deletion loops, which can only be repaired by MMR.

We know that MMR defects are associated with microsatellite instability because we can readily detect such events from analyzing the tumor DNA. The single germ-line mutation in an MMR gene in an individual with HNPCC does not result in MMR deficiency. It is a second hit in the remaining functional allele in a tumor that eliminates MMR and results in such high levels of alteration in the notoriously unstable microsatellites that it can be detected directly. However, this microsatellite instability reflects a much broader genomic instability, known as a mutator phenotype, which helps the tumor cell to acquire new attributes for unhindered growth (see also Chapter 6). This explains the predisposition to cancer of MMR genetic defects in HNPCC. Indeed, genomic instability and predisposition to cancer are hallmarks of MMR defects, more so than in the case of NER defects. This is best illustrated by the MMR-deficient mouse models that have been generated. MMR-defective mice, as a consequence of the inactivation of any of the MutS or MutL homologs, develop both lymphomas, a common spontaneous tumor in normal mice at advanced age, and intestinal tumors. These mice usually die of cancer as early as 6 months of age[332].

To study mutations other than microsatellite instability, MMR-defective mice were crossed with other mouse strains harboring bacterial reporter genes that can be recovered from the mouse genome and tested in *E. coli* for mutations (see Chapter 6 for a detailed explanation of such mouse models). Such reporter loci are more similar to gene sequences than microsatellites and therefore a better marker for overall genome instability. Using such hybrid mouse models it was found that this high susceptibility to cancer of MMR-defective mice is associated with dramatically increased levels of genomic instability. For example, in the absence of any mutagenic treatment, mice lacking PMS2 showed a 100-fold elevation in mutation frequencies in all tissues examined, as compared with the wild-type or heterozygous animals[333]. Most of these mutations were deletions and insertions of one nucleotide within mononucleotide repeat sequences, which is consistent with the role of MMR in the excision of mismatches caused by slippage during replication.

As we have seen in the case of NER, it is often informative to model human disease-causing mutations in the mouse, rather than generating complete knockouts. Indeed, in many cases these disease-causing mutations do not completely eliminate gene function, but only affect part of it. MMR mouse lines have been generated that harbor specific mutations causing HNPCC in humans. Analysis of cells from such knockin mice revealed that whereas MMR knockouts lacked both mismatch repair and the apoptosis response, an MSH2-specific mutation inactivated only the former. Whereas the mutant MSH2–MSH6 complex retained normal mismatch recognition and, hence, can presumably still signal apoptosis, it lost its capacity to initiate the next stage in MMR. To retain the apoptotic response is obviously advantageous because mutant cells that fail to be repaired by MMR can be eliminated. This is in keeping with an observed significantly longer lifespan of these mice compared with the complete MMR-knockout mice. Nevertheless, eventually they also succumbed to cancer (about 6 months later) and their cells also displayed increased mutation rates[334]. Interestingly, this separation of the repair function from the DNA-damage signaling function of MMR would rule out the futile repair hypothesis described in section 4.2.6.

MMR-deficient mice show increased predisposition to cancer but not premature aging. In this respect, it is of interest to compare the MMR-deficient mouse models with the previously discussed NER models. Whereas a deficiency in NER can also predispose to cancer (as, for example, in XP), several NER mutants are not cancer-prone. Some, including mice with a mutational defect in the *XPD* gene, which in humans causes trichothiodistrophy, are even resistant to cancer[335]. There are two possible explanations for this difference. First, as we have seen, MMR-defective mice lack an apoptotic response, an important barrier against cancer. In NER-deficient mice this response is not only retained, but likely plays a major role in reducing cancer in some of these mutants and promoting some of the degenerative aspects of aging (Chapter 5). Second, as we have seen, MMR defects can result in very high levels of genomic instability, much higher than what has been observed in most NER-deficient mice. It is likely that this increased mutagenesis, which mainly involves small mutations, is the cause of the greatly elevated cancer risk. This is because it increases the chance of mutations in genes that are critical for cancer development or progression. However, point mutations may contribute less to the non-cancer-related, degenerative aspects of aging. As we will see later in the book, it is possible that this non-cancer component of the aging phenotype is caused by other types of random genome alteration, such as large genome rearrangements. Such events are introduced into the genome as a consequence of erroneous repair of DNA double-strand lesions. It is to the repair of this type of DNA damage that I will now turn.

4.3.4 REPAIR OF DNA DOUBLE-STRAND LESIONS

DNA double-strand lesions are forms of DNA damage that affect both strands simultaneously at opposite sites that are sufficiently close together to prevent the use of excision repair. The model lesion for the types of repair that are usually employed to overcome double-strand lesions is the DNA double-strand break DSB; that is, when two or more breaks are formed in opposite strands of DNA within about 10–20 bp of each other. DSBs are induced by ioniz-ing radiation or ROS, and can also arise as a consequence of replication-fork collapse—when encountering DNA damage or difficult-to-resolve secondary structures. The best example of the latter is the passing of a replication fork through a SSB, which will then be converted to a DSB on one of the sister chromatids. DSBs are also generated deliberately, as part of the process of V(D)J recombination in B and T lymphocytes, to generate antigen-binding diversity in the immunoglobulin and T-cell receptor proteins. In this process DSBs are generated at specific sites by a nuclease composed of the RAG1 and RAG2 proteins. After these controlled, site-specific rearrangement events, the repair of the resulting DSBs employs the same non-homologous end-joining (NHEJ) process used by the cell to repair randomly induced, so-called illegitimate, DSBs (see below).

DSBs and double-strand lesions in general are highly toxic and their correct repair is critical for the cell to survive. Incorrectly repaired or persistent DSBs can cause chromosomal aberrations, such as deletions, insertions, and translocations, the spontaneous accumulation of which may play a causal role in age-related cellular degeneration and death. Most DSBs are likely to be repaired by either HR or NHEJ. Of these two processes, HR is generally thought to be error-free since it employs a homologous DNA molecule, usually a sister chromatid, as template to overcome the break. As will be described in more detail below, NHEJ merely joins the ends of breaks without the option of distinguishing one end from another when a cell suffers from more than one DSB at the same time.

As for all DNA-repair pathways, there are strong interactions between DNA-damage sensors and the repair of DSBs, in this case especially ATM. The DSB signaling cascade, involving ATM, ATR, DNA-PK, and other proteins, leading to the activation of various downstream substrates, such as p53, has already been discussed. Although the details regarding the coordination between DNA-damage sensors and the various DNA-repair executive branches are not clear yet, components of DNA-damage signaling are known to be involved in repair per se. The role of ATM and especially DNA-PK in activating NHEJ has already been mentioned. In addition, ATM phosphorylates active participants in HR, including BRCA1. The DNA-damage checkpoints are also thought to be responsible for the p53-dependent elevation of the levels of deoxyribonucleotides in the nucleus to facilitate the DNA synthesis steps of DSB repair[336]. Most importantly, DNA-damage sensors are thought to play a major role in facilitating access of the repair enzymes to the sites of damage by reorganizing chromatin structure. We have seen that this is true for the excision-repair pathways and it applies equally well to DSB repair. As already discussed, chromatin remodeling as a prelude to DSB repair is triggered in mammalian cells by phosphorylation of H2AX by ATM or DNA-PK. Finally, if all else fails, the damage signaling pathways may lead a cell into apoptosis or cellular senescence. Since DSBs are so toxic, it is likely that they contribute significantly to both spontaneous apoptosis events and the accumulation of senescent cells over a lifetime in a given mammalian tissue. As will be discussed in Chapters 5 and 7, apoptosis and cellular senescence are two of the three major cellular endpoints contributing to aging, neoplastic transformation being the third. Unfortunately, we do not yet know how a cell decides to enter apoptosis or cellular senescence. It is possible that, similar to the decision to abandon repair, it is the amount of damage that dictates whether a cell should die or enter a state of permanent replication arrest.

Below I will briefly discuss the two major pathways for repairing DSBs and the physiological consequences associated with their action or lack of action. For more detailed discussions of the repair of double-strand lesions in general, including the highly toxic interstrand cross-links, the reader is referred to several excellent recent reviews[337–340].

4.3.4.1 Repair by homologous recombination (HR)

Repair by recombination is probably the most ancient form of repair in the living world. It has been proposed that sexual reproduction, which is essentially HR contributing to the generation of genetic diversity, arose very early in evolution as a way of overcoming damage in the genome[165]. A model for an early form of recombinational repair has been presented by Michael Cox (Madison, MI, USA)[164], which is essentially a form of postreplication repair, applicable to either an RNA or DNA genome, that requires an homologous double-stranded nucleic acid from elsewhere.

Most of our knowledge of HR as a mechanism to overcome DSBs stems from work with bacteria and yeast. This has resulted in the elegant DSB-repair model for meiotic recombination by Szostak *et al.*[341]. The pathway is obviously well conserved and critically important in mammals. The mechanism of the repair of DSBs by HR in mammalian cells is essentially similar to the Szostak model or the ancient form of HR proposed by Cox. Before strand invasion, the first major step in HR, the ends of the DSB undergo resection in the 5' to 3' direction by the 5' to 3' exonuclease activity of the MRN complex (see section on DNA-damage sensors). This results in 3' single-stranded tails, the targets for a complex of proteins originally discovered in yeast and known as the RAD52 group. Mammalian homologs of all the factors in this group have been described. From the molecules in this group, it is the RAD51 protein that acts as the mammalian counterpart of *E. coli* RecA by forming a nucleoprotein filament that coats the single strands and invades the exchange partner of the damaged DNA. Once identified, the homologous sequence is subsequently used as a template for DNA synthesis to extend the single-strand DNA tails. The strands are sealed to the correct parental strand by DNA ligase. The resulting Holliday junctions are resolved by resolvases to yield two intact DNA molecules. This leads to crossover and non-crossover products, depending on the mechanism of resolution. As we will see later, the protein helicase encoded by the BLM gene, which is defective in the human disease Bloom syndrome, plays an important role in the resolution process.

There is still considerable debate about the details of HR, which may not always be the same and could also differ from species to species. The entire process, which involves many different actors, including the above-mentioned BLM helicase and the breast cancer-susceptibility proteins BRCA1 and BRCA2, is concentrated in nuclear foci. These foci can be visualized under the microscope after fluorescent staining with antibodies against one or more of the participants.

In principle, HR is error-free, since it uses a clean copy of the damaged DNA molecule as a template. The ideal template in this respect is a sister chromatid, which is why HR is thought to be particularly active in repairing DSBs that are induced during the S/G_2 phase of the cell cycle. However, HR could also use the homologous chromosome as its exchange partner during the G_1 phase of the cell cycle, when a sister chromatid is not present. This requires searching around in the nucleus for the homolog, a cumbersome process which may explain why G_1 cells prefer to repair DSBs through NHEJ (see below).

Another reason to avoid the homologous chromosome as template for HR is the possibility that mutations arise as a consequence of crossing-over. In the absence of crossing-over the repaired segment will simply acquire the donor sequence, a form of nonreciprocal transfer called gene conversion. With crossing-over, however, large parts of a chromosome are exchanged with a substantially increased risk of loss of heterozygosity (LOH); that is, the loss of the single functional allele when the other is already inactive due to a pre-existing mutation. This is a major reason to avoid crossing-over when resolving Holliday junctions. Candidate proteins for resolving Holliday junctions in mammalian cells (and possibly also in yeast) without crossover products include the protein encoded by the BLM gene, a member of the RecQ family (see below).

In spite of the penalty, mitotic recombinations do occur as demonstrated[342], indicating that HR does occasionally use the homologous chromosome as template, thereby generating crossover products. However, they would probably occur much more frequently were it not for its apparent suppression by MMR components on the basis of sequence differences between the potential exchange partners. This effect of increased genetic distance has been demonstrated by a comparison of progeny from crosses between strains of mice with significantly diverged nucleotide sequences; in progeny of such strains, mitotic recombination is hardly detectable[343] (see also Chapter 6).

The impact of mitotic recombination on spontaneous levels of genomic instability during aging of mammals is unclear. In budding yeast, however, a dramatic increase in LOH events during replicative aging of this organism has been reported[151]. This phenomenon is unrelated to the accumulation of extrachromosomal rDNA circles in yeast mother cells (see Chapter 2), which appeared to cause replicative aging in at least some strains of yeast and may also arise as a consequence of illegitimate recombination events. The possible relationship of these two forms of genomic instability in yeast to the aging process will be discussed extensively in Chapter 7.

Another disadvantage of HR repair in G_1 involves the possibility of a DSB occurring in any of the large number of repeat elements in the mammalian genome. In the absence of a conveniently lined-up sister chromatid, this could easily lead to the selection of a repeat family member on another chromosome as exchange partner. This would result in chromosomal translocations. When the exchange partner is a repeat family member on the same chromosome this would result in deletions, a potential mechanism for the appearance of extrachromosomal circular DNA in aging yeast mother cells. Whereas in mammals there is no evidence that extrachromosomal circles accumulate with age as in yeast, extrachromosomal DNA is common and possibly also a consequence of erroneous recombinational repair[152] (see also Chapter 6). To limit such instability, HR is restricted to late S and G_2, but it is nevertheless possible that a significant portion of the chromosomal aberrations as they are observed in human and animal cells are a result of such illegitimate recombinations.

Finally, in an error-prone variant of HR DSBs are repaired in a pathway that does not use the sequence information from a sister chromatid or homologous chromosome. This mechanism is called single-strand annealing. Like HR, single-strand annealing requires the presence of DNA sequence homology, in this case in the form of complementary sequences on both sites of the break. The process begins very similar to HR by resecting the 5' ends of the break. This resection exposes complementary regions within the 3' overhangs, usually repeat sequences. In mammalian cells such repeats are of course frequently present. Annealing takes place at these repeats with the loss of the flap-overhangs (by a FEN1-like nuclease). Loss of the sequence between the repeats is the inevitable result of single-strand annealing, which is therefore always mutagenic.

Apart from DSBs, HR is also involved in repairing the highly toxic interstrand cross-links[339]. Interstrand cross-links are induced by chemotherapeutic agents, such as mito-mycin C, and cisplatin, but also by natural compounds in plants as well as endogenous agents formed during lipid peroxidation. To effectively remove interstrand cross-links, different repair pathways have to act in concert. Repair of interstrand cross-links in *E. coli* has been well characterized and is based on incisions by the NER enzymes UvrABC, followed by HR. Alternatively, a combination of NER and translesion synthesis is employed. As expected, in yeast and mammalian cells the situation is somewhat similar, but more complicated. In mammalian cells it is unknown how the interstrand cross-links are recognized. Also the precise role of the NER enzymes in mammalian cells is unclear. However, Chinese hamster ovary cells with defects in the endonuclease ERCC1–XPF are exquisitely sensitive for agents that induce interstrand cross-links. Moreover, ERCC1-knockout mice have a much more severe phenotype than other NER mutant mice. They are not only UV-sensitive, but display premature aging symptoms and live only for a few weeks (for a more detailed discussion of this mouse model, see Chapter 5). Like the ERCC1–XPF mutant Chinese hamster ovary cells, cells from these mice are very sensitive to mitomycin C and other agents that induce interstrand cross-links. There is no evidence that XPG, the other endonuclease active in NER, plays a role in interstrand cross-link repair. The role of ERCC1–XPF in repair of interstrand cross-links is unclear and it may be based on its function in removing the non-homologous DNA tails during HR.

Defects in interstrand-cross-link repair have been implicated as the cause of Fanconi anemia, a rare, autosomal recessive disorder, with developmental abnormalities, progressive bone-marrow failure, and a greatly increased risk of cancer[344]. The disorder is characterized and diagnosed by the sensitivity of the patient's cells to agents inducing interstrand cross-links. Fanconi anemia is a genetically heterogeneous disorder with as many as 12 complementation groups identified, corresponding to defects in distinct genes involved in the genome-maintenance pathway represented by this disease. Some protein products of the identified genes have been shown to functionally or physically interact with BRCA1, RAD51, and the MRN complex. Indeed, one protein, FANCD1, was identified as BRCA2, a human breast cancer-susceptibility gene. Like BRCA1, this gene has been implicated directly in the process of HR repair (see above). The function of most of the Fanconi anemia proteins is still unclear, as is the exact nature of the condition, which is also characterized by symptoms of oxidative stress.

4.3.4.2 Non-homologous end-joining (NHEJ)

In yeast, HR is the preferential pathway to repair DSBs, although NHEJ also occurs. This is different in mammals, in which NHEJ is extensively employed. This difference in preference is likely to be due to the greatly increased time mammalian cells *in vivo* spend in G_1 or G_0. Indeed, NHEJ seems to be especially important in differentiated cells; early in development, HR is probably more important. In mammals, NHEJ is also used to repair the RAG-generated DSBs in the process of V(D)J recombination. This can only occur by NHEJ, as indicated by the observation that inactivation of each of the six NHEJ genes in humans or mice prevents B- and T-cell maturation, resulting in severe combined immunodeficiency (SCID)[345].

The main player in NHEJ is the Ku heterodimer with its 70- and 80-kDa subunits, known as Ku70 and Ku80, respectively. Ku80 is sometimes also referred to as Ku86. As we have seen, Ku is a part of DNA-PK, which consists of the Ku heterodimer and DNA-PK$_{CS}$, the catalytic component with kinase activity upon binding to DNA ends. Ku, which is basically a ring, binds to the DSB by threading its ends. In this way it holds them together, probably protects them against nucleolytic attack, and also serves as the recruiting platform for the repair proteins, including DNA-PK$_{CS}$. The role of the latter in NHEJ is not exactly clear. We have already seen that this giant protein kinase of the PIKK family, like ATM and ATR, plays a role in DNA-damage sensing and signaling, especially in signaling the apoptotic response. It also plays a role in telomere maintenance and it is activated by bacterial DNA as part of the innate immune response. In NHEJ its function must be specific for vertebrates since they are the only organisms in which it has been identified.

Two molecules of DNA-PK$_{CS}$ on each side of the break may assist Ku to hold the DNA ends together. The structures of these ends, however, are not amenable to a simple ligation (as is also usually the case with SSBs; see above under BER). They may have single-stranded overhangs and damaged bases or sugar moieties that require processing. It is likely that the protein called Artemis plays an important role in resolving the various DNA end modifications[346]. Artemis binds to DNA-PK$_{CS}$ and is phosphorylated. This stimulates and extends the nuclease activity of Artemis, so that the protein becomes capable of cutting away protruding single-stranded regions at DNA ends and creating double-stranded structures that are good ligase substrates. Other factors, such as the Werner syndrome protein (WRN) and the previously discussed MRN complex may play a role. The polymerases needed to fill the gap are those belonging to the X family. These enzymes, such as polμ, polλ, or the terminal deoxytransferase (TdT), are apparently utilized depending on the template. DNA ligation is subsequently carried out by ligase IV in a complex with XRCC1.

NHEJ is active throughout the cell cycle in all vertebrate tissues and only competes with HR in the S and G_2 phases. In contrast to HR, NHEJ is error prone. Part of the necessary processing of DNA ends before they can be religated is resection. We have seen this happening in HR, which is not a problem since it relies on its identical sister chromatid to retrieve the lost information. In NHEJ, however, such resection leads to the permanent loss of between 1 and 20 nucleotides each time a DSB is repaired. Realizing the potential impact of such a continual loss of small amounts of genomic sequence over the lifetime of an organism, Michael Lieber (Los Angeles, CA, USA) has made an attempt to estimate its consequences in terms of the loss of gene activity of an average cell[347]. Based on his estimate of 5–10% of all cells harboring a DSB at all times and a repair time of 24 h, he calculated that at the end of a human life each cell would have 2300 imprecise repair sites distributed throughout the genome. In a genome with 50 000 genes (we now know that this is probably less) and only 5% of the genome functional (a very low estimate; Chapter 3), he reached the not inconsiderable number of 430 genes being adversely affected.

A second potential source of errors associated with NHEJ is its lack of a mechanism to distinguish DNA ends corresponding to different DSBs arising in the same cell at the same time. It is likely that NHEJ is responsible for random integration of transforming DNA in transgenesis via sporadic DSBs in the chromosomes. It has been demonstrated that by increasing the number of DSBs in a cell from one to two the frequency of translocations is increased by at least 2000-fold[348]. The situation will be worse with more DSBs in a single cell. Using γH2AX as a marker for multiple DSBs induced by a linear track of α particles, Aten *et al.*[349] observed congregation of these breaks into small clusters in G_1-phase cells. This supports the notion that distant DSBs can be juxtaposed, which would greatly increase the generation of translocations. Whereas we may assume that Ku would be quick enough to capture the ends of each DSB before they can float apart, the promiscuity of NHEJ could lead to chromosomal translocations, especially when multiple DSBs are juxtaposed. Interestingly, whereas one might think that inactivation of NHEJ would prevent such translocations (DNA ends cannot ligate to one another by themselves) its absence in the mouse has been associated with increased chromosomal aberrations[350]. It is possible that in such a situation HR functions as a back-up, inaccurate in G_1 as described above, or single-strand annealing of complementary single strands from different chromosomes. Finally, non-classic forms of end-joining could be active that have thus far been elusive.

4.3.4.3 Effect of DSB repair defects in animals

As for all genome-maintenance pathways, mouse models have been made in which HR or NHEJ genes were inactivated. The results indicate that deficiencies in critical HR genes are usually fatal, but those in genes involved in NHEJ generally not. For example, inactivation of key genes involved in the crucial strand-invasion step of HR, for example, *RAD51 BRCA1*, and *BRCA2*, is embryonically lethal. This lethality is understandable in view of the role HR plays in repairing the DSBs that must arise when the replication fork meets a SSB. By contrast, most gene deficiencies in core NHEJ genes, for example *Ku70, Ku80*, and *DNA-PK$_{CS}$*, result in viable mice. This lack of lethality must be due to the relative scarcity of non-dividing cells during development. Indeed, the animals do suffer from SCID (due to the defect in V(D)J recombination) and they show multiple symptoms of accelerated aging (see Chapter 5).

4.3.5 ANCILLARY SYSTEMS

In addition to the main DNA-damage checkpoint and repair pathways discussed above, a number of systems have been identified for resolving particular problems related to DNA damage and genome instability. These systems often closely interact with the DNA-repair systems per se; that is, those that act to remove lesions, such as the excision-repair systems. Here I will discuss three of such ancillary systems: telomerase, RecQ helicases, including WRN, and DNA topoisomerases. Inactivating mutations in genes participating in these systems in the mouse result in phenotypes resembling accelerated aging, similar to the situation for NHEJ and NER.

4.3.5.1 Telomere maintenance

As discussed in Chapter 3, mammalian telomeres are composed of TTAGGG repeat arrays bound by a complex of proteins and organized in a so-called T loop (Fig. 4.7a). This prevents attack by DNA-repair enzymes, which would otherwise recognize these chromosome ends as DSBs. In contrast to bacteria, which have a circular genome, eukaryotes are unable to complete lagging-strand synthesis during replication (see Chapter 3 for a detailed description). As a consequence, 50–200 bp of telomeric repeats are lost with each cell division. The telomerase reverse transcriptase (TERT) maintains telomere length by copying a short template sequence within its intrinsic RNA moiety (TERC), running 5' to 3' towards the end of the chromosome. It extends the 3' tail, after which normal lagging-strand synthesis generates the opposite strand (Fig. 4.7b). This process is controlled by regulator proteins, some of which (TRF1, TIN2) inhibit telomerase-mediated lengthening to prevent unrestrained elongation of telomeres.

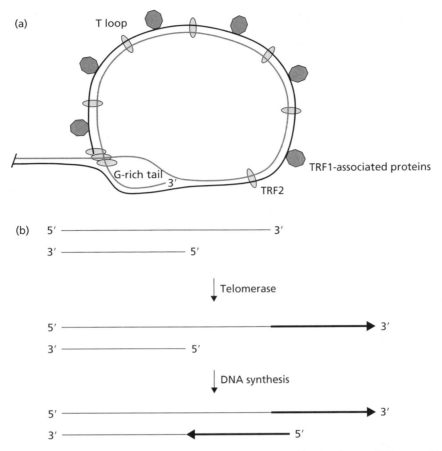

Fig. 4.7 (a) The telomeric ends of chromosomes are organized in the form of T loops. (b) Action of telomerase as it corrects the effects of the end-replication problem.

Both telomeres and telomerase are important genome-maintenance systems. The T-loop structure prevents NHEJ from fusing different chromosomes together and telomerase prevents telomeres from becoming so short that they can no longer form T loops. Telomeres that shorten to the point that they become dysfunctional cause replicative senescence by activating p53. Inactivation of the p53 pathway allows cells to bypass replicative senescence and proliferate until they reach a stage called crisis, which is characterized by widespread chromosomal instability (with its characteristic telomeric end-to-end fusions), apoptosis, mitotic catastrophe, and senescence, with the rare emergence of immortal cell clones all of which have acquired a mechanism to re-stabilize their telomeres. The mechanism of such genomic instability is not entirely clear but NHEJ is the most likely culprit[351]. Inactivation of the p53 pathway is enough to bypass senescence due to telomere dysfunction.

Normal human cells do not express telomerase at all, except embryonic tissues, germ-line tissues (testes and ovaries), and progenitor cells in tissues with high cell turnover, such as bone marrow and intestinal crypts. As mentioned above, in the majority of human cancer cells telomerase activity is easily detectable. Such cells exhibit stable telomere lengths upon extended propagation in culture and do not undergo replicative senescence. In mouse cells, the situation is different in the sense that most laboratory strains have very long telomeres and *TERT*, the gene encoding the telomerase reverse transcriptase component, is widely expressed at low levels in adult tissues, with greatest abundance during embryogenesis and in adult thymus and intestine[352]. The continued availability of telomerase in mice could be a major factor determining its short lifespan as compared to humans. Cancer, a major aging-related disease, is kept at bay much longer in humans than in mice (in absolute time), possibly because of the strict regulation of telomerase as a major gatekeeper to prevent immortalization. The difference in telomere length and telomerase regulation also explains, at least in part, why senescence in mouse cells is fundamentally different from senescence in human cells[353].

In humans, even a partial loss of telomerase function already causes serious disease. This is illustrated by the human inherited disorder dyskeratosis congenita, an autosomal dominant form of which is caused by mutations in one of the two copies of the gene that encodes TERC, the RNA part of telomerase. These patients suffer from abnormalities in the production of cells in the blood and in the gut, and have poor wound healing, early baldness, hypogonadism, and lung fibrosis. Not surprisingly, they generally die early. From these symptoms it can be derived that the cells affected are progenitor cells in tissues with a high cell turnover, which is very similar to the situation in *Terc*-deficient mice. In such animals symptoms manifest only after several generations because telomeres of most laboratory strains of mice are so much longer than those of humans. Interestingly, whereas the short telomeres in *Terc*-deficient mice are associated with increased cancer incidence as a consequence of the genomic instability, some tissues and cell types seem to suffer less from cancer as compared with normal control mice. For example, late-generation *Terc*-null mice are resistant to skin tumorigenesis[354]. Reduced cancer has also been observed in mice deficient in both telomerase and p16 (INK4a)[355]. This may reflect slow tumor progression in these tissues due to high levels of tumor cell death as a consequence of their short telomeres, which would be in keeping with the need of most tumors to stabilize telomeres, which is most easily accomplished by induction of telomerase (see above). It is important to notice that these mice have intact DNA-damage responses. Hence, whereas the increased genomic instability accelerates tumor formation, the constitutive activation of DNA-damage response factors, such as p53, will greatly impair their progression by inhibiting cellular proliferation. Impairment of cellular proliferation through short telomeres will also adversely affect stem-cell reservoirs, which may explain the premature appearance of multiple aging-related phenotypes in *Terc*-null mice. The antagonism between cancer and aging and its possible implications for the causes of normal aging will be discussed further in Chapters 5 and 7.

Clearly, in tissues with high cell turnover telomerase is critically important. However, it is not the only system that can extend and maintain telomeres. As first discovered in yeast[356], *est1* mutants (lacking telomerase) can adapt to telomere erosion by two mechanisms termed type I and type II survival. Type I survivors require RAD51, a sure sign that HR is involved[357]. Indeed, in mammals, the so-called alternative lengthening of telomere (ALT) pathway was first identified in replicatively immortal, tumor-derived cell lines that do not express telomerase[358]. Dividing tumor cells cannot do without a mechanism to maintain their telomeres and generally manage to switch telomerase on. If telomerase expression cannot be restored, the ALT pathway is an alternative. Its activity can be recognized by the accumulation of so-called ALT-PML bodies, a variant of the PML bodies mentioned in Chapter 3, in this case consisting of DNA-repair proteins co-localized at telomeres. Interestingly, HR may play a role in normal telomere maintenance as well. This is suggested by the presence of RAD51D, one of the five RAD51 paralogs, at telomeres of normal mouse cells[359]. Its absence was shown to cause chromosomal aberrations, including typical end-to-end fusions. Such chromosomal instability was also observed in telomerase-negative immortalized human cells in which RAD51D was suppressed by siRNA[359].

Many different DNA-repair proteins have been localized at telomeres, including Ku, DNA-PK_{CS}, BRCA1, BRCA2, and the MRN complex, as well as the RecQ proteins WRN and BLM (see below). It is possible that these repair factors play a role in maintaining telomeres, perhaps in association with telomere-specific proteins. This is somewhat paradoxical since many of these proteins are supposed to repair broken DNA molecules, presumably including the ends of uncapped chromosomes[360]. In yeast, proteins of the NHEJ system may be intimately involved in maintaining telomere length, as indicated by the shortened telomeres in cells lacking either of the two Ku proteins. Also in mammals there is evidence for DNA-repair proteins playing a role in normal telomere maintenance, but the details of this involvement are still incompletely understood. Alternatively, telomeres could merely be a storage site for DNA-repair proteins, similar to nucleoli and PML bodies.

Whereas we now know that repair proteins are located at telomeres and may even play a role in telomere maintenance, the recognized telomere cap proteins, such as the telomere repeat factors TRF1 and TRF2, are clearly the most important in this respect. As already mentioned, TRF1, together with its interacting partners TRF2, TIN2, TANK 1, and TANK 2, are regulators of telomere length[361]. Some of these factors negatively control telomere length by promoting a telomeric architecture that limits the ability of telomerase to access telomeres. For example, dominant negative forms of TIN2 or TRF1 can extend telomere length[362]. The importance of such regulation is illustrated by lack of viability of TRF1-deficient mice[363]. Interestingly, this embryonic lethality could not be rescued by telomerase, suggesting that it is not due to uncapping or attrition of telomeres.

TRF2 has mainly been implicated in the formation of the T-loop structure and in preventing repair enzymes from recognizing telomeres as DNA DSBs. After TRF2 deletion in mouse cells, the telomeres lose the 3' overhang and are processed by the NHEJ pathway. Overexpression of TRF2 in the mouse skin was found to resemble the NER disorder XP, with keratinocytes from these animals hypersensitive to UV and DNA cross-linking agents[364]. Skin cells of these mice were also found to have short telomeres and loss of the

G-strand overhang. It was suggested that the increased TRF2 raises the activity of XPF (with which it is known to interact, as with many other repair proteins) at telomeres, possibly at the cost of XPF activity elsewhere in the genome, which may explain the increased UV sensitivity.

Evidence is now emerging that whereas general DNA-repair proteins are important for telomere maintenance, telomere-specific proteins may be important for genome maintenance in general. Interestingly, TRF2 was recently found to quickly and transiently localize at DSBs induced by a laser microbeam at all possible locations in the genome[365]. This suggests a general role for TRF2 in genome maintenance, possibly in preventing premature action of DNA-repair enzymes until the correct repair complex is assembled.

How important is telomere maintenance in the context of maintaining the genome overall? As I will discuss in more detail in Chapter 6, telomere length measurements, mainly in white blood cells, have revealed that telomeres shorten during human aging. However, age-adjusted telomere length is highly variable, possibly because of heritable factors and/or disease. This considerable individual heterogeneity and the overlap in telomere lengths between young and elderly individuals renders any correlation weak and the significance unclear. Nevertheless, a relationship between enhanced telomere attrition, reduced cancer, and a more pronounced manifestation of non-cancer, degenerative symptoms of aging can be rationalized. Telomere shortening protects us from cancer by impairing tumor cell proliferation, through cell-cycle arrest or by activating replicative senescence and apoptosis. However, this would also result in a loss of functional cells, which in turn can be expected to lead to age-related organ dysfunction. Adverse effects of telomere shortening may be especially prominent in stem-cell reservoirs. Indeed, it has been demonstrated that telomere shortening (in *Terc*-deficient mice) inhibited mobilization of epidermal stem cells out of their niche in hair follicles, whereas telomerase overexpression promoted stem-cell mobilization[366]. The latter is apparently not related to telomere lengthening but a reflection of another TERT functional pathway[367]. On the other hand, overexpression of TERT in mouse hematopoietic stem cells had no effect on the transplantation capacity of these cells; irrespective of telomerase activity, such cells could be transplanted for no more than four generations[368]. Of note, the situation could be very different in humans due to a difference in telomere biology between the two species. Nevertheless, it is possible that whereas telomere attrition is the main barrier of replicative capacity of human cells in culture, it may not necessarily exert a major influence on cells *in vivo*.

4.3.5.2 RecQ helicases and DNA topoisomerases

Members of the RecQ helicase family occur in organisms varying from prokaryotes to mammals and are named after the RecQ protein originally identified in *E. coli*. In mammals there are five RecQ helicases, defects in three of which have been associated with genomic instability, cancer, and premature aging in humans in the form of the heritable, segmental progeroid disorders Bloom syndrome, Werner syndrome, and Rothmund–Thomson syndrome[369]. The RecQ homologs share a central helicase domain, with only the Werner syndrome protein WRN harboring an exonuclease activity as well. It is also Werner syndrome, and to a lesser extent Rothmund–Thomson syndrome, in which premature aging is especially prominent; Bloom syndrome is mainly characterized by excessive tumor formation (see Chapter 5 for a more detailed discussion of segmental progeroid syndromes). Most of what we know about RecQ helicases in mammals is derived from the properties of the BLM and WRN proteins.

RecQ homologs are able to catalyze unwinding of many different DNA substrates, such as forked DNA, Holliday junctions, and triple and tetraplex DNA (see Chapter 3). Whereas all the probably multiple roles of RecQ homologs in genome maintenance are not yet known, they are generally thought to be ancillary factors in replication, recombination, or repair by resolving secondary structures. The best-characterized RecQ protein in this respect is BLM. As established by Ian Hickson (Oxford, UK) and collaborators, one function of BLM, in concert with topoisomerase IIIα (TOPIIIα; see below) is to resolve recombination intermediates containing double Holliday junctions by a process they called double Holliday junction dissolution[370]. This function is apparently specific for BLM since WRN and other RecQ homologs cannot substitute for BLM in these dissolution reactions. Human Bloom-syndrome cells show a characteristically elevated level of spontaneous exchanges between sister chromatids and homologous chromosomes, which indicates increased HR. BLM's resolution of recombination intermediates avoids crossover events during HR. It was proposed that this would protect organisms with large genomes, containing repetitive sequences, against mutagenic genome rearrangements as a consequence of exchange between non-sisters. The danger of such erroneous forms of HR in generating LOH events or other types of genomic mutation has already been mentioned in the section on HR. Since Bloom-syndrome cells are not hypersensitive to ionizing radiation, indicating that BLM is not essential for the repair of DSBs, it was proposed that crossover suppression through this mechanism was especially important in the repair of daughter strand gaps arising during replication of a damaged template. As discussed earlier, this postreplication mechanism for damage avoidance relies on HR.

Although none of the other RecQ family members is able to substitute for BLM in suppressing crossing-over through Holliday-junction dissolution, they may all be involved in decreasing genomic instability during some form of DNA transaction. This would explain the increased genomic instability observed in cells from Bloom syndrome and Werner syndrome as well as the high cancer incidence. It should be noted that mutant cells from both Werner syndrome and Bloom syndrome are viable and the existence of the human syndromes shows that impairment of RecQ functions is not incompatible with life, but merely increases spontaneous levels of genomic instability and cancer. However, in mice complete BLM deficiency is embryonically lethal[371,372] and it is likely that this is true also in most human cases. Indeed, Bloom syndrome is much rarer than expected from the number of heterozygotes. So it is likely that only some allelic combinations allow for pre- and postnatal survival. WRN-deficient mice show virtually no phenotypic differences from their wild-type counterparts, unless they also harbor null mutations in *Terc*, *Atm*, or *p53* (see Chapter 5).

Whereas BLM and WRN display a vast overlap in function *in vitro*, there are significant differences *in vivo*. Of the two syndromes, Bloom syndrome is the most severe, with death usually before the age of 30. Werner syndrome patients die about 20 years later, mostly of cancer and cardiovascular disease. Cancer in Werner syndrome develops much later than in Bloom syndrome and is restricted to tumors derived from mesenchymal cells. Cells from Werner syndrome do not show the greatly increased rate of sister chromatid exchange as Bloom-syndrome cells. There is evidence that the WRN protein is required for maintaining telomere function during replication. Cells lacking WRN have been demonstrated to suffer from a loss of telomeric sequences from a single sister chromatid, which was explained by a defect in lagging strand synthesis[373]. Telomere loss was dependent on the helicase function of WRN alone and could be counteracted by expression of telomerase.

WRN has an exonuclease function that BLM lacks. It is possible that the exonuclease activity is needed to process telomeric DNA. Another difference between the two RecQ homologs is that WRN, but not BLM, plays a role in NHEJ. This can be derived from observations that WRN binds to the Ku70-Ku80 component of DNA-PK, which stimulates the exonuclease, but not the helicase activity of WRN[374]. There is evidence that WRN acts in optimizing the repair functions of NHEJ and HR[375]. Nevertheless, at least in mice, its role in these processes cannot be that important since WRN-null mice do not suffer from similar problems as other mouse models in which a NHEJ or HR core protein has been inactivated.

RecQ proteins have many binding partners among DNA-repair and -replication proteins, as can be expected from their role in assisting in the seamless performance of these processes. A most frequently found interaction is with DNA TOPIII, which is a type I topoisomerase. Type I DNA topoisomerases can change the topological status of the DNA by introducing a DNA SSB and then transfer another DNA strand through this break. By doing this they change the degree of supercoiling of the DNA. Physical interaction between Sgs1, the yeast RecQ helicase, and BLM with TOPIII has been described for yeast and human cells, respectively. Deletion of the TOPIII homolog in yeast (TOP3) results in slow growth, hyper-recombination, genomic instability, impaired sporulation, and increased sensitivity to genotoxic agents[376]. Interestingly, this phenotype is suppressed by deletion of Sgs1, the yeast RecQ helicase, indicating that Sgs1 creates a deleterious topological substrate that needs to be resolved by TOP3[377]. In vertebrates, there are two isoforms of TOPIII, termed α and β. The human BLM protein functionally interacts with TOPIIIα. In mice, deletion of TOPIIIα is embryonically lethal. By contrast, mice lacking TOPIIIβ are viable and grow to maturity with no apparent defects. However, once adults these animals show a reduced lifespan and display lesions in multiple organs resembling premature aging (see Chapter 5). It is conceivable that this effect is due to impairment of a RecQ family member rather than a direct consequence of the lack of TOPIIIβ. Thus far, there is no evidence that WRN interacts with TOPIII. However, it has been demonstrated that the WRN protein physically and functionally interacts with TOPI[378].

4.3.6 DNA REPAIR AND CHROMATIN

Nucleosomes are severe obstacles for processing of DNA lesions and must be disrupted to allow repair. This has been studied most intensively for NER, especially after UV damage. Since the chromatin remodeling events necessary for disruption and restoration of nucleosomal structure associated with DNA repair are likely to depend on the type of lesion and repair pathway involved, we are still far from a comprehensive insight into these processes and their impact on genome stability in aging organisms.

As we have seen, chromatin structure contains epigenetic information essential for genome functioning, which is encoded in the specific patterns of DNA methylation and histone modification. Are there chromatin-repair systems and what are the potential consequences of incomplete or erroneous restoration of higher-order DNA structure? One important player that has thus far been identified is the chromatin assembly factor CAF-1[379]. This multimeric protein specifically deposits histones onto DNA that has been subject to DNA-repair synthesis. This is very similar to chromatin assembly following DNA replication, which also requires the CAF-1 complex (Chapter 3). As in replication, CAF-1 recruitment depends on its interaction with PCNA. It has been demonstrated that CAF-1 is recruited to sites of NER (or SSB repair), but not in cells that are deficient in repair, such as XP cells. Yeast deleted for the gene encoding CAF-1 is highly sensitive to double-strand DNA-damaging agents[380].

It has been suggested that CAF-1 can also act as a chaperone in mediating the correct re-assembly of the histones displaced during repair. Indeed, another candidate chromatin repair protein is the histone chaperone anti-silencing function 1 (ASF1). It is unclear as to how and where this protein acts, but it is known to enhance the nucleosome-depositioning activity of CAF-1. In yeast, asf1 mutants are sensitive to DNA damage, suggesting its involvement in maintaining chromatin structure after repair[381].

Whereas it is unknown whether aberrant or incomplete chromatin repair can lead to loss of genome integrity through epigenetic changes, evidence for both incomplete restoration of DNA-methylation patterns after repair and methylation changes with age has been obtained (Chapter 6). Methylation patterns are established during embryogenesis and remain largely unchanged in adult cells[382]. Genome-wide demethylation occurs before embryo implantation followed by *de novo* global remethylation (by DNMT3a and

DNMT3b *de novo* DNA methyltransferases) after implantation, before organ development. This allows gene-specific methylation patterns to develop, which determine tissue-specific transcription repression, including repression of either the maternal or paternal allele of imprinted genes. Once established during early life, methylation patterns are replicated during mitosis by the enzyme DNMT1, a maintenance DNA methyltransferase. However, methylation patterns are not completely stable. Gradual hypomethylation of the mammalian genome occurs with age in most tissues as well as aberrant hypermethylation in the promoter regions of genes (see Chapter 6). Interestingly, it has been demonstrated that after UV irradiation of non-dividing cells in culture restoration of the methylation patterns after the resynthesis step of NER was slow and incomplete[383]. This could explain the observed gradual demethylation associated with the aging process. It is unknown whether DNMT1 is also involved in replicating methylation patterns during DNA repair, but this is likely since evidence for an early accumulation of this enzyme (but not the *de novo* DNA methyltransferases) at sites of DNA damage has been obtained[384]. The observed incremental methylation of CpG islands with age of human colon[385] could be due to aberrant *de novo* methylation.

4.3.7 PRE- AND POST-GENOME MAINTENANCE

Naturally, systems for the prevention of DNA damage or the mitigation of its adverse effects are critically important parts of genome maintenance. A detailed discussion of the full repertoire of such systems is beyond the scope of this book. Here I will only briefly discuss the main systems for preventing oxidative DNA damage as well as the physiological buffering of the phenotypic penetrance of deleterious mutations.

4.3.7.1 Antioxidant defense

In view of their abundance as normal by-products of metabolism, ROS, such as singlet oxygen, superoxide, peroxyl radicals, hydroxyl radicals, and peroxynitrite, are considered as probably the main source of spontaneous DNA damage[386]. However, ROS are also used by many cell types in normal processes, for example, by macrophages as part of their ability to defend the body against intruding microorganisms. To prevent ROS from rising to excessive levels, cells are equipped with a variety of antioxidant defense systems. Such systems include the enzymes superoxide dismutase (SOD), catalase, or glutathione peroxidase and a variety of dietary antioxidants. Antioxidants are compounds that protect cells against the damaging effects of ROS. The main dietary antioxidants are vitamin E, β-carotene, and vitamin C. An imbalance between antioxidant defense and ROS results in oxidative stress, leading to cellular damage.

Oxidative stress has been causally linked to many diseases, including cancer, atherosclerosis, ischemic injury, inflammation, and neurodegenerative diseases, such as Parkinson's disease and Alzheimer's disease. As we have seen, oxidative stress has been implicated as the main causal factor in aging and antioxidant defense is therefore considered critically important in longevity assurance. However, the results obtained after inactivating antioxidant defense systems, such as SOD or catalase, or increasing their activity through overexpression of these enzymes or by supplementing antioxidants to the diet, are conflicting. Full or partial inactivation of antioxidant defense systems does not generally result in premature aging, which is in contrast to the aforementioned mice with inactivated DNA-repair systems (see also Chapter 5). Overexpression of antioxidant defense genes, such as SOD and/ or other ROS-scavenging enzymes, has been reported to increase lifespan in fruit flies[140], but in the mouse there are few obvious beneficial effects. Indeed, in my own laboratory we have generated mice overexpressing both SOD1 and catalase as part of large, almost 100-kb bacterial artificial chromosomes, providing most if not all the normal regulatory *cis*-acting control elements. Whereas the activity of these enzymes was upregulated about 2-fold in all tissues analyzed[387], the first results of lifespan studies do not indicate that they live significantly longer (A. Richardson, personal communication).

In humans, clinical trials of antioxidant supplementation have failed to show benefit with respect to disease outcome and sometimes adverse effects were observed. However, these results are controversial because there is evidence that plasma levels of antioxidants are correlated with decreased disease risk[388]. In mice, no effect of long-term dietary supplementation with antioxidants on the pathological outcome or on mean and maximum lifespan has been observed[389].

4.3.7.2 Mitigating the effects of mutations

Although in general DNA mutations are irreversible their effect can sometimes be buffered. An example of such physiological buffering is the action of molecular chaperones, such as the heat-shock protein Hsp90[390,391]. Hsp90 assists with the maturation of many key regulatory proteins. Like other chaperones, Hsp90 recognizes and transiently binds hydrophobic residues often found in incompletely folded proteins. Chaperones, therefore, prevent improper protein interactions, which is especially important under conditions that promote protein unfolding and aggregation, such as environmental stress. Interestingly, this makes them critically important for aging, which has been associated with increased aggregation of disease-related proteins, such as in Huntington's disease, Alzheimer's disease, and Parkinson's disease. The action of chaperones camouflages the adverse effects of polymorphic variants (in the germ line) or accumulated somatic mutations that would normally result in protein-folding defects. Hence, these protein-quality control mechanisms act as effective genetic buffer or capacitor systems against adverse molecular consequences of aging, such as the accumulation of somatic mutations. We have already seen in Chapter 2 that overexpression of heat-shock genes increases the lifespan of nematodes and flies, which illustrates the importance of these systems in aging.

The repertoire of buffering mechanisms is surprisingly diverse and does not only include mutations in the client protein itself, but also in other proteins as well as in gene-regulatory regions. However, all such proteins become dependent on the chaperone and compromising the function of the capacitor—for example, through a mutation or increased environmental stress—would uncover the hidden mutation load. Although Hsp90 is highly abundant and can be further induced by heat stress, it can be overwhelmed when more and more proteins are destabilized.

4.4   Genome maintenance and aging

A key element in current theories of how we age is the accumulation of somatic damage. The decline in the efficacy of natural selection in protecting the soma after the age of first reproduction is the most likely ultimate cause of age-related cellular degeneration and death. Since the first replicators, genome maintenance has been the most important somatic maintenance system, preventing untimely death of its individual carrier and maintaining its genetic integrity, while simultaneously providing the variation on which evolutionary success depends. Together, the various pathways for processing natural damage to DNA and regulating information transfer provide a balance between individual stability and evolutionary diversity. The possible role of genome-maintenance systems in the control of aging and longevity has been recognized almost since the discovery of DNA repair. At that time, the first somatic mutation theory of aging had already been formulated by Leo Szilard[392]. Alterations in the genome of somatic cells and the possibility that they are a primary causal factor in aging will be discussed extensively in the following chapters. Here I will briefly consider the role of genome maintenance in controlling the rate and extent of such changes and its possible relationship to aging and lifespan.

4.4.1   GENOME MAINTENANCE AS A FUNCTION OF AGE

The possible decline of genome maintenance, especially DNA repair, with age and its relation to species-specific lifespan were major topics of aging research, especially during the 1970s and 1980s. Testing these hypotheses was plagued by difficulties encountered in measuring DNA-repair capacity and interpreting the results obtained. From a technical point of view it is clear that there is no easy way to obtain an integral measure for the multitude of DNA-repair activities as they take place in an intact organism. Indeed, even an assessment of one or few DNA-repair activities in a given tissue can be fraught with error. This is simply due to the fact that we know far too little about the mechanisms of repair in intact tissues or cells taken directly from the situation *in vivo*. The most reliable way of measuring the capacity of an organism to repair damage in DNA is to analyze the removal of known lesions induced through treatment with relatively low, physiological doses of a genotoxic agent. In my laboratory we have treated rodents with benzo(*a*)pyrene or *N*-acetyl-2-aminofluorene and measured the removal of the main lesions from liver and other tissues in young and old animals[393,394]. Whereas significantly lower levels of removal were observed in the old animals, the difference was only small. Similarly, we have analyzed the capacity of primary rat fibroblasts, derived from young or old animals, to perform DNA-repair synthesis after UV irradiation. Only a very small decline in repair synthesis in cells derived from the old animals was observed[395].

More recent work has focused on the analysis of BER as a function of age, in view of the importance of this pathway in processing oxidative DNA damage caused by ROS. ROS are now considered as a major contributing factor to the various degenerative aspects of aging. BER activity, measured as the ability of nuclear extracts from various tissues to repair BER substrates, such as uracil, in synthetic duplex DNA, was found by Cabelof *et al.* to significantly decline with age in various tissues of rodents: brain, liver, spleen, and testes[396]. This reduction by 50–70% in BER activity correlated with decreased levels and activities of DNA polβ, as well as with increased spontaneous mutation frequencies. Similar results were obtained by Intano *et al.*, who specifically looked at germ cells, but in a later study compared brain and liver extracts as well[397]. Interestingly, these workers found that the age-related BER defect in germ-cell nuclear extracts could be restored by the addition of AP endonuclease, suggesting that this enzyme is rate limiting rather than DNA polβ, at least in this tissue[398]. These investigators have now also analyzed other BER substrates and their results indicate an age-related decline in BER not only using G–U mismatches as substrates, but also for abasic sites. However, they did not observe a decline with 8-oxoG (C. Walter, personal communication). To some extent, these findings are in conflict with results obtained by Imam *et al.*[399], who observed a significant age-related decrease in the *in vitro* repair of uracil, 8-oxoG and 5-hydroxycytosine in mitochondrial extracts from various regions of the mouse brain. However, in this study nuclear extracts were also tested, without evidence for an age-dependent change in uracil repair[399].

Using a synthetic DNA duplex with a 1–4 nucleotide gap in one of the strands, evidence has been obtained that neuronal extracts from old rats are deficient in completing gap repair, a measure for BER, possibly due to a deficiency of DNA polymerase β and/or DNA ligase[400]. Interestingly, also using synthetic DNA templates and neuronal extracts, these same workers provided evidence for an age-related decline in NHEJ activity[401].

Overall, therefore, there is evidence that NER and BER activities and possibly NHEJ activity decline with age, albeit sometimes only marginally and with some discrepancies between the different studies. The possible consequences of the age-related decline in BER activity are still unclear. In all these studies BER activity was measured as glycosylase activity in nuclear or mitochondrial extracts. It should be realized that these assays are difficult to carry out in a reliable, quantitative manner. In addition, glycosylase activities in protein extracts are not the same as BER activities in cells or *in vivo*. Even if these results faithfully reflect BER activities *in vivo*, it is possible that any observed decreases reflect switches in the organism's utilization of repair pathways or other adaptations to an altered situation driven by the consequences of the aging process. An age-related decrease in cell-proliferative activity comes to mind as a possibility, which may require less BER activity than at a young age.

Hanawalt[402], Tice and Setlow[403], as well as my own group[404] reviewed this field in the 1980s and early 1990s, and basically came to the conclusion that there is no evidence for a drastic decline in DNA repair during aging. This is not surprising in view of the critical importance of genome maintenance for the cell. It should be noted that a decline in DNA repair is not necessary to explain an accumulation of alterations in the somatic genome. Since by nature genome-maintenance systems are imperfect, one would expect such alterations to accumulate. However, in the presence of declining repair activities such an accumulation would be expected to be exponential rather than linear.

Apart from the absence of a general marker for DNA-repair capacity, it is also very difficult to interpret higher or lower levels of particular repair processes. Genome-maintenance pathways necessarily act in the short-term interest of the cell or the individual. A high activity of a repair pathway could be beneficial for the survival of a cell population affected by genotoxic stress. However, this may be at the cost of reduced genomic integrity, for example, when the repair pathway is error-prone. At old age, therefore, organisms may actually benefit from somewhat reduced repair activities. Along the same lines, an easily triggered DNA-damage response could lead to increased apoptosis rates, which may not be a problem at early age. However, at later age, such increased genome maintenance may cause organ dysfunction due to loss of functional cells. It is because of these considerations that changes in DNA-repair activities are not always easy to interpret.

It is possible that DNA-repair activities decline with age as part of a more general decline in enzymatic function. On the other hand, when it is true that aging is associated with increased genotoxic stress one would probably expect an increase in DNA repair in response to the increased damage. Both possibilities may actually be true, as can be illustrated by the apoptosis response, a major component of genome maintenance. In an important series of experiments, Yousin Suh, then in Seoul, South Korea, first observed an association between the resistance of rat liver to tumor formation after treatment with the direct-acting genotoxic agent methyl methanesulfonate and the robust apoptosis response of this tissue[405]. This was in contrast to the brain, which is highly susceptible to tumor induction by this agent, but displays a lack of apoptotic response. She then tested the apoptotic response in liver to methyl methanesulfonate in rats at old age and found this to be diminished[406]. Hence, whereas at young age the animals responded to the challenge of this mutagen with a robust apoptosis response, underscoring the importance of this system in getting rid of heavily damaged cells to prevent cancer, at old age this defense system was clearly not functioning well. Without treatment, however, she observed a trend towards an increased level of apoptotic cells in liver of old rodents, which confirmed earlier observations by others[407]. Hence, it is possible that a system becomes both less effective at old age in responding to a challenge, but increasingly active during aging due to increased spontaneous genotoxic stress. Subsequent findings by others of a significant reduction in mean apoptotic response to γ radiation with increasing age in human peripheral blood lymphocytes[408] suggests that such a decline may be a general phenomenon associated with the aging process in mammals.

4.4.2 DNA REPAIR IN STEM CELLS

Whereas DNA damage in somatic cells can have numerous deleterious consequences, of which malignant transformation and cell death are the most critical, damage accumulation in stem cells would severely constrain the capacity of such cells to replenish organs. Indirect evidence, such as the increased expression levels of DNA-repair genes in stem cells[409], suggests that genome-maintenance capacity in stem cells is generally higher than in normal somatic cells. The first direct evidence that stem cells do have increased capabilities to maintain their genome integrity comes from the laboratories of Peter Stambrook (Cincinnati, OH, USA) and Jay Tischfield (Piscataway, NJ, USA). These investigators have been mentioned previously with regard to their discovery that LOH as a consequence of mitotic recombination is a frequent spontaneous event in cells *in vivo*. Using the *Hprt* and *Aprt* selectable marker genes as a measure for mutations (described in more detail in Chapter 6), they compared mouse embryonic stem (ES) cells with mouse embryonic fibroblasts[410]. The results indicate a dramatically lower spontaneous mutation frequency in the stem cells compared with the embryonic fibroblasts (Fig. 4.8). Mutation frequency

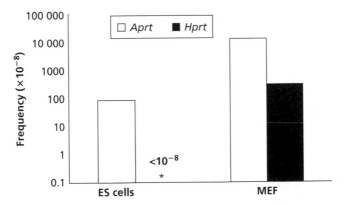

Fig. 4.8 Comparison of mouse embryonic stem (ES) cells and mouse embryonic fibroblasts (MEF) for their spontaneous mutation frequency at the *Aprt* and *Hprt* loci. Spontaneous mutation frequency in MEF is two orders of magnitude higher than in ES. Note that the *Hprt* mutation frequency in these cells is so low that it is not detectable (*). Courtesy of P. Stambrook, Cincinnati, OH, USA.

in the embryonic fibroblasts was similar to that in adult fibroblasts or lymphocytes. Interestingly, whereas in both cell types LOH was the predominant type of mutation, mitotic recombination appeared to be suppressed in the stem cells. Of note, extended culture led to mutation accumulation in the ES cells, which underscores the fact that even in these cells genome maintenance is not infallible. Whereas it is possible that in this case mutation accumulation occurred as a consequence of the high oxygen concentration in standard culture medium, it is likely that even under optimal conditions stem cells would eventually lose genetic integrity. This issue is critically important for the potential of stem cell therapy in delaying or reversing aging-related functional degeneration (see Chapter 8).

If we could know what makes stem cells superior in maintaining the integrity of their genome then strategies could be designed to improve genome stability further. This is especially important since there are no obvious strategies to interfere in genome-maintenance pathways. Since most forms of genome maintenance are based on enzymatic pathways, often interconnected with other pathways or enzymatic functions, it is not easy to make improvements simply by increasing the expression of one or few genes, which may often lead to adverse side effects. As we have seen in a previous section, attempts to enhance the efficiency of repair of endogenous damage by overexpressing in mammalian cells single enzymatic activities, such as the *MGMT*-encoded O^6-alkylguanine alkyl transferase, have been partially successful. However, overexpression of some BER enzymes is detrimental rather than beneficial with respect to protection from spontaneous mutagenesis. The same is true for other pathways, such as NER[411].

A relatively simple explanation for the reduced mutation frequency in ES cells would be an increased apoptotic response to spontaneous DNA damage. This would diminish the chance of errors and therefore increase genome stability, albeit at the expense of the number of surviving cells. As suggested by Stambrook, Tischfield, and co-workers[410], this increased apoptosis, as well as a reduction in mitotic recombination, could be regulated through an increased mismatch-repair activity.

Some time ago, John Cairns suggested that somatic stem cells might be protected against the accumulation of DNA replication-induced mutations through the selective retention of the 'old' template strands[412]. This would ensure that errors made during replication would not stay with the stem cell but pass to the non-stem daughter cells. Experimental evidence that such a mechanism exists has come from work by Chris Potten (Manchester, UK) for stem cells in the gut[413], and more recently from Tom Rando (Stanford, CA, USA) for muscle satellite stem cells (T. Rando, personal communication). These investigators used different types of label, incorporated during replication, to demonstrate the arrangement of sister chromatids in these cells through successive divisions.

Obviously, template retention as a strategy to avoid mutations in stem cells can only work when cells are prevented from using DNA-repair mechanisms based on recombination, because this would involve the mixing of old and new strands. We have already seen that mouse ES cells display dramatically lower rates of mitotic recombination, possibly due to increased mismatch-repair activities. In turn this could explain the much higher sensitivities of stem cells to DNA damage, which readily results in apoptosis[414].

Whereas it is therefore possible that embryonic and adult stem cells (as well as germ cells) are better equipped to maintain the integrity of their genome, this is not a guarantee for indefinite stability. Indeed, as we will see in the next chapter genome alterations in such cells do occur and their frequency may increase with age. This brings us to the next question in this section: is there any evidence that variations in the efficacy of genome-maintenance systems are related to the lifespan of a species?

4.4.3 GENOME MAINTENANCE AND SPECIES-SPECIFIC LIFESPAN

Apart from investigating the possibility that DNA-repair systems decline in effectiveness at old age, thereby promoting age-related genome alterations, workers in the field of aging have explored the possibility that genome-maintenance capacity has been a critical factor in the evolution of species-specific maximum lifespan. Genetic factors must underlie the sizable differences in maximum lifespan between species (see also Chapter 2). The rapid evolutionary changes that have led to a dramatic increase in longevity during the evolution of primates have sometimes been considered as evidence that maximum lifespan must be controlled by a small number of genes[184]. Indeed, there has been an increase in maximum lifespan of about 2-fold between humans and chimpanzees since our last common ancestor, about 5 million years ago. This period was considered too short to account for changes in many genes at the known rate of DNA-sequence change. This seems a reasonable argument, since a comparison of the recently completed genome sequence of the chimpanzee with that of humans quickly indicates that they are about 99% identical[415]. In general, there is only about a 1.2% nucleotide difference between humans and chimpanzees. However, I already mentioned the long-known, dramatic karyotypic differences even between closely related mammals, and more detailed comparison between the genomes of human and chimpanzee indicated significant structural variation, involving as much as 90 Mb of insertions and deletions and approximately 35 million single-nucleotide changes[416]. Many of these differences will affect regulatory elements. Whereas a substantial amount of this variation may reflect random genetic drift, it seems premature to assume that the adaptive evolution since the divergence of humans and chimpanzees could not have affected a substantial number of different genes and gene families. Hence, whereas it is possible that longevity in all mammals is controlled by one or a few of the same gene families, this cannot be derived from a perceived small number of genetic differences due to a short period of evolutionary time. More detailed functional analysis of completely sequenced genomes of closely related animals may ultimately offer the key to resolving this issue.

The assumption that longevity-assurance genes as they came to be called might be few, and possibly restricted to some basic cellular defense mechanisms, almost immediately focused attention on DNA repair as one of several likely candidates for universal lifespan control. This prediction was intimately linked to the theory that aging is caused by the adverse effects of accumulating genome alterations, which are suppressed by genome-maintenance systems. We have already seen that the correlation originally found by Hart and Setlow between the amount of UV-induced unscheduled DNA synthesis in skin fibroblasts and the maximum lifespan of seven different species has become difficult to interpret because of the inability of rodent cells to induce DDB2, which is essential for the repair of UV-induced CPDs and possibly other lesions. Whereas the possibility cannot be excluded that this deficiency has more to do with such animals being nocturnal than with their short lifespan, the fact remains that through DDB humans have an improved NER as compared with short-lived rodent cells. It remains to be seen whether other NER proteins also differ in inducibility or basal expression levels between cells and tissues among species with different lifespan. Apart from the previously mentioned correlation between PARP activity and species-specific lifespan there has been continuous interest in potential differences between short-lived rodents, still our favorite model system for aging and human disease, and humans with respect to genome maintenance and genome instability. In this respect, there are a number of intriguing differences related to control of DNA damage.

As mentioned earlier in this chapter, rodent cells express much higher levels of PARP or DNA-PK than human cells, lack p53 induction of DDB2, and do not have the same strict control mechanisms as human cells for inducing DNA repair and other stress-response systems upon DNA damage. And as we have seen, rodent cells lack a strict telomere barrier against cell immortalization. Such lower stringency of DNA-damage control might lead to an increased mutation load in cells and tissues of short-lived as compared to long-lived organisms.

Other evidence suggesting that longer-lived species do have superior genome-maintenance systems as compared to shorter-lived species can be derived from the varying rates of mammalian DNA sequence evolution. For example, the rate of sequence change in the germ line along evolutionary lineages, as determined from comparisons of mitochondrial and nuclear genes, is quicker in rodents than in artiodactyls (hoofed mammals), which is faster than in primates. Interestingly, in what is called the hominid slowdown, lineage-specific rates in primate evolution decline from apes to monkeys and humans, with humans (with the longest lifespan) having evolved the slowest. Whereas it has been suggested that these differences in rates of molecular evolution are due to differences in DNA-repair efficiency, it should be realized that also in this case mere correlations will not conclusively demonstrate deficient DNA repair in short-lived species. This can only be derived by comparing cells from long- and short-lived species directly for their capacity to maintain both genome integrity and viability upon genotoxic stress.

Mouse cells, for example T cells from the spleen, appeared to have a similar spontaneous mutation frequency *in vivo* as human T cells, as indicated by studies examining chromosomal aberrations or mutations at the *Hprt* locus, although such results are often difficult to interpret due to extensive inter-laboratory variation (see Chapter 6 for a more detailed discussion). Both species also accumulate mutations in these cells during aging at about the same rate. Therefore, on a chronological time scale, mice accumulate mutations during aging at a much faster pace than humans do; that is, over a period of about 3 years instead of 100 years. It is possible that this difference reflects a much higher quality of genome maintenance in human cells.

In principle one could examine mouse and human primary cells in culture, side by side, for their resistance to spontaneous or induced genome alterations over longer periods of time. To my knowledge, such experiments have never been done. A fair comparison would be difficult in view of the different behavior of mouse and human cells in culture which may involve not only different genome-maintenance capacities but also distinct ways in which genome-maintenance pathways are utilized. One important difference that makes it highly likely that primary cells from short-lived rodents have an inferior genome-maintenance machinery is the much higher probability of a mouse cell becoming neoplastically transformed than a human cell. This is clear in the intact organism, with tumors arising much earlier, on a chronological time scale, in mice than in humans. Similar to the increase in the number of genome alterations with age, increased tumor incidence in mice and humans is a function of biological age and displays an identical pattern (although the types of tumor differ). This again would point towards better genome-maintenance systems in humans, since DNA mutational alteration is generally considered as the main causal factor in cancer. However, other factors that might contribute to this increased likelihood of mouse cells becoming cancer cells *in vivo* cannot be ruled out. For cultured cells the situation may be easier to interpret.

In 1963, George Todaro and Howard Green published a paper in the *Journal of Cell Biology* that would become very influential[417]. At the time, it was well known how to bring human fibroblasts into culture and 2 years earlier Hayflick and Moorhead had published their landmark paper describing the finite lifespan of human fibroblasts *in vitro*[418]. Todaro and Green described the behavior of mouse cells, which when placed in culture appeared to be very different from human cells. They obtained their cells from mouse embryos and demonstrated that proliferation of such cells depended on the inoculation density. At low density cells could multiply through several transfers, but after about eight generations lost the ability to divide, somewhat similar to Hayflick and Moorhead's finite lifespan of human cells, albeit at a much quicker pace. At higher densities, however, mouse cells showed a much slower rate of proliferation decline. These populations quickly overcame the dip in proliferation rate and established themselves as immortal cell lines, most of which harbored a mutation in at least one p53 allele. This indicates that spontaneous mutations readily accumulate in such cells[419].

Spontaneous immortalization never occurs in primary populations of human cells. Such cells always become senescent, which is essentially an irreversible process due to telomere attrition (see above). The process of establishment of mouse cells in culture is accompanied by changes in their karyotype. Already in an early stage tetraploid cells arise, which start dominating the culture when this has become a permanent cell line. At that time, abnormal chromosomes start to appear, a sign of genomic instability. Although primary human fibroblasts occasionally also undergo karyotypic changes[420], their level of genomic instability is very low as compared to mouse cells established in culture.

Whereas it is tempting to conclude from the above that mouse cells are both less viable in culture and easily transformable, due to a lower capacity to repair damage in their DNA, the question is where this damage comes from. As already mentioned, mouse cells have very long telomeres and in contrast to human cells their telomerase activity in culture is not repressed. Hence, what causes mouse cells in culture to show all the outward signs of senescence as early as after eight population doublings, followed by immortalization and transformation, whereas human cells can easily double their population 60 times before their telomeres become so short that a senescence response is induced, without ever undergoing spontaneous immortalization? This issue has been resolved by a series of experiments in the laboratory of Judy Campisi. In the recent past, all cells used to be cultured under atmospheric oxygen, which is about 20%. This is much higher than what cells under physiological conditions experience, which is not much higher than about 3%. Campisi realized that mouse cells might not be able to sustain the increased DNA damage resulting from such high levels of oxygen. The results of her experiments[353] clearly indicated that only mouse embryonic fibroblasts, cultured at 20% oxygen, underwent the reduction in proliferation originally described by Todaro and Green. At 3% oxygen senescence was not observed and it is likely that these cells gradually developed into immortal cell lines similar to mouse cells cultured at high density. High density might increase viability and proliferative activity by increased production and secretion of growth and survival factors. Indeed, in my own laboratory we compared rat skin fibroblasts and cultured them at high or low density. Whereas the cells at high density showed no signs of proliferation inhibition, the cells at low density underwent senescence very rapidly[277].

All together, these results indicate that rodent cells are highly susceptible to culture conditions with cell density and oxygen concentration identified as key determinants of cell fate. Whereas it is not exactly known how cell density determines whether or not cells become senescent or established as immortal cell lines, the role of oxygen appears to be more straightforward. Oxygen is a major source of DNA damage, which likely activates a senescence response, similar to what has also been observed in normal human fibroblasts treated with ionizing radiation or hydrogen peroxide. In such cases, telomere attrition is unnecessary to cause human cells to senesce. It is now apparent that many forms of stress can induce a senescence response[421]. For example, normal cells senesce when they overexpress certain oncogenes, such as activated components of the Ras-Raf-mitogen-activated protein kinase (MAPK)/extracellular-signal-regulated kinase (ERK) kinase (MEK) signaling cascade, after telomere uncapping (as distinct from attrition) through the inhibition of TRF2, or in response to epigenetic changes to chromatin organization. This strongly indicates that senescence is a response to DNA damage threatening to overwhelm the DNA-repair machinery. This response is mediated by the p53 pathway, with the p16/RB pathway also playing an as yet incompletely understood role.

The fact that mouse cells respond so much more readily to such damage than human cells is likely to be yet another indication of their relative inferiority with respect to the quality of their genome-maintenance systems[422]. This is in keeping with observations made in the Campisi laboratory of mouse fibroblasts accumulating far more oxidative DNA damage than human cells (significantly reduced by culturing cells at 3% oxygen) and results obtained by Pankaj Kapahi (Novato, CA, USA) that mouse and rat cells are much more sensitive to a variety of agents producing oxidative stress[423].

4.4.4 SUMMARY

In summary, whereas striking reductions in genome-maintenance capacity in mammalian cells or tissues at old age are not supported by the evidence, modest declines in repair activities have been found, both in the repair of DNA adducts and in enzymatic steps that may reflect the activity of core pathways. Still uncertain are the effects of such changes on genome integrity and cell viability of the organism during its lifespan. In principle, decreasing DNA-repair activities could lead to an exponential increase of genome alterations with age. However, it is also possible that requirements for DNA-damage processing alter with age. For example, a decline in cell-proliferative activity may reduce the utilization of certain repair pathways. On the other hand, an increase in DNA-damage production or a decrease in antioxidant defense systems would require increased DNA-repair activities. Not enough solid data on the activity patterns of the various genome-maintenance systems as a function of age are available for general conclusions to be drawn. Indeed, our knowledge of genome-maintenance activities as they take place in different organs and tissues is far too limited to allow any conclusions as to the potential relevance of age-related alterations.

Also incompletely understood is the relationship of genome maintenance with lifespan. Whereas direct evidence has been slow to emerge and side-by-side comparisons of cells in culture are still sparse, it is likely that a long-lived species, such as humans, has a greater capacity to maintain its genome than a short-lived species, such as mice. A general relationship of genome-maintenance proficiency with species lifespan would indicate a causal role of genome alterations in age-related cellular degeneration and death. However, very limited information is presently available. What has recently become a major topic of research is the effect on lifespan of heritable mutations that inactivate or weaken specific genome-maintenance pathways. This will be discussed in the next chapter.

5 Genome instability and accelerated aging

Aging remains poorly defined at the mechanistic level. This is true in spite of the spectacular progress that has been made in recent years by studying genetic mutations altering lifespan and the onset of age-related characteristics. As we have seen in Chapter 2, these studies have revealed that over a range of model organisms, from yeast to nematodes and fruit flies, all the way up to mice, mutations downregulating activities of growth, reproduction, energy metabolism, or nutrient sensing significantly increase longevity. It is possible that the beneficial effects of reducing these activities, which have thus far been exclusively observed in laboratory populations, stem from a decrease in the generation of somatic damage and/or the upregulation of mechanisms that protect against such damage. This would be in keeping with current theories of aging as an adverse effect of the life-history strategies of multicellular organisms. Indeed, whereas the mechanisms underlying age-related degeneration and death in these and other species still need to be uncovered, a consensus as to why and how we age has begun to emerge.

Recapitulating the essence of what has already been explained in more detail in Chapter 2, it is now generally accepted that the time-dependent decrease in fitness in most multicellular organisms is non-adaptive; that is, it is not controlled by a purposeful genetic program similar to the control of development. Aging provides no specific advantage to an individual animal and most researchers now accept that age-related degeneration and death is ultimately due to the greater relative weight placed by natural selection on early survival and reproduction than on maintaining vigor at later ages. This decline in the force of natural selection at later ages is largely due to the scarcity of older individuals in natural populations owing to mortality caused by extrinsic hazards. Because resources are limited, somatic maintenance and repair are optimized to the effective lifespan of the population, not maximized to help those fortunate enough to survive the hazards of their environment to live a healthy life for extended periods of time. High extrinsic mortality diminishes the chances to reproduce at later ages. Selection would then favor those genotypes that help shift the allocation of resources to early growth and reproduction. By contrast, low extrinsic mortality would permit reproduction at later ages, shifting the balance towards increased somatic maintenance, thereby increasing lifespan. This trade-off in life-history strategy is known as the disposable soma theory. First proposed by Tom Kirkwood[9], this theory not only provides a rationale for why we age, but also predicts the nature of its proximate cause: the accumulation of unrepaired somatic damage. This theory, which is now supported by a large body of evidence, also explains the similarities in symptoms of aging, both within and across species, and the apparent universality of genetic pathways of life extension across different phyla. Indeed, rather than being programmed to age, animals are programmed to survive long enough to reproduce, possibly by using highly conserved cellular defense systems against somatic damage common to all or most species. Such damage may come from the environment, for example radiation or infectious agents, but also from inside the organism, for example normal by-products of metabolism.

Somewhat paradoxically, the focus in the science of aging is now more on the mechanisms that help to delay aging and extend lifespan than on aging itself as a complex phenotype. Indeed, it is often argued that extension of lifespan postponing all known markers of the aging process is the only acceptable way to study how we grow old. However, this concept is of doubtful validity in the face of a lack of model systems that can help us to better understand aging phenotypes, how they come about and how they interact with each other and with genetic and environmental factors. An approach based solely on animal models of extended lifespan is especially problematic since all of them are inbred laboratory strains, with the possibility that they represent artifacts difficult to rule out (see Chapter 2). Hence, it is far from certain that the information obtained will ever translate into practical interventions to prevent, ameliorate, or treat human aging. It is also counter to established practice of creating disease models by exaggerating its phenotypes, for example, through genetic manipulation or chemical treatment. In fact, based on the almost general consensus that aging is caused by the accumulation of unrepaired somatic damage, it seems relatively straightforward to generate models for aging based on the inactivation or weakening of somatic maintenance systems. Such models exist and they are almost all based on human and mouse mutants harboring defects in genome maintenance. In this chapter I will discuss such models, their molecular basis, the possible nature of the pro-aging processes underlying the phenotypes observed and their validity in representing genuine aging processes as they take place in normal animals.

5.1 Premature aging

The first models for accelerated aging were natural human mutants. This is not surprising in view of the century of clinical observations on subjects of our species. Thanks to this enormous reservoir of knowledge clinical practicioners recognized over 100 years ago accelerated aging in a number of people suffering from life-shortening genetic defects[424]. These so-called segmental progeroid syndromes, characterized by the premature appearance

Table 5.1 Some of the best-known human segmental progeroid syndromes

Syndrome	Incidence (per live birth)	Inheritance	Age of onset (years)[a]	Mean lifespan (years)	Gene/abnormality	Defect
Hutchinson–Gilford	<1/1 000 000	*De novo*	2	13	*LMNA*	Nuclear stability, transcription?
Werner	<1/100 000	Autosomal recessive	25	50	*WRN*	DSB repair
Cockayne	~1/100 000	Autosomal recessive	5	20	*CSA, CSB, XPD, XPG*	Transcription-coupled excision repair
Ataxia telangiectasia	~1/60 000	Autosomal recessive	10	20	*ATM*	Response to DNA damage
Down's	~1/1000	*De novo*	40	60	Trisomy 21	Antioxidant defense?

[a]The age when accelerated aging becomes apparent, as distinct from age at diagnosis.

of multiple signs of normal aging, were described by the medical community well before the discovery of DNA and are therefore not biased towards a DNA-based hypothesis of aging. So it is remarkable that so many of these syndromes are caused by a defect in genome maintenance. Table 5.1 lists the best-known human segmental progeroid syndromes, which we met when discussing genome maintenance in Chapter 4. At the top of the list are Werner syndrome[425] and Hutchinson–Gilford progeria syndrome (HGPS)[426]. Werner syndrome is caused by a defect in a gene that is a member of the RecQ helicase family[427]. The affected gene, *WRN*, encodes a RecQ homolog whose precise biological function remains elusive, but is important for DNA transactions, probably including recombination, replication, and repair (Chapter 4).

HGPS is caused by a defect in the gene *LMNA*, which through alternative splicing encodes both nuclear lamins A and C[428]. Nuclear lamins play a role in maintaining chromatin organization (Chapter 3). Less striking segmental progeroid syndromes include ataxia telangiectasia, caused by a heritable mutation of the gene *ATM*, a relay system conveying DNA-damage signals to effectors[429], Cockayne syndrome, and trichothiodystrophy, diseases based on defects in DNA repair and transcription[430], and Rothmund–Thomson syndrome, like Werner syndrome based on a heritable mutation in a *RecQ* gene[431]. There is evidence that each of these genes, when defective, can also lead to aging symptoms in the mouse, sometimes in combination with other gene defects[432] (see also below).

The discovery that human segmental progeroid syndromes are almost without exception based on heritable defects in genome maintenance is unlikely to be a coincidence. Indeed, there is no other gene family associated with accelerated aging. It seems therefore logical to consider this as support for the idea that genome instability is a major cause of aging. As we have seen, the disposable soma theory predicts that aging is caused by the accumulation of unrepaired somatic damage[55]. The fact that defects in genome maintenance lead to premature aging, but not or much less so defects in other somatic maintenance systems, strongly suggests that DNA damage is the most important type of age-accumulated damage and is a likely cause of many aging-related phenotypes. This would be in keeping with the unique role of the DNA of the genome in transferring genetic information from cell to cell and from generation to generation and its lack of a back-up template. This is in contrast to proteins, which, at least in principle, can be easily replaced by using the corresponding gene as a template. Indeed, the maintenance of genomic DNA is of crucial importance to survival because its alteration by mutation is essentially irreversible and has the potential to affect all downstream processes. The rationale for adopting genome-maintenance mutants as models for studying aging is therefore strong.

Because humans are often unsuitable experimental models of aging a logical approach would be to model genome-maintenance defects in the mouse. The mouse is a good model system for studying human aging for several reasons. First, mice are positioned close to humans on the evolutionary scale. Second, their relatively short lifespan and small size permits extensive lifespan studies on an economic basis. Third, mouse genetics has closely emulated the progress in human genetics and is now almost equally powerful. Fourth, although in mice a full phenotypic characterization of aging is still far from the systematic catalog of signs and symptoms of old age presently available for humans, the species ranks a solid second, with rapid improvements underway. Indeed, in the wake of the current explosion in genetically engineered mouse models, major coordinated efforts have emerged to obtain standardized and comprehensive databases for morphologic, biochemical, physiologic, and behavioral characteristics of various mouse strains (e.g. the mouse phenome project; www.jax.org/phenome). More recently, patterns of age-related pathology of the mouse have been made computationally accessible using ontologies or controlled vocabularies of terms (see Chapter 1). Brent Calder, in my own laboratory, has designed an ontology for studying phenotypes in the aging mouse in the context of other clinical, biochemical, and molecular variables. He called this system MPHASYS for Mouse Phenotype Analysis System, which represents an integrated platform for studying the systems biology of aging (http://mphasys.info). Such progress greatly facilitates the development and use of mouse models of accelerated aging.

Before I discuss the validity of accelerated aging as a phenotype in the mouse, the first mouse model of premature aging, the so-called senescence-accelerated mouse, or SAM for short, deserves some attention. SAM mice are a collection of ill-defined mouse lines derived from the notoriously short-lived AKR/J mice[433]. Much of the criticisms that have been raised with regard to the SAM model apply to all models of accelerated aging; that early diseases, unrelated to basic mechanisms of aging, also lead to a reduced lifespan[434]. Such diseases typify genetically inbred mouse strains, which have been selected for early fertility and growth. This issue will be discussed below. More importantly for this discussion, because the genetic factors that lead to premature aging in the SAM model are not defined, the model does not point us towards a potential cause of normative aging, which is in contrast to models based on genome-maintenance defects. Of note, an approx. 5-fold increase in mutation rate at the locus encoding hypoxanthine phosphoribosyltransferase (*Hprt*; a measure for genomic instability; see Chapter 6) was observed in SAM mice up until 6 months, compared with control animals over the same age range[435]. The increased mutation frequency paralleled the phenotypic features of aging in this SAM line, which also develop around this time. It is not impossible, therefore, that relative defects in genome maintenance also underlie premature aging in SAM mice.

5.2 Validity of accelerated-aging phenotypes

Aging exceeds all human diseases in complexity and is the only example of generalized biological dysfunction. Its effects become manifest in all organs and tissues, it influences an organism's entire physiology, impacts function at all levels, and increases susceptibility to all major chronic diseases. Nevertheless, typical symptoms of aging, often similar across species, can and have been defined.

Mouse models of accelerated aging have been criticized in the past based on the argument that many of the degenerative phenotypes associated with aging could result from a variety of interventions and need not necessarily involve the same causes that underlie natural aging. Whereas this is a valid argument, it should be realized that the use of model systems to study natural phenomena is a generally accepted approach in biology. For example, human cancer has been studied extensively using laboratory rodents subjected to treatment with a variety of genotoxic agents. Whereas we were all well aware of the fact that natural human cancers were normally not caused by such treatments, this approach nevertheless allows us to obtain valuable information about the etiology of this disease. Of note, rodent models for studying human cancer are based on the rationale that cancer is caused by DNA damage, hence the use of DNA-damaging agents to generate these model systems. Nevertheless, many if not most of the early cancer models would fail a simple validation, with morphological attributes of the model not matching those of the human disease. Likewise, models of accelerated aging suffer from this problem, which can be addressed by more careful modeling based on increased information on the etiology of the adverse phenotype and the pathways affected. For example, the most recent series of mouse models for human cancer has been greatly improved, based on highly specific genetic alterations, known to increase human susceptibility to cancer[436].

Although in both humans and mice cancer incidence increases exponentially with age, the tumor spectrum in the two species differs significantly, with sarcomas and lymphomas predominant in the mouse and epithelial cancers predominant in older humans[437]. Likewise, the spectrum of non-cancer, degenerative age changes in mice and humans is not exactly the same, which always needs to be kept in mind when using these models[438]. Moreover, whereas cancer as a phenotype is generally undisputed, aging has also more diffuse degenerative characteristics. For humans the degenerative and disease-related characteristics of the aging process are best described in textbooks of pathology or geriatrics, such as Robbin's *Basic Pathology* (Saunders, 2004) and *Principles and Practice of Geriatric Medicine* (Wiley, 2006). For rodents there are a number of useful papers, chapters and edited books[439–442] (see also our own mouse pathology ontology; http://mphasys.info). As far as I know there is no literature on global comparisons of aging phenotypes in humans and mice. This would also be difficult in view of the wide variety of laboratory strains, each with its own unique genetic background.

Progeroid genotypes are associated with an early onset of some, but not all, characteristics of senescence and must therefore be interpreted with caution. Together with my colleague, Paul Hasty (San Antonio, TX, USA), I formulated the following loose criteria that help to identify genuine mouse mutants of accelerated aging: (1) the phenotype should present after development and maturation are complete; (2) the phenotype should be demonstrable in control populations at a more or less similar point in their survival curve; and (3) the genetic alteration should accelerate multiple aging phenotypes[438]. None of these criteria is written in stone. Indeed, accelerated aging can occur even before development is complete, as in the case of HGPS. Such cases, however, are more difficult to recognize as authentic models of aging, and may not be as valid as those that exhibit aging phenotypes after maturation. It is also easily imaginable that a genetic alteration accelerates certain symptoms of aging much more than expected on the basis of biological age; that is, the point of occurrence on the survival curve is earlier and/or the symptom is more severe. Such so-called exaggerated aging would be expected if, rather than a quantitative, chronological masterswitch, the mutation would affect only one critical pathway for somatic maintenance leading to severe imbalance of the survival network.

In Chapter 2 I mentioned that the single-gene mutations that increase lifespan of worms, flies, and mice may do so through the upregulation of cellular defense systems, including DNA repair and antioxidant defense. Genes implicated in the control of such survival responses are those belonging to the FOXO and SIRT families, which have been demonstrated in nematodes and fruit flies to control downstream targets of the pro-longevity mutations affecting energy metabolism, nutrient sensing, reproduction, or growth[443]. Downregulation of these effector genes could then conceivably lead to an acceleration of all possible aging phenotypes. In mammals, the *FOXO* genes are diverse, and knocking them out in the mouse has not revealed overt premature aging phenotypes, possibly with the exception of *Foxo3a*-null female mice, which showed age-dependent infertility and abnormal ovarian follicular development[444,445]. Likewise, of the seven SIRT homologs there is not much evidence that their inactivation leads to premature aging. Mice with inactivated *SIRT1*, the most closely related to the yeast *Sir2* gene, were small, exhibited developmental defects, and showed only infrequent survival in the postnatal period[446]. More recently, ablation in the mouse germ line of another *Sir2* homolog,

SIRT6, was shown to cause abnormalities—after only 3 weeks—that resemble some aspects of aging, such as thinning of the skin due to loss of subcutaneous fat and signs of osteoporosis[447]. However, the premature aging phenotype in these mice was not wholly convincing. Interestingly, cells from these mice suffered from impaired proliferation and increased genomic instability, possibly as a consequence of a defect in BER, which thus far has never been associated with accelerated aging.

Hence, neither *FOXO*- nor *SIRT*-knockout mice display a convincing premature aging phenotype, which might have been expected if these genes truly are master regulators of cellular defense and stress responses. It is of course possible that a progeroid phenotype will become visible by downregulating rather than completely ablating the activity of these genes.

As discussed in Chapter 2, the mutations that lead to increased longevity in nematodes, flies, or mice are likely to do so only at the cost of some selective disadvantage, sometimes already under laboratory conditions[122,123]. For some of the mouse longevity mutants, such as the growth hormone deficient Ames dwarf mice, fitness costs are readily apparent in the form of infertility and hypothyroidism[448]. However, for another long-lived mouse mutant discussed in Chapter 2, the p66^{SHC} mutant, initially no selective disadvantage was obvious[449]. We now know that these mice are more susceptible to cold and are therefore not likely to survive in the wild for very long (P. Pelicci, personal communication). Another possible disadvantage in the wild would paradoxically be the loss of its free-radical-generating capability. Indeed, p66^{SHC} functions to increase cellular ROS to initiate cellular destruction as part of our defense system against infectious agents. This potential disadvantage would be masked in the p66^{SHC}-mutant mice because they are housed in a pathogen-free environment. Hence, whereas it is often assumed that long-lived mutants have no price to pay, this is unlikely to be the case. Apart from the phenotypes only manifest in the wild, we also lack the detailed phenotypic comparisons to know whether longevity-conferring mutations do so by retarding all possible symptoms of aging equally. In fact, the relationship between symptoms of aging and longevity is often unclear. Therefore, long-lived mutants are also likely to have a segmentally rather than a comprehensively delayed aging phenotype. Hence, based on what we now know of the phenotypes of mouse models of accelerated and delayed aging, the concept of master regulator genes to control the rate of aging is doubtful.

Based on the above, generating animal models for human aging through genetic manipulation or pharmacological intervention based on our increased knowledge of what causes human aging seems a valid approach. However, there are certain caveats. The first one involves the lack of mouse models recapitulating all possible aging symptoms in an accelerated fashion. One could argue that only gene defects that accelerate all possible symptoms of normal aging are likely to truly intervene in its core mechanisms. I have already argued that this is unlikely, not only for models of accelerated aging but also for models of delayed aging. At any rate, it is difficult to ensure whether a mutant animal does or does not display a full complement of aging characteristics as compared with their littermate controls. First of all, in a typical population of humans or mice, there is not a single individual displaying all possible symptoms of aging. Whereas it is indeed very likely that human progeroid disorders are in fact segmental[424], the number of patients is often too small to exclude the possibility that other aging-related phenotypes are also accelerated. In mice, the situation is seriously confounded by the use of inbred strains and the variation in genetic background. It is very well possible that a particular phenotype that is associated with normal aging in one genetic background is not observed in others. However, as mentioned above, the expectation of mutants accelerating or decelerating all aspects of normal aging is probably unrealistic. I have already alluded to the lack of evidence for pacemaker genes orchestrating all the critical aspects of the fundamental mechanisms of normal aging. Because of its very nature as a byproduct of evolution, the postmaturational changes considered as aging are ill-defined, with major confounders in the form of genetic, environmental, and stochastic variation. It seems highly unlikely that the many different individual phenotypes of aging, which may be related and/or interact with each other, are controlled by one or few genes. Therefore, it is not to be expected that single-gene mutations accelerate all aging symptoms characteristic for the normal population. Indeed, if DNA damage is the single most important cause of aging, the hundreds of genes directly involved in its metabolism immediately refute the possibility of a simple genetic control center specifying each and every aging-related phenotype. Mutations in only one of these genes may cause one or a few aging symptoms, but will never accelerate all possible signs of aging that are characteristic for the population under study.

On the other hand, we are confronted with the fact that for the long-lived worm, fly, and mouse mutants there is abundant evidence that multiple cellular defense systems can be upregulated in concert, to increase stress resistance and survival in a diverse range of organisms[52]. Whereas we do not know whether increased survival is controlled in the same way, by the same mechanism, in all these mutants in different species, it is possible that in some way they all reflect the selective advantage of temporarily delaying reproduction and maximizing survival in times of famine, awaiting better times. As mentioned above, there is no evidence that the gene families that potentially control these responses, *FOXO/daf-16* and *SIRT/Sir2*, serve as master regulators, since when inactivated they do not produce accelerated aging phenotypes. There is also not much evidence that an increase in the signaling pathways inhibited by the longevity mutations, such as the IGF-1 axis, leads to accelerated aging. Growth hormone-overexpressing mice have been found associated with at least some symptoms of accelerated aging, but the overall picture is not convincing[450]. A more intriguing example is the gene *KLOTHO*. Overexpression of this gene in mice extends lifespan, possibly because the Klotho protein functions as a circulating hormone that binds to a cell-surface receptor and represses intracellular signals of insulin and IGF-1[451]. Inactivation of the *KLOTHO* gene in the mouse germ line had previously been found to result in multiple symptoms of human aging after 3–4 weeks, which could therefore be ascribed to the upregulation of the IGF-1 axis. Whereas, therefore, the possibility cannot be excluded that premature aging phenotypes can be obtained by promoting rather than opposing activities related to growth, reproduction, energy metabolism, and nutrient sensing, much more convincing patterns of premature aging can be found on the basis of mutations in genome-maintenance pathways. As we shall see, such mutants are so widespread that it is logical to conclude that DNA damage is a major proximate cause of aging and that only direct interference at that level uncovers accelerated aging phenotypes (see below). Any upstream interference would probably have too many side effects or require a much more sophisticated type of genetic engineering than we currently have available.

A second caveat in using mouse models of accelerated aging is the potential misinterpretation of a life-shortening effect of early diseases as a sign of the acceleration of fundamental mechanisms of normal aging. It is often argued that mouse models with a mutation that accelerates aging are simply showing some disease that has nothing to do with normal aging. Based on the aforementioned evolutionary basis of aging one could say that all genetic variations that exclusively manifest after the age of first reproduction are by definition part of the aging phenotype. How likely is it that a random germ-line change would give rise to a phenotype that can superficially be diagnosed as aging; that is, adversely affects postmaturation fitness? The first results of mouse ethyl nitrosourea mutagenesis programs, in which the effects of many different randomly induced germ-line mutations are assessed, suggest that aging phenotypes are rare. The phenotypes observed in ethyl nitrosourea mutagenesis screens are to a large extent developmental and very few if any have been defined as aging-related[452]. In addition, very few mouse knockouts exhibit aging phenotypes. Thus, genetic alterations rarely impact aging. In view of the convincing array of aging phenotypes in the genome-maintenance mutants discussed, the point is probably moot, but it remains useful to point out the difference between a disease and an aging phenotype.

Consider the well-known genetic diseases cystic fibrosis and sickle cell anemia. These genetic alterations reduce lifespan by impairing the function of a vital system independent of normal aging. The negative impact of both these diseases is apparent very early in life and neither disease is observed in the elderly. On the other hand, atherosclerosis and its sequelae, such as coronary and cerebrovascular disease, is an age-related disease process. Similar examples to the one used by Williams to illustrate the effect of a pleiotropic aging gene, promoting late-life arteriosclerosis through enhanced calcium deposition in the bones (a selective advantage at young age)[76], could be found for diabetes, osteoporosis, cancer, and other diseases of the aged. Although none of these diseases occur exclusively in the elderly, the high incidence in this segment of the population is compelling and it is very difficult to argue that they are not genuine aging phenotypes (see also Chapter 7). Hence, in interpreting accelerated aging symptoms it is critically important to correctly diagnose the phenotype. On this basis it should be possible to differentiate between genuine aging phenotypes and phenotypes related to disorders unrelated to aging.

It is conceivable that a response to stalled transcription at sites of DNA damage is responsible for early aging in the *Xpd* mutant mice. In this scenario, a defective TFIIH results in stalled transcription that decreases gene activity and leads to a mild accumulation of DNA damage. Cells respond by undergoing apoptosis, which may cause the early organismal senescence in these mice. Indeed, increased apoptosis and activated p53 have been observed in liver from both young and old *Xpd* mutant mice (Y. Suh, personal communication). In the absence of Xpa alone (that is, with a fully functional Xpd protein), transcription from a damaged DNA template can still take place, possibly because the amount of spontaneous damage subject to NER is low. (BER, not NER, is the main pathway for removing the most abundant forms of spontaneous DNA damage, such as oxidative lesions.) In combination with the Xpd mutation, however, the complete absence of Xpa would exacerbate the suboptimal performance of TFIIH, leaving the DNA lesion exposed for a greater period of time; this would result in a further decline in gene activities and an enhanced response. Thus, the premature aging symptoms in the *Xpd* mutant mouse may be primarily the result of a cellular response to impaired TFIIH at the site of a spontaneous DNA lesion, rather than the accelerated accumulation of DNA damage or mutations. Indeed, mutation frequencies at a *lacZ* reporter locus in this mouse model were not elevated as compared to control animals of the same age[325].

As mentioned above, in spite of their severe symptoms of premature aging, *Xpd* mice show only a modest reduction in lifespan. This may be due, at least in part, to some unexpected features in these mice that appear associated with caloric restriction, a condition usually linked to retardation of age-related pathology and extension of lifespan (Chapter 2). These features include a reduced incidence of cancer, cataract, and inflammation[335]. Global gene-expression analysis of the liver of these mice, using microarrays, indicated reduced IGF-1 signaling and decreased energy metabolism. This was accompanied by reduced plasma IGF-1 levels (Y. Suh, personal communication). A reduction of IGF-1 signaling, which as we have seen is generally associated with increased lifespan, was also found in the aforementioned *SIRT6*-defective mice[447]. The attenuation of IGF-1 signaling in mice with DNA-repair defects may reflect a compensatory mechanism to limit the onslaught of spontaneous DNA damage[457]. Indeed, reducing IGF-1 signaling might be a general mechanism to cope with increased stress, including genotoxic stress, similar in nature to the mechanism underlying increased lifespan as a consequence of dampening IGF-1 signaling in nematodes, flies, or mice (Fig. 5.1). Activation of this protective pathway in the Xpd-defective mice could explain why these mutants live as long as they do in spite of their severe premature aging phenotypes.

Another NER-defective mouse model showing symptoms of premature aging involves the gene *ERCC1*. The ERCC1–XPF complex forms an endonuclease, which is required for the 5' incision to remove the damage-containing oligonucleotide during NER, but is also essential for interstrand cross-link repair. Hence, *Ercc1*-knockout (*Ercc1*$^{-/-}$) mice are deficient in GG-NER, TC-NER, and interstrand cross-link repair. These mice show a severe phenotype, including runted growth, progressive neurological abnormalities, kyphosis,

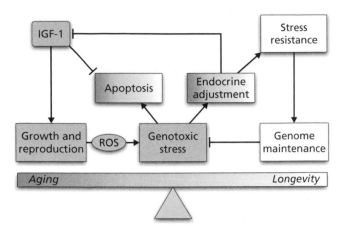

Fig. 5.1 The balancing act. Increased genotoxic stress resulting from DNA-repair defects may cause a compensatory metabolic shift towards reduced IGF-1 signaling, thereby lowering the production of ROS. Shaded areas: pro-aging; white areas: pro-longevity.

a short lifespan of about 3 weeks, and liver and kidney dysfunction[455]. At the cellular level the *Ercc1* defect leads to accelerated nuclear polyploidization. The combination of a knockout allele with a truncated *Ercc1* allele (*Ercc1*$^{-/m}$), resulting in a protein lacking the last seven amino acids, delays the onset of the premature aging phenotype and extends the maximal lifespan to about 4–6 months[455]. Indeed, by generating a conditional knockout mouse model for this gene (in which the gene can be inactivated at will at any time), Jan Hoeijmakers and co-workers succeeded in postponing aging in these mice even further (J. Hoeijmakers, personal communication). Similar to the *Xpd* mutant, the premature appearance of aging symptoms is not caused by the defect in NER (although this may contribute), but is due mainly to the defect in the removal of the highly toxic interstrand cross-links. In the hypomorphic *Ercc1* mutant mouse an increased level of genomic instability at a *lacZ* reporter locus was observed at 4 months of age[325]. Whereas this mouse model may ultimately appear as a valid model of accelerated aging, the fact that severe symptoms already appear before development is complete make it less suitable as a model for aging phenotypes that normally appear only in adult animals. However, what this model appears to show is that the severity of the premature aging phenotype varies with the different gradations of gene inactivation.

5.3.2 THE IMPORTANCE OF DNA DOUBLE-STRAND-BREAK REPAIR

DSBs in DNA are highly toxic lesions that can be created through a variety of mechanisms, including effects of ROS. Explained in Chapter 4, DSBs are repaired by either of two

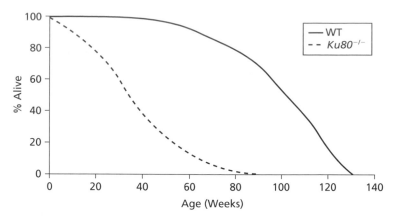

Fig. 5.2 Survival of homozygous *Ku80*-defective mice as compared to their littermate controls. Re-drawn with permission from ref. 458.

mechanistically distinct DNA-repair pathways, homologous recombination (HR) or non-homologous end-joining (NHEJ)[338]. A key factor of DSB repair by NHEJ is the DNA-end-binding Ku70–Ku80 heterodimer. As first shown by Paul Hasty (San Antonio, TX, USA), mice harboring a null mutation in the *Ku80* gene have a significantly shorter lifespan and display a range of premature aging phenotypes[458]. This mouse model is one of the few in which age-related phenotypes have been compared with those of their littermate controls in a side-by-side study in an identical environment (same air, same food, same bedding, same cage). This is extremely important because aging phenotypes in the mouse are not necessarily the same in humans and can have different causes. In this study a phenotype was only considered a premature aging phenotype when exactly the same phenotype was also observed during normative aging in the wild-type littermate controls. I will use this study as an example to demonstrate the validity of the premature aging phenotype in mice with defective DNA repair.

Figure 5.2 shows the survival curve of the *Ku80*-mutant mice, indicating a significantly shorter lifespan than their littermate controls. One of the most prominent aging-related phenotypes occurring early in the mutants is kyphosis (also called lordokyphosis), the lateral curvature of the spine that is so frequently observed at old age, both in mice and humans (Fig. 5.3). In humans the causes of kyphosis are not necessarily the same as in mice. Kyphosis in the *Ku80*-mutant mice is likely due to osteoporosis because histology showed the older mutant and control bones to exhibit osteopenia (thinning of the bone and reduced trabeculae). Figure 5.4 shows growth-plate closure, a well known age-related phenotype that in humans is part of maturation, not senescence. Mice are different from humans in that growth plates do not close until well after maturation. Here, growth plates close much earlier in *Ku80*-mutant mice than in control mice. Figure 5.5 illustrates skin atrophy; a well described age-related phenotype in both mice and humans. Again, skin

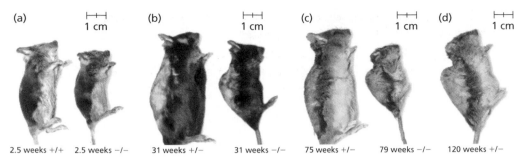

Fig. 5.3 Kyphosis in *Ku80*-defective mice as compared with their littermate controls at the same age. Note that at 120 weeks, when all mutants have died, the aged normal mouse shows kyphosis similar to the mutant of 79 weeks. Re-printed with permission from ref. 458.

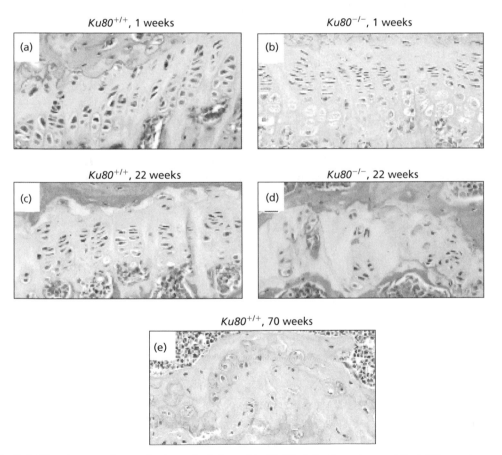

Fig. 5.4 The closure of growth plates (epiphyses) in *Ku80*-defective mice and their littermate controls. Note that whereas in the *Ku80*-null mice the number of chondrocytes is reduced, with the cartilage becoming ossified and the parts fused (d), this is not yet the case in the control animals of the same age (c). Courtesy of P. Hasty.

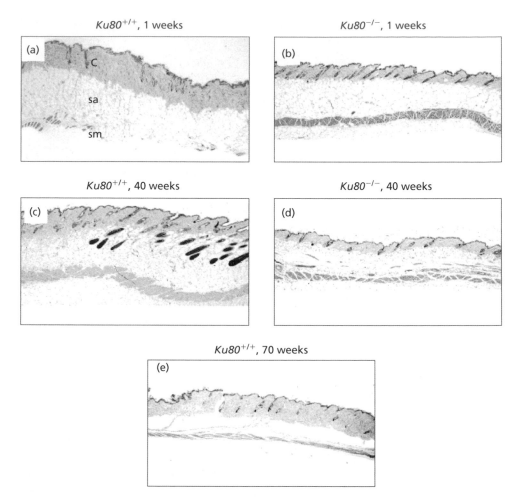

Fig. 5.5 Skin atrophy occurs much earlier in the *Ku80*-defective mice than in their littermate controls, as revealed by a comparison of a section of skin. Whereas at 1 week the skin of the mutant (b) is the same as that of the control animal (a), at 40 weeks of age all subcutaneous elements, including collagen (marked by c in panel a), adipose (sa), and skeletal muscle (sm), are atrophied in the mutant (d). In the controls this same phenotype is only observed at 70 weeks. Courtesy of P. Hasty.

atrophy was observed earlier in *Ku80*-mutant mice than in their littermate controls. In addition, various other aging-related phenotypes were observed in *Ku80*-mutant mice, well before they appeared in their littermate controls (forms of liver degeneration, reactive immune responses)[458]. Even though these phenotypes occur earlier in *Ku80*-mutant mice than in the controls, they all occur at about the same point in their biological lifespans (the latter half of their survival curve). Despite these similarities there are also differences; most obvious is the difference in cancer incidence. Tumor frequency at similar points on the survival curve is much lower in *Ku80*-mutant mice than in their littermate controls[458].

The most straightforward explanation for the accelerated aging phenotypes in *Ku80*-null mice is increased genomic instability resulting from erroneous or inefficient repair of DNA DBSs in the absence of NHEJ. Such increased genomic instability would trigger apoptosis and interfere with normal cell growth and tissue regeneration. Whereas increased genomic instability would be expected to promote tumor initiation, the presence of an intact p53 checkpoint is likely to suppress tumor progression through an apoptotic or senescence response. This scenario is supported by actual observations. *Ku80*-mutant mouse cells display growth impairment, increased susceptibility to apoptosis and replicative senescence, and a marked increase in chromosomal aberrations, including breaks, translocations, and aneuploidy[350]. Results from my laboratory indicate increased genomic instability at a *lacZ* reporter locus in liver, and especially spleen, at just 5 months of age (see also Chapter 7).

Hence, whereas the inability of the *Ku80*-mutant mouse to repair DNA damage leads to excessive genome rearrangements, the resulting increase in cell death or dysfunction becomes manifest as impaired proliferation and regeneration, resulting in diminished cancer and accelerated age-related organ and tissue degeneration. This situation may be very similar for two other mouse models with defects in DSB repair: mice with defects in DNA-PK$_{CS}$—the catalytic subunit of the Ku70–Ku80 complex—and mice harboring a hypomorphic mutation in the *BRCA1* gene, a major player in HR. In DNA-PK$_{CS}$-null mice the situation resembles the *Ku80*-null model: the absence of intact NHEJ promotes genomic instability leading to impaired tissue growth and regeneration[459]. Complete loss of BRCA1 is embryonically lethal and the same is true for the homozygous BRCA1 hypomorph, lacking exon 11. However, the homozygous hypomorph can be completely rescued in a p53-heterozygous background. In a p53-homozygous mutant background these mice exhibit a high incidence of cancer. In the p53-heterozygous background BRCA1 hypomorphic mice exhibit a long list of premature aging phenotypes and a significant reduction of lifespan, probably caused by the activity of the remaining p53 allele, which is triggered by genomic instability to prevent normal cell proliferation[460].

Thus far, there is no evidence that inactivation of Ku80's partner in NHEJ—Ku70—is causing accelerated aging[461]. However, *Ku70*-null mice have never been studied as an aging cohort over longer periods of time in parallel with their littermate controls. Such studies are now underway and should soon reveal if there really is a difference between the *Ku80* and *Ku70* mutants. It is possible that each of these two key players in DSB repair has other, tissue-specific functions which confound their role in suppressing aging.

5.3.3 WRN, ATM, AND TELOMERES IN MICE AND HUMANS

Several mouse models of premature aging involve defects at the interface of telomere maintenance and DNA DSB repair. The first of these models that should be mentioned is the WRN-defective mouse. Whereas in most of the above cases the mouse mutants of genome instability were generated primarily for studying cancer, the WRN gene defect was modeled on its own account as a model of accelerated aging[462,463]. Human patients with Werner syndrome prematurely exhibit signs of senescence including atrophic skin, graying and loss of hair, osteoporosis, malignant neoplasms, diabetes, and shortened lifespan[464]. Furthermore, a greatly increased frequency of genomic rearrangements has been reported in peripheral blood lymphocytes from these patients[465]. Mice with *Wrn* deleted exhibit no obvious phenotype, suggesting redundancy with another RecQ helicase[462,463]. However, the fact that *Wrn*-mutant mice do not recapitulate Werner syndrome as seen in humans does not diminish the potential importance of the human phenotype with regard to aging or the importance of mouse models. Similar cases, in which the mouse model failed to recapitulate the human phenotype, have been observed many times for genes that suppress cancer in humans. After a closer look, the proteins involved were found to perform remarkably similar biochemical functions with the same physiological significance in both species. For example, the retinoblastoma protein (pRb) is a tumor suppressor that in humans prevents the formation of tumors in the retina. However, in Rb-deficient mice, retinoblastoma does not occur. Instead, adenomas develop in the intermediate pituitary gland. However, retinoblastomas will develop in Rb mutant mice after reducing the expression of an Rb family member, p107[466]. Thus, the difference in phenotype between mouse and human is simply due to different levels of expression of genes in the Rb family. Therefore, tumor-suppressor genes perform remarkably similar functions in mice and humans at both the biochemical and physiological levels.

An important confounder of modeling human cancer and aging in the mouse involves telomere instability. As we have seen in Chapters 3 and 4, telomeres are the nucleoprotein complexes that occur at the ends of eukaryotic linear chromosomes. In the mammalian genome they consist of several kilobases of repetitive DNA sequences (TTAGGG) that attract a number of sequence- and structure-specific binding proteins. These chromosomal caps prevent nucleolytic degradation and provide a mechanism for cells to distinguish natural termini from DNA DBSs, which would otherwise signal DNA damage, resulting in cell-cycle arrest, senescence, or apoptosis. As we have seen, cells inevitably lose telomeric DNA with each cell division, which can be countered by elongation mechanisms. The most important of these mechanisms is telomerase, a reverse transcriptase composed of a protein component and an RNA (complementary to the telomeric single-stranded overhang), that can synthesize telomeric DNA directly onto the ends of chromosomes[467].

Telomerase-null mice have been made by ablation of the RNA component of the enzyme, Terc[468]. The enzyme is less important in mice than in humans because mice have very long telomeres. This is thought to be the reason that mice are more prone to sarcomas and lymphomas whereas humans are more prone to epithelial tumors[437]. Indeed, late generations (the fifth to eighth generations) of telomerase-knockout mice display a more human-like spectrum of tumors[469]. Progressive telomere erosion in Terc-null mice of somewhat earlier generations has been found to be associated with premature aging symptoms, but these represented far from a full spectrum of classical pathophysiological symptoms of aging. In these mice, age-dependent telomere shortening and increased genetic instability were associated with shortened lifespan as well as a reduced capacity to respond to stresses such as wound healing and hematopoietic ablation[470].

As mentioned above, the *ATM* gene, when defective, causes a segmental progeroid disorder in humans. ATM stimulates cell-cycle responses to the highly toxic DNA DSBs. One of its targets is actually p53, which can also be activated via ATM-independent mechanisms. ATM likely plays a role in telomere maintenance, possibly through direct binding to short telomeres, phosphorylation of telomere-binding proteins, or some other form of regulation of other telomere proteins[257]. Premature aging in ataxia telangiectasia is not that obvious[471] and in its mouse counterpart progeroid symptoms are even less prominent[472] unless they are bred into a *Terc*-deficient background. Against the background of eroding telomeres, the double null mice exhibited a general growth and cell-proliferation defect, probably causing the extensive organ dysfunction observed in this mouse model[473]. The cause of the problems in these mice is likely to be increased genomic instability and an elevated rate of apoptosis. This is typically observed in the *Atm* and *Terc* mutants separately, but greatly accelerated by the combined defect. It results in diminished stem-cell reserve in several organ systems and this impairment in regenerative capacity results in premature aging, exemplified by kyphosis, reduced muscle and fat mass, and several other aging-related phenotypes. Hence, in the combined *Terc/Atm* mutants the accelerated aging phenotype is much more convincingly expressed than in each mutant separately.

Similar to the *Atm* mutants, bringing the *Wrn* defect into a *Terc*-deficient background dramatically uncovered an array of premature aging phenotypes reminiscent of the human syndrome[474]. After four to six generations, during which the initially long telomeres of these mice become progressively shorter, mice were obtained of shorter lifespan displaying premature aging symptoms at 12–16 weeks. Symptoms included hair loss, cataract formation, hypogonadism, osteoporosis, type II diabetes, and kyphosis, which are all typical also for the patients with Werner syndrome. Since patients with Werner syndrome display high levels of spontaneous genome instability in their peripheral blood lymphocytes, bone-marrow metaphases in the late-generation double knockout mice were examined and showed similar, marked genomic instability. Increased genomic instability was also observed in embryonic fibroblasts from these mice as well as a reduction in their replicative lifespan. A marked increase in apoptosis rates was observed in intestinal crypt cells. Whereas not prominently cancer-prone, the earlier-generation $Terc^{-/-}$ $Wrn^{-/-}$ mice succumbed to osteosarcomas and soft-tissue sarcomas, typical for human patients with Werner syndrome, but not for mice. The later-generation mutants (generations 4–6) may live too short for these tumors to manifest[474].

As pointed out in Chapter 4, WRN (like ATM) plays a role in telomere maintenance, and the conclusion could easily be drawn that the *Wrn* mutation mainly reinforces the premature phenotype reported to be associated with the ablation of *Terc* alone. Conceivably, in mesenchymal tissues, which are mainly affected in the *Terc/Wrn* double knockouts, intact WRN is able to compensate for the lack of telomerase. It may be able to do that through its role (like ATM) in HR, a repair process known to be involved in the maintenance of eukaryotic telomeres[475]. Ultimately, the cause of the accelerated aging phenotypes is likely to be very similar as the one discussed for the *Terc/Atm-*, *Ku80-*, and *Brca1*-defective mice, namely increased genomic instability resulting in impairment of cell proliferation. The same would probably apply to a third accessory DNA-repair-defective mouse model, the TOPIII β-deficient mouse model.

It is known that topoisomerases interact with RecQ helicases[476]. Even though there is virtually no phenotype in Wrn-mutant mice, deletion of the helicase domain of Wrn increases the frequency of spontaneous mutations in T lymphocytes and inhibits proliferation of embryonic fibroblasts from these mice. Both cellular phenotypes are exacerbated by inhibition of TOPI[462]. The Wrn defect also increases tumor formation in a p53-defective background, much more than the p53 defect by itself[463,477]. The combined defect also results in tumors, such as sarcomas, which do not normally develop in a p53-null mouse model.

Because of the increased sensitivity to topoisomerase inhibition of cells from the Wrn-defective mice (which has also been demonstrated for lymphoblastoid cells derived from patients with Werner syndrome[478]), it is possible that the Wrn defect is to some extent caused by impaired interaction with topoisomerases. This is supported by the observation that mice with a topoisomerase defect—TOPIIIβ-null mice—develop normally but have a reduced lifespan and display a premature aging phenotype[479]. As pointed out in the previous chapter, work with yeast suggests that the TOPIIIβ phenotype is influenced by RecQ helicase activity. It is conceivable, therefore, that ablating the *TOP3B* gene may impair the RecQ family members causing the symptoms of aging, which Wrn inactivation alone, at least in the mouse, cannot accomplish.

5.3.4 NUCLEAR STRUCTURE

Attention was drawn to the importance of maintaining nuclear organization as a longevity-assurance system once it was demonstrated that HGPS is caused by *de novo* mutations in the gene *LMNA*, encoding both lamin A and lamin C through alternative splicing[428]. A *de novo* mutation activates a cryptic splice site, effectively resulting in a protein with a 50-amino-acid deletion near the C-terminus of lamin A. This deletion allows farnesylation, but prevents proteolytic cleavage of the prelamin A to generate the final product. The resulting mutant lamin A acts in a dominant fashion (patients are heterozygous) and the aberrant protein is called progerin. Whereas mouse models accurately mimicking the human defect are now being generated in different laboratories, two existing models deserve some discussion. The first is a null mouse for the metalloproteinase gene *Zmptse24*, encoding an enzyme involved in the proteolytic processing of prelamin A. Apart from nuclear abnormalities, this model shows growth retardation, alopecia, bone fractures, muscle weakness, and early cardiac dysfunction[480]. More recently, accelerated aging in these mice has been linked to p53 signaling activation, supporting the concept that hyperactivation of the tumour suppressor p53 may cause accelerated aging[481].

The second model was originally aimed at mimicking a mutation in *LMNA* that in humans normally results in Emery–Dreifuss muscular dystrophy (EDMD). (Heritable mutations in lamin A/C or lamin-binding proteins cause various diseases other than HGPS. For example, mutations in emerin, a lamin-binding protein, or *LMNA* cause EDMD, and mutations in *LMNA* can also cause Dunnigan-type partial lypodystrophy. The mutations that cause these diseases are non-overlapping and it is unclear as to how different mutations in the same gene can cause different diseases.) The homozygous EDMD mutation in the mouse causes death in 4–6 weeks. The mice develop severe growth retardation and a number of other symptoms reminiscent of HGPS[482]. In neither of the two mouse models has genomic instability, cell growth rate, or apoptosis rate been studied thus far. In interpreting these results, it should be kept in mind that HGPS is an early-onset segmental progeroid syndrome that occurs before reproductive maturation. Hence, it is unrealistic to expect an adult-onset type of premature aging.

Is it possible to ascribe HGPS to a defect in genome maintenance? Lamins at the nuclear periphery and throughout the nucleoplasm are thought to maintain nuclear shape. However, they may also contribute to tissue-specific gene expression through their role as key elements in nuclear architecture[214]. Regions of chromatin appear to be anchored to the lamina and, at least *in vitro*, lamins have been demonstrated to bind directly to chromatin. As pointed out in Chapter 3, the actions of a gene are not only influenced by its position in the one-dimensional DNA sequence, relative to regulatory elements, but also by its particular location in the nucleus. Such position effects may influence not only transcription, but also replication, repair, and recombination. For example, when DNA undergoes more than one DBS the spatial proximity of the broken ends is positively correlated with the probability of illegitimate joining. Hence, it is of the utmost importance to faithfully maintain lamin organization, defects in which are likely to result in genomic instability. Indeed, it was recently found that *Zmpste24*-deficient mouse embryonic fibroblasts as well as cells from HGPS patients show increased DNA damage and chromosome aberrations and are more sensitive to DNA-damaging agents[483]. It is conceivable that as a consequence of the dominant progerin or unprocessed prelamin A the distorted cell-nuclear architecture will no longer faithfully support the proper assembly of DNA-repair foci.

5.3.5 MITOTIC-SPINDLE CHECKPOINT AND THE ROLE OF ANEUPLOIDY

During the process of cell division the spindle-assembly checkpoint (also called the mitotic checkpoint; see Chapter 3) allows every chromosome to send a stop signal, preventing cells from entering anaphase until each chromatid is properly attached to the microtubules. Defects in this checkpoint provoke chromosome mis-segregation and aneuploidy (gain or loss of chromosomes), which can have adverse functional consequences, including cell death and cancer. For example, defects in different components of the mitotic checkpoint have consistently been observed in cancer cells, characterized by chromosomal instability[484]. Recently, a mouse model was made based on partial inactivation of the gene *Bub1b*, encoding a spindle-assembly checkpoint protein BubR1. These mice, *Bub1b*[H/H](H for hypomorph), which express BubR1 at 11% of normal levels, have a median lifespan of only 6 months and display a host of premature aging symptoms, including kyphosis, cataracts, thinning of the skin, reduced wound healing, infertility, and muscle atrophy[485]. They develop aneuploidy (as measured in splenocytes) from the age of 2 months onwards that increase as the mice age further. There is a strong correlation between the degree and severity of the aneuploidy and the onset and progression of the aging phenotypes. Increased apoptosis in several tissues from these animals at 1 year of age was not found. However, senescence-associated β-galactosidase activity was high compared with control mice. This observation of increased cellular senescence *in vivo* was further supported by an accelerated rate of senescence of embryonic fibroblasts cultured from these mice, which also correlated with the degree of aneuploidy observed in these same cells. Hence, the progressive aneuploidy was assumed to be the underlying cause of both senescence *in vitro* and premature aging symptoms *in vivo*. Interestingly, BubR1 expression declines in tissues of normal mice with age[485]. Hence, it is conceivable that increased aneuploidy, which has been observed in aging animals, contributes to normal aging as well.

A somewhat unexpected finding in this study was the virtual lack of spontaneous tumors in the mutants, in spite of the fact that aneuploidy is one of the hallmarks of cancer. In this respect, it is possible that the severity of the aneuploidy, while initially promoting the genesis of cancer cells, eventually prevents further tumor progression, similar to the situation in the aforementioned *Atm/Terc* mutant mice. This could be a direct effect of ongoing aneuploidization or the result of increased cellular responses activated by an aneuploidy checkpoint.

More recently, the same research group reported similar results for mice haploinsufficient for both *Bub3* and *Rae1*, two other genes involved in mitotic-checkpoint control in mammals[486]. Living much longer than the *Bub1b*[H/H] mice, the *Bub3*[+/−] *Rae1*[+/−] mice display similar premature aging characteristics between 15 and 27 months. Also, these mice displayed enhanced cellular senescence rather than apoptosis, which could be the underlying cause of premature aging. Indeed, aneuploidy in the compound mutant mice appeared to be higher than in the $Bub1b^{H/H}$ mice in spite of the much shorter lifespans of the latter. Aneuploidy was also present in the individual *Bub3*- and *Rae1*-haploinsufficient mice, neither of which displayed signs of early aging. However, it is difficult to draw any definite conclusions in the absence of complete lifespan studies, as pointed out above, especially since aneuploidy was only measured in splenocytes.

5.3.6 DEFECTS IN REPLICATION OF MITOCHONDRIAL DNA

Another example of a mouse model with a defect in genome maintenance especially designed to test the hypothesis that genomic instability is involved in aging is the *polgA* proofreading-deficient mouse model. POLG, the catalytic core of the mtDNA polymerase γ (polγ), is a mtDNA polymerase encoded in the nuclear genome. Mutations in both nuclear and mitochondrial DNA have been demonstrated to accumulate with age in a tissue-specific pattern[487,488]. Hence, in the *polgA*-deficient mouse one would expect to see accelerated mutation accumulation in the mitochondrial genome as well as the premature appearance of aging-related phenotypes. Both predictions turned out to be correct. Apart from a 3-5-fold increase in mtDNA mutations in brain, heart, and liver, and respiratory chain dysfunction in the heart, a host of premature aging symptoms was observed in homozygous mutant animals, including osteoporosis, lordokyphosis, alopecia, a growth defect, reduced subcutaneous fat, hypertrophy of the heart, and anemia[489]. Lifespan was reduced by about 50%. Very similar results were obtained in a somewhat later, independent study[490] with a mouse expressing another proofreading-deficient version of POLG. Interestingly, the enhanced mitochondrial mutation load in these 'mtDNA mutator mice' was already established at 2 months and did not dramatically increase thereafter. This suggests that, once present, somatic mutations can continue to exert adverse physiological effects even without any further dramatic increases.

Naturally, one would expect that the mutations and the premature aging phenotypes arise through increased ROS production as a consequence of respiratory-chain dysfunction. In turn, this would further increase ROS production, leading to more mutations, and so on. This is called the vicious-cycle hypothesis, which will be discussed in more detail in the next chapter. A causal role of ROS generation and mitochondrial damage in aging is supported by the recent demonstration in mice that overexpression of a catalase transgene targeted to mitochondria significantly extends lifespan (20% increase in both average and maximum lifespan)[491]. However, the mtDNA mutator mice were found not to affect ROS production and there was no evidence for increased oxidative stress. Instead, accumulation of mtDNA mutations was found to be associated with increased apoptosis[490].

Whereas it is tempting to interpret the association between increased mtDNA mutations and premature aging as observed in the *polg*-deficient mice as strong evidence for a causal role of mtDNA mutations in aging it should be noted that the mutation frequencies observed in these mice were more than an order of magnitude higher than typical levels in aged humans. Also the pattern of affected tissues in these mice was different from normally aged humans or mice, in which very high mtDNA mutation frequencies are observed in muscle, heart, and brain. The types of mutation were different too, since in normally aged tissues deletions are a major component of the spectrum (see Chapter 6). Hence, ironically the results with the *polg* mouse models may argue against a causal role of mtDNA mutations in aging, at least as far as point mutations are concerned[492]. Interestingly, results obtained with transgenic mice expressing a dominant mutant variant of the mouse Twinkle protein suggest that the accumulation of deletion mutations in the mitochondrial genome may also not be causally related to aging[493]. Such mutational variants of this nuclear-encoded mitochondrial replicative helicase are known to cause multiple mtDNA deletions (and disease) in humans and do so also in mice. These so-called deletor mice do not show premature aging, which was interpreted to indicate that subtle accumulation of mtDNA deletions and progressive respiratory-chain dysfunction are not sufficient to accelerate aging. Whereas muscle atrophy and mitochondrial dysfunction was observed in these mice, their lifespan was normal. Taken together, these results suggest that mtDNA mutations, including deletions, are by themselves not sufficient to cause aging.

5.3.7 THE p53 ANTI-TUMOR RESPONSE AS A CAUSE OF AGING

By now the reader will probably have noticed that many if not all of the mouse mutants thus far discussed share not only a defect in some aspect of genome maintenance accompanied by symptoms of premature aging, but also a series of enhanced cellular responses that appear to converge on p53, the so-called guardian of the genome. In fact, defects in the gene *TP53*, encoding this cell-cycle control protein, can also cause premature aging in mice. The p53 tumor suppressor is a transcription factor that controls a network of genes that regulates responses to DNA damage[494]. The p53 protein inhibits cell-cycle progression, facilitates DNA repair, and activates both cellular senescence and apoptosis pathways[421]. Two mutant alleles of *TP53* have been described which, in combination with a wild-type allele, result in multiple symptoms of aging and a shortened lifespan. The p44 allele, a naturally occurring splice variant first described in 1987, lacks the first transactivation domain[495]. The other mutant allele, termed M, is an accidentally introduced deletion of the 5' region of the *TP53* gene, lacking both transactivation domains[496]. Both mutants display prominent signs of premature aging, but the two animal models differ, mainly in the severity of the symptoms. In the M mutant mice, in which expression of the variant allele is very low, the first symptoms of aging only become apparent at 18 months. The p44 transgenic mice, which express the variant allele at a much higher level, already show aging-related mortality as early as 5 months. Hence, severity is likely to be correlated with the expression level of the mutant allele. In the p44 model it has been demonstrated that with increasing p44 expression symptoms become much more severe. Interestingly, an increase in p44 was found to alter IGF-1 signaling, by increasing both plasma IGF-1 levels and IGF-1 receptor levels in tissues[495]. The relevance of this finding is not completely clear, but it should be noted that the p53 pathway is interconnected with the IGF-1/Akt pathways (as well as with the TOR pathway; see Chapter 2) in their joint role of sensing growth factors, nutrients, oxygen, and stress[497].

Both models display early osteoporosis and kyphosis, major forms of aging-related pathology in both mice and humans. The p44 mice show early fertility loss, to some extent caused by a breakdown of the reproductive axis (H. Scrable, personal communication). This loss of the hypothalamic pituitary gonadal axis is very similar to the situation in natural human and mouse aging[498]. This loss of fertility is not observed in the M mutant, but again, the symptoms in that model are generally much less severe. Both mouse models suffer from a typical aging-related redistribution of fat, resulting in loss of subcutaneous fat. Other typical aging-related phenotypes in these mice are skin problems and various forms of atrophy. Surprisingly, in both mutants cancer incidence was lower than in the controls. Of note, mice expressing additional copies of wild-type p53 under the control of its own promoter do not show signs of premature aging, but do show lower cancer incidence[499]. In this case, the normal p53 gene dosage is minimally increased—by one additional copy—resulting in an increased response to DNA damage without affecting basal levels of p53. By contrast, in the two accelerated aging mutants, p53 may be contitutively activated through an interaction of the mutant p53 allele with the wild-type allele. This apparently results in postnatal growth impairment of the animals, possibly caused by a reduced cell proliferation rate as observed in cultured fibroblasts from these animals[495]. It is also supported by the observation of a more rapid accumulation of senescent cells in tissues of the $p53^{+/M}$ mice, as compared with control animals[500]. This was based on staining for senescence-associated β-galactosidase (SA-β-gal), the most widely used biomarker for senescent cells and originally discovered by Judy Campisi and co-workers[501]. However, it is unclear what such a marker, defined as β-galactosidase activity detectable at pH 6.0, would indicate in liver cells normally not proliferating. It is possible that SA-β-gal under such conditions reflects a stress response. Of note, increased rates of apoptosis (programmed cell death) in these two mutant mouse models have not been observed (H. Scrable and L. Donehower, personal communication). Indeed, the hyperactive p53 protein most likely causes its effects through increased cell-cycle arrest, leading to a general inhibition of cell proliferation. This may explain many of the premature aging phenotypes. For example, impairment of osteoblast proliferation, a likely natural cause of osteoporosis in the elderly, may cause the increased, premature osteoporosis in these mutant mice.

Overall, therefore, constitutively activated p53 activity appears to promote many aging-related phenotypes through the inhibition of normal regenerative processes which are essential for adult animals to survive and maintain organ function. This may be the case, not only in the two *TP53* mutants just discussed, but also in a host of other mouse models harboring defects in genome maintenance. Indeed, in such animals increased genotoxic stress would keep p53 continuously activated. The role of *TP53* as a pro-aging gene is schematically depicted in Fig. 5.6. As the guardian of the genome, p53 is supposed to inhibit cell growth and proliferation, but probably only occasionally and transiently, after a burst of spontaneous DNA damage. Whereas cancer is also a major aging-related phenotype, constitutively activated p53 would be expected to greatly suppress tumor formation due to its inhibition of normal cell-proliferative activities. This is exactly what is often observed in mice with defects in genome maintenance. It is conceivable that in normal mice p53 responses naturally increase with age due to the aforementioned increased load of DNA damage. This would then result in very similar phenotypes as in the p53 mutants, but later in life, as part of the normal aging process. Below I will discuss the possibility that the cellular response to DNA damage, rather than genome instability per se, could be the cause of aging.

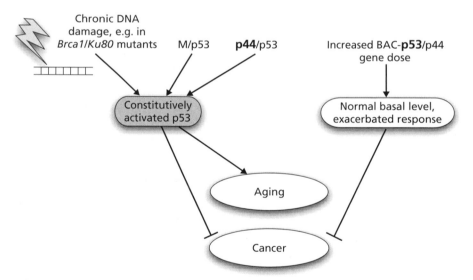

Fig. 5.6 Schematic representation of the putative role of p53 as a pro-aging factor. Increased copy number of the gene for p53 in the context of its normal regulatory sequences (as part of a bacterial artificial chromosome (BAC) clone) causes its exacerbated response, but not constitutive activation. Constitutive activation of p53 (shading) can be accomplished through chronic DNA damage (as in various mouse mutants harboring genetic defects in genome maintenance), or through the M or p44 allele in the context of the normal p53 gene product. Note that in the p44 transgenic mice the ratio of p53 to its variant is altered in favor of the latter, as denoted by the relative letter sizes.

5.4 **Conclusions**

In this chapter I reviewed a range of genetically engineered mouse models, harboring specific defects in genome maintenance, and exhibiting a common array of aging-related phenotypes, even though the mutated genes represent a diverse range of pathways. As discussed, the most convincing models undergo normal development and then prematurely exhibit a set of common aging features. In such cases it is difficult to maintain that these mutants do not faithfully mimic aspects of the normal aging process. At this point in time, the question is no longer whether the premature aging phenotypes in these mice are genuine, but rather why different genetic defects in some aspect of the complex network of genome maintenance all lead to very similar endpoints in aging. Of note, genome maintenance is the only system in which such a wide variety of defects leads to such a recognizable pattern of phenotypic changes.

How can so many clinical endpoints of aging undergo acceleration as a consequence of genetic defects in a set of functional pathways that superficially have no relationship to the original function that is impaired; for example, bone metabolism, functional skin mass, neuronal activity, and growth? As already alluded to, this observation points to a key role of the network of genome-maintenance genes as the last line of defense against aging by removing damage induced by metabolism-related ROS.

The key consideration here is that aging is characterized by progressive system dysfunction as a consequence of the loss of functional cells, ultimately driven by the accumulation of DNA damage. This could reflect two major underlying processes. First, organ and tissue dysfunction could be caused by cellular responses to DNA damage, including apoptosis and cellular senescence. Second, cells could degenerate by a process of intrinsic aging, possibly driven by increased genomic alterations. Whereas the former would mainly result in various forms of tissue atrophy and loss of regenerative capacity, increased genomic instability at old age is likely to contribute to the well documented exponential increase in cancer during aging. Indeed, hyperplastic and neoplastic lesions can comprise as much as 50% of all pathological lesions observed at old age[439]. However, an increased load of genomic mutations, including sequence alterations and methylation changes, could also lead to a progressive increase in the number of dysfunctional cells, without causing them to immediately die or become transformed. We can call such cells senescent cells and they may include permanently arrested cells, as a consequence of replicative senescence[421] (for a discussion of the phenomenon of cell senescence and its implications for aging *in vivo*, see Chapter 7). Similar to increased cell death, an increased fraction of dysfunctional cells would be expected to lead to organ dysfunction.

As we have seen, in all mouse models of premature aging the genetic defect leads to increased impairment of growth and cellular proliferation, either through cell-cycle arrest, or other elicited DNA-damage responses, such as replicative senescence and apoptosis. In many of these mutants we also see an increase in mutations; irreversible changes in DNA sequence due to misreplication or misrepair. This increased genomic instability promotes the emergence of cancer cells but is so dramatic that it ultimately represses tumor progression, due to increased anti-cancer responses, such as apoptosis. Indeed, deletion of p53 rescues the premature replicative senescence phenotype often observed in cells from mice with defects in genome maintenance[502]. This suggests that many of the premature aging phenotypes in mice with defects in genome maintenance are caused by cellular responses to DNA damage, including the activity of a p53-dependent cell-cycle checkpoint (Fig. 5.6).

Are these same cellular responses, identified in the mouse models as a likely cause of their accelerated aging, also responsible for normal aging? There can be no doubt that loss of cells per se contributes to normal aging and that various forms of atrophy are common among older mice or humans. Hence, part of normal aging is likely to be also due to the adverse effects of cellular responses to DNA damage. However, normal aging is not generally associated with the dramatic forms of growth reduction, tissue atrophy, and loss of organ functional mass as observed in many of the models of accelerated aging. For example, the symptoms of aging in the *Xpd* mouse model listed in Table 5.2 are much more severe than would be expected from their biological age as derived from the survival curve. Indeed, these data were obtained by complete histopathology at the end of life. Hence, one would expect to see an incidence and severity level very similar to the control animals, but this is not the case. Hence, it is possible that accelerated aging in many of these models is mainly caused by excessive cellular responses to genomic stress, in contrast to normal aging, which is more likely due to a gradual accumulation of persistent DNA damage or mutations. Indeed, the bursts of DNA damage necessary to evoke, for example, an apoptosis response, must be quite rare during normal aging, as can be derived from the extremely low number of apoptotic cells in tissues of normally aged animals. Whereas even very low numbers of apoptotic cells at a given time point may still correspond to a significant cell turnover, this is not necessarily a bad thing. Indeed, caloric restriction has been found associated with slight increases in apoptosis[407], which may point towards a healthy cell turnover to eliminate defective cells so that they can be replaced by new ones from the stem-cell reservoirs. Only when apoptosis levels are excessive, as is the case in several of the mouse models just discussed, are organ and tissue cellularity likely to become adversely affected. As will be discussed in Chapter 7, there is no convincing evidence for excessive cell loss or depletion of stem-cell reservoirs at old age. Nevertheless, at this point in time it is not possible to conclude that normative aging is not based on cell loss; it just seems unlikely.

As discussed extensively in the next chapter, normal steady-state levels of DNA damage, such as 8-oxoG, are extremely low[503] and therefore also unlikely to directly cause aging-related degenerative defects. However, DNA damage can be turned into mutations by erroneous processing of DNA damage. Evidently, the frequency of such events is low, as exemplified by the normally low frequency of neoplastic transformation. However, it is likely that many mutations of different kinds (from point mutations to large genome

rearrangements and aneuploidy) are needed to generate a tumor cell. Most of such mutations do not involve actual genes, but gene-regulatory sequences. Indeed, rather than a
catalog of useful genes interspersed with functionless DNA, each chromosome is now
viewed as a complex information organelle with sophisticated maintenance and control
systems. In this concept of a genome, each part has a function, even its non-protein-coding
parts. Such a holistic view of the genome would assign a variety of functions to non-coding
DNA (e.g. structural maintenance, gene regulation). The continuous process of irreversible
alteration of the DNA sequence organization of the genome would lead to a mosaic of
cells with each individual cell in a tissue bearing a different pattern of genomic scars[504].
Rather than the inactivation of unique genes with unique functions, which has thus far
been given primacy, a scenario of an initially tolerable loss of genome structural integrity
is more likely. This could lead to increased loss of function, highly variable from cell to
cell, thus contributing to increased organ dysfunction. Hence, whereas age-related systems dysfunction in premature aging may predominantly involve cellular responses
evoking proliferation impairment and programmed cell death, normal aging could
mostly be due to the gradual accumulation of persistent genomic or epigenomic alterations leading to a progressive increase of dysfunctional cells. Both scenarios would result
in organ and tissue dysfunction. Figure 5.7 schematically depicts the hypothetical causes

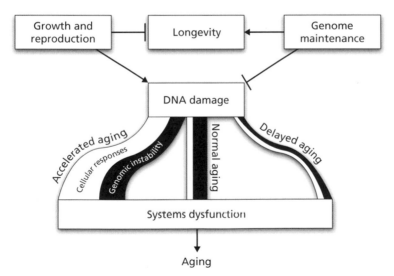

Fig. 5.7 DNA damage, the adverse by-product of normal growth and metabolism, may be a
major driver of the aging process. Whereas in accelerated aging cellular responses impairing cell
proliferation (apoptosis, cellular senescence) as a consequence of defects in genome maintenance are likely to be mainly responsible for the aging phenotype, in wild-type animals or
mutants showing delayed aging, cellular responses may be much less important and aging
mainly the result of cell-functional decline.

of aging as driven by DNA damage for (1) accelerated aging, mostly due to adverse effects of cellular responses to DNA damage; (2) normal aging, with a reduced role of cellular responses, and (epi)genomic alterations as the main cause of cell and organ dysfunction; and (3) delayed aging with reduced cellular responses as well as reduced accumulation of irreversible, (epi)genomic alterations.

It would be premature to consider DNA damage as the sole driving force of aging in all multicellular organisms. However, based on the model systems now available, it accounts for a significant number of the observations that have been made. Of note, DNA damage as the original driver of a universal process of aging has some inherent logic. After all, damage to nucleic acids is the most ancient example of damage accumulation in the living world. Ever since the first replicators, genetic damage posed both a fundamental problem and an opportunity for living systems. A problem, because genetic damage essentially prevents the perpetuation of life, since it interferes with replication (and transcription); an opportunity, because it allows the generation of genetic variation through errors in the replication or repair of a damaged template, thereby facilitating evolutionary change. Genetic stability in somatic cells of metazoa has become part of the trade-off between the allocation of scarce resources to either reproduction or somatic maintenance and is not expected to be maximized. Whereas the concept of DNA damage as a main driver of aging is intuitively attractive, evolutionary logic dictates that other causes of aging would have emerged over evolutionary time (Chapter 2). To judge whether genome instability is likely to play a causal role in aging it will be necessary to assess the rate and severity of genome alterations in normal individual animals during their lifetime. This is the topic of the next chapter.

6 The aging genome

Spontaneous instability of the nuclear genome of somatic cells has been considered as a possible explanation for aging since the 1940s when the biological effects of high-energy radiation were first systematically studied in mice[505]. These studies were part of the Manhattan project and undertaken when it was realized that the personnel involved in the production of radioactive isotopes needed protection. Hence, an experimental basis was needed to establish tolerance levels. It was found that small periodic doses of fast neutrons or γ rays resulted in premature aging, as indicated by the acceleration of typical end-of-life pathology, such as generalized atrophy and lymphomas. Shortening of lifespan was found to be a function of the daily dose. It is not surprising, then, that after the discovery of the DNA double helix in 1953, the first somatic mutation theories were formulated by physicists. They reasoned that since radiation was known to induce mutations in DNA, aging could be the result of life-long exposure to low, natural levels of background radiation[392,506]. At the time, not much was known about the mechanisms underlying mutation induction. DNA damage still remained to be discovered and nothing was known about the relentless stream of DNA chemical lesions inflicted continuously on the cells of the body. Nevertheless, the logic behind the general idea that somatic mutations are the cause of aging was already then compelling.

Instability of the nuclear genome is a broad concept. Distinctions should be made between DNA damage, DNA mutations, and epigenomic DNA alterations. DNA damage includes chemical alterations in DNA structure, leading to a non-informative template; that is, a structure that can no longer serve as a substrate for faithful replication or transcription. As a consequence of DNA damage mutations can be introduced, for example, by mishandling of the damage by repair or replication processes. DNA mutations are heritable changes in genomic DNA sequence or in their organization, which are transmitted to daughter cells or to offspring (when they occur in germ cells). Mutations can vary from point mutations, involving single or very few base pairs, to large deletions, insertions, duplications, inversions, and translocations. In organisms with multiple chromosomes, DNA from one chromosome can be joined to another and the actual chromosome number can be affected. Epigenomic alterations are changes in patterns of DNA modification, such as methylation patterns or the histone code, which are heritable and can influence patterns of gene expression without altering the sequence of base pairs.

In a strict sense, changes in histone proteins that would alter the information that can be retrieved from the genome are not genome alterations but alterations in chromatin, the ordered complex of DNA and protein that forms the chromosomes. In this chapter I will refer to all three types of alteration collectively as genome alterations. Epigenomic profiles, once established, are only partially stable and can vary substantially as compared with the more static DNA-sequence code. This has to do with the role of epigenomic changes in orchestrating various cellular activities, as we have seen in Chapter 3. Furthermore, the epigenome also varies in a stochastic manner, similar to the random alteration in the genome's DNA sequences. Because epigenomic patterns are thought to be erased during maturation of the germ line and then re-programmed, they are generally considered not to contribute to the transmission of information from one generation to the next. However, there is some evidence that this is incorrect and that there are cases of meiotic inheritance of epigenomic states[507].

As discussed in the previous chapters, genome instability as a potential cause of aging is based on the need to provide for evolutionary diversity. The price to pay for such diversity is the continuous threat of population extinction when the mutation load of the germ cells becomes too high. Sexual reproduction has greatly facilitated the maintenance of this precarious balance by its capacity to generate individuals with lighter mutation loads through recombination. Relatively small, asexual populations are unable to do that and, as a consequence, tend to lose mutation-free genomes (due to genetic drift) and inevitably suffer from loss of viability[73]. Such fates and their similarities to aging have been discussed in Chapter 2 for protozoa and for *E. coli*. While this explains the advantage of recombination and therefore sexual reproduction, fitness loss as a consequence of mutation accumulation is not limited to asexual organisms. Indeed, reduced fitness due to mutation accumulation in the germ line has been demonstrated, under certain conditions, in *Drosophila*[508] and *C. elegans*[509]. In a somewhat simplistic manner one could argue that what is true for germ cells should be true for somatic cells, especially since the evolutionary theory of aging would not predict that any adverse effects associated with high mutation loads become subject to natural selection as long as they occur at late age, after the reproductive period (Chapter 2). Somatic cell populations have only limited opportunity to lighten the burden of their mutation loads through apoptosis or senescence because this would lead to cell loss and organ dysfunction.

In spite of the logic of ascribing aging to mutation accumulation, analogous to the generation of genetic variability in the germ line, some serious criticisms against the hypothesis were expressed almost from the very beginning. In 1962, John Maynard Smith critically reviewed the possible causes of aging according to the information then available[510]. Many of the arguments presented in this lucid paper are still absolutely valid and contributed greatly to our understanding of aging. However, Maynard Smith's discussion of somatic mutations as a possible cause of aging in *Drosophila* clearly reflects the lack of insight at that time as to how mutations arise. The key issue, which also obscured the interpretation of the results obtained with the haploid and diploid wasps of the genus

Habrobracon (Chapter 2), is DNA damage and the cellular responses to it. For example, Maynard Smith argued that there are reasons for expecting a single concentrated dose of radiation to produce more mutations than a smaller dose spread over a longer time. We now know that the opposite is true due to the higher toxicity associated with the acute dose, which will prevent many mutations from reaching fixation. The ignorance of DNA damage as the intermediate between the mutagen and its mutational consequences has plagued a correct evaluation of the somatic mutation hypothesis ever since.

The emergence on the stage of DNA damage immediately raises the question of whether this type of genome alteration can itself be a cause of aging, or its molecular consequences in the form of mutations; that is, changes in DNA-sequence content. In addition, the recent realization that epigenomic modifications should also be considered as genuine changes in the genetic code introduces yet another potentially pro-aging type of genome alteration. Like DNA-sequence alterations, epigenomic changes are now considered as causes of cancer. The question addressed in this chapter involves the types of genome alteration that occur spontaneously in organisms, and their frequencies in aging cells and tissues.

6.1 **DNA damage**

Apart from errors made during information transfer per se, the ultimate source of permanent genetic changes, including DNA mutations and epigenomic changes, is the continuous introduction of chemical lesions in the DNA by a variety of environmental and endogenous, physical, chemical, and biological agents (Fig. 6.1). In contrast to DNA mutations, the induction of DNA damage is a frequent event. As discussed in Chapter 4, there are tens of thousands of spontaneous lesions inflicted on the genome of a typical cell on a single day. These lesions need to be repaired quickly to prevent rapid cellular degeneration and death. DNA damage has drawn ample attention in the past because of its causal relationship with cancer. This has led to an extensive research effort to identify the critical lesions causally related to the various types of cancer in humans. Whereas the early focus was on environmental agents, it was later realized by Larry Loeb[511] (Seattle, WA, USA), Bruce Ames[245] (Berkeley, CA, USA), and others that endogenous genotoxicants were probably much more important in causing the majority of spontaneous tumors. The focus of these studies was on humans and rodent model systems, which explains why most of what we know about DNA damage is from work with mammalian systems. Before discussing the evidence for DNA damage persisting with age and its potential functional impact, I will briefly review the different sources of DNA damage, focusing on mammalian genomes.

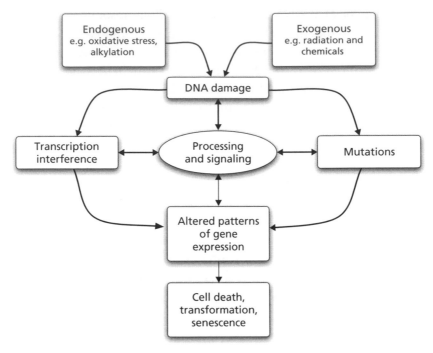

Fig. 6.1 Induction of DNA damage from endogenous and exogenous sources during aging. Subsequent processing of the damage by DNA repair and DNA-damage signaling systems can result in a number of molecular and cellular endpoints, including mutations, altered gene expression, and cell death. Note that DNA damage itself is generally not an endpoint, but continuously induced and repaired.

6.1.1 POTENTIAL SOURCES OF DNA DAMAGE

6.1.1.1 Endogenous sources

Most data on DNA-damage induction, from either endogenous or environmental sources, are estimates, mostly based on experiments with isolated DNA; quantitative data on the spontaneous induction of endogenous DNA damage in cells and tissues of higher animals are for the most part still lacking, with the exception of some relatively recent results on the presence of oxidative lesions, such as 8-oxoG (Table 6.1). Much of what we know of spontaneous DNA damage, including the enormous magnitude of the problem, derives from the pioneering work of Tomas Lindahl (South Mimms, Herts, UK)[512]. Lindahl also greatly contributed to our current insights into the main pathway for repairing the vast majority of spontaneous DNA damage—base excision repair (BER)—and his work will be extensively cited below.

Water Under physiological conditions—at the normal human body temperature of 37°C in aqueous solution—the N-glycosylic bond between the purine or pyrimidine base

Table 6.1 Estimated numbers of DNA lesions induced in human cells each day

Source	Lesion	Estimated number of lesions induced per cell/day[a]	Reference
Spontaneous hydrolysis	SSBs	20 000–40 000[b]	515
	AP sites	10 000[b]	513
	Deamination	100–300[b]	517
Oxidation	8-oxoG	27 000[c]	503
	Thymine glycol	270[d]	531
Methylation	N^7-methylguanine	4000[b]	543
	N^3-methyladenine	600[b]	543
	O^6-methylguanine	10–30[b]	543
Glucose	Glucose adducts	3[b]	541
Sun exposure	Pyrimidine dimer/6–4 photoproduct	60 000–80 000[e]	552
Smoking	PAHs	100–2000[f]	545–547
Coke ovens	BaP diol epoxide	7000–70 000[g]	553
Radon	SSBs	2[h]	556

AP site, apurinic or apyrimidinic site; BaP diol epoxide, benzo(a)pyrene diol epoxide; PAH, polycyclic aromatic hydrocarbon.
[a] Calculated on the basis of 1.2×10^{-10} nucleotides per mammalian cell.
[b] Based on experiments with isolated DNA.
[c] Based on the half-life of 8-oxoG in mouse brain.
[d] Based on data obtained from urine of humans.
[e] Number of pyrimidine dimers induced during 1 h of sun exposure.
[f] Level of DNA adducts found in individuals who smoke about 20 cigarettes per day.
[g] Level of DNA adducts found in individuals working for several years in factories with coke ovens.
[h] Calculated number of SSBs induced per year based on the estimated level of background radiation (see text).

of DNA and its deoxyribose is relatively labile and can easily be broken, resulting in apurinic or apyrimidinic sites (AP sites). From experiments in which the release of labeled bases was measured in double-stranded *Bacillus subtilis* DNA, incubated at different temperatures, Lindahl and Nyberg[513] estimated that at the normal body temperature of 37°C about 10 000 depurinations are induced per mammalian cell per day. Apyrimidinic sites are formed at a significantly lower rate. In isolated DNA at 37°C AP sites are spontaneously converted into SSBs in a few days[514]. This number may even be higher. Based on measurements from Crine and Verly[515] of the spontaneous degradation of isolated T7 phage DNA *in vitro* at 37°C (using alkaline sucrose-gradient centrifugation), Saul and Ames[516] calculated that about 20 000–40 000 breaks per cell are induced per day. According to Crine and Verly most of the breaks found were generated by the spontaneous destruction of deoxyribose moieties or direct hydrolysis of phosphodiester bonds and not by depurination.

Other forms of water-induced DNA alterations include the deamination of cytosine to uracil and, to a smaller extent, the deamination of adenine and guanine to hypoxanthine and xanthine, respectively[517,518]. As discussed in Chapter 4, uracil, hypoxanthine, and xanthine can be removed from cellular DNA enzymatically by the specific glycosylases of the BER process, resulting in AP sites. Less frequently occurring heat-induced DNA modifications are caused by hydrolytic reactions such as the opening of the imidazole ring of adenine, hydration of pyrimidines, and conversion of deoxyguanosine to deoxy-neoguanosine[519].

At present only limited information is available as to the actual formation of SSB and AP sites in tissues of higher animals *in vivo*. Since DNA *in vivo* is tightly packed together with proteins, the induction of damage might be quite different from the situation *in vitro*. In 1998 Nakamura *et al.* reported results obtained with a novel, highly sensitive assay for AP sites, indicating a steady-state level of 23 000 of such lesions per cell in a human B lymphoblastoid cell line[520]. To put this in perspective, such a number for one of the most frequently occurring lesions corresponds to only four AP sites per 10^6 nucleotides.

Oxygen A mechanism contributing to the second, major endogenous source of DNA damage is the generation of oxygen free radicals associated with a variety of normal cellular processes. Important classes of such free radicals, also termed reactive oxygen species (ROS), are the active oxygen species superoxide (O_2^-), singlet oxygen ($^1O_2$), and the hydroxyl radical (OH$^\bullet$). A major source of ROS is the sequential reduction of oxygen to water as part of the process of oxidative phosphorylation, our main energy generator (Fig. 6.2). However, ROS are also by-products of enzyme systems, such as the cytochrome P450 systems involved in the detoxification of toxic chemicals, and xanthine oxidase, a key enzyme in purine catabolism[521]. The reaction of ROS with cell membranes can initiate a radical chain reaction, leading to the oxidation of unsaturated lipids present in these membranes and resulting in lipid radicals, such as lipid-peroxy radicals, fatty acid hydroperoxides, and cholesterol epoxides[522]. ROS are also generated through the oxidation of small molecules, such as thiols, hydroquinones, and flavins[523]. Whereas ROS can clearly damage a variety of vital cellular constituents, such as proteins, membrane lipids, carbohydrates, and nucleic acids, they are also functionally important in a variety of signaling processes. In the innate immune system, polymorphonuclear leukocytes or granulocytes are activated for phagocytosis, and these cells produce ROS and reactive nitrogen species in amounts sufficient to kill invading bacteria. ROS have been proposed as regulators of apoptosis[524]. A possible critical factor in this process is p66[SHC], the inactivation of which in the mouse decreases the incidence of aging-associated diseases, such as atherosclerosis,

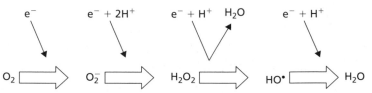

Fig. 6.2 Generation of free radicals during the univalent reduction of oxygen to water.

and prolongs lifespan[449,525], as we have seen in Chapter 2. p66[shc] is a redox enzyme that generates mitochondrial ROS (hydrogen peroxide) as signaling molecules for apoptosis[526].

Hence, there must be a balance between ROS formation and the various antioxidant systems that defend cells against the adverse effects of ROS. Such systems include simple antioxidants, such as vitamins E and C, β-carotene, uric acid, and numerous other so-called free-radical scavengers that can react with ROS. Probably even more important are a variety of enzyme systems that can deactivate ROS. They include superoxide dismutase (SOD), catalase, and glutathione peroxidase, each of which has several representatives. Together, antioxidant systems function as a first-line defense against oxidative DNA damage. It is for this reason that in Chapter 4 they have been categorized as a genome-maintenance system.

Much knowledge concerning free-radical reactions with DNA has been derived from studies with radiation (e.g. γ rays, X-rays, and UV radiation). The transfer of radiation energy to cellular compounds, such as water, results in the formation of ROS. Indeed, OH$^•$ radicals are responsible for 70–80% of the SSBs induced by γ rays[527,528]. In DNA of cultured cells or in isolated DNA, γ ray-induced radicals give rise to a variety of oxidative lesions, including SSBs, DSBs, AP sites, and cross-links, but also to several modified bases such as 8-oxoG, thymine glycol, and 5-hydroxymethyluracil[529,530].

There is both indirect and direct evidence that small, oxidative base damages are formed in cells of higher animals *in vivo*. The modified bases mentioned above, and also thymidine glycol, have been found in the urine of rats and humans, which has been interpreted as an indication for the continuous induction and removal of these and other DNA-oxidation products in mammalian tissues[531,532]. Interestingly, the amount of both thymine glycol and thymidine glycol per kilogram of body weight found in the urine of short-lived animals (rats and mice) is higher than that in the urine of long-lived animals (monkeys and humans), suggesting a higher induction of oxidative DNA damage in the short-lived species[533]. This would be in accordance with the higher rate of mitochondrial ROS production per unit of body weight in short-lived mammals than in long-lived ones, such as humans[534], which in turn would be expected to lead to an increased formation of oxidative damage. According to calculations made by Saul *et al.*[532], about 1000 of these DNA-oxidation products (thymine glycols, thymidine glycols, and 5-hydroxymethyluracil) are induced in each human cell per day (Table 6.1), whereas about 15 000 of such lesions are generated per day in the genome of a rat cell.

More direct evidence for the presence of oxidative DNA damage in cells *in vivo* has been obtained through the development of sensitive assays, based on high-performance liquid chromatography (HPLC) and gas chromatography. There are at least 20 known products of DNA oxidation, with 8-oxoG likely to be the most frequent one, comprising perhaps 5% of the total[535]. As we have seen in Chapter 4, this lesion base pairs preferentially with adenine rather than cytosine and this generates transversion mutations after replication. Whereas initially very high levels of 8-oxoG in animal tissues were reported (as many as three per 10^5 dGs), it is now clear that most of these lesions were artifacts of the experimental procedure to measure them[536,537]. Increased precautions to prevent oxidation at various stages of sample preparation have now led to the assumption that the real level is likely to be no more than 1–10 8-oxoGs per 10^7 dGs. Indeed, at the laboratory of Arlan Richardson (San Antonio, TX, USA) steady-state levels of this lesion of only several hundred per cell were found in various mouse tissues[503] (see also below). Interestingly, as originally discovered in laboratory of Bruce Ames, the level of 8-oxoG is an order of magnitude higher in mtDNA than in nuclear DNA[538], which is not unexpected since mitochondria are a major source of ROS. This was comfirmed by the group of Richardson.

Sugar Glucose reacts non-enzymatically with proteins and lipids *in vivo*, chemically forming covalently attached glucose-addition products and cross-links termed advanced glycation end products (AGEs)[539]. AGE formation is increased in situations with hyper-glycemia (e.g. diabetes mellitus) and is also stimulated by oxidative stress. AGEs are the products of the Maillard reaction, a complex set of non-enzymatic protein browning and cross-linking reactions. AGEs can react with DNA to form specific lesions[540]. However, glucose and other reducing sugars can also react non-enzymatically with the amino groups of DNA bases, which can lead to the formation of DNA adducts and mutations. The frequency of these lesions is unknown, but most likely only a fraction of the above-mentioned AP sites and oxidation products[541].

Methyl donors As also mentioned in Chapter 4, a potentially important source of endoge-nous DNA damage is *S*-adenosyl-L-methionine (SAM), the normal intracellular methyl-group donor[542,543]. Non-enzymatic transmethylation from *S*-adenosyl-L-methionine to DNA occurs at a slow rate. This reaction can lead to the formation of N^7-methylguanine, N^3-methyladenine and small amounts of O^6-methylguanine. The methylation of guanine and adenine bases causes a further destabilization of the N-glycosylic bond, resulting in an increased spontaneous cleavage and the formation of AP sites[513]. It has been estimated that about 600 N^3-methyladenine residues per day are generated in a human cell through non-enzymatic methylation from *S*-adenosyl-L-methionine[512]. This is a toxic lesion, which can be efficiently removed through BER, as we have seen.

6.1.1.2 Exogenous sources

Although endogenous sources of DNA damage now appear to be the most important, various exogenous sources should not be ignored. The extent of exposure of human indi-viduals to such sources is, of course, determined predominantly by lifestyle, occupation, and place of residence (Table 6.1).

The human diet contains a great variety of natural mutagens and carcinogens, such as polycyclic aromatic hydrocarbons (PAHs), aflatoxin B1, and nitrosamines[544]. These agents or their metabolites can react with DNA, thereby forming several types of lesion, including SSBs, DSBs, and bulky adducts. Cigarette smoke also contains various carcinogenic compounds, including benzo(*a*)pyrene (BaP). DNA adducts induced by BaP have been detected—by using the ^{32}P-postlabeling assay (see below)—in placental tissue, bronchus, and lung of heavy smokers[545–547]. My former mentor, Paul Lohman, then in Rijswijk, The Netherlands, developed highly sensitive assays for specific DNA lesions on the basis of mono-clonal antibodies. One such antibody, against the reaction product of BaP diol epoxide with guanine, was able to detect DNA adducts in the bronchial cells of heavy smokers[548].

Like other non-polar, chemically unreactive compounds, BaP needs to undergo meta-bolic activation to yield more reactive forms that can then interact with nucleophilic centers in DNA[549]. Another well-known example of such a compound is N^2-acetyl-2-aminofluorene (2-AAF), an aromatic amine. Metabolic activation occurs through the action of specific metabolizing enzymes. These enzymes function to protect the cell against the toxic effects of such compounds by converting them into water-soluble, excretable forms. Unfortunately, in some cases they yield electrophiles, which attack DNA (and other macromolecules with nucleophilic centers). The best known metabolizing system is the aforementioned cytochrome P450 system. The actions of these enzymes, which involve oxygenation, produce ROS as yet another genotoxic by-product. Since we have already seen that ROS are also a by-product of ionizing radiation, it is often very difficult to know the exact source of the effect of environmental genotoxic agents. For example, Erik Mullaart, a student in my laboratory at the time, demonstrated that after oral adminis-tration of BaP to rats the main source of DNA damage is not the carcinogen–DNA adduct, but oxidative damage resulting from the free radicals that are a by-product of the metabolization[550]. Other indirect effects of chemical carcinogens that act through ROS are caused by the binding of carcinogens to the cell membrane. This results in stimulation of the arachidonic acid cascade, which elicits an oxidative burst and disturbs the membrane structure. In this way active oxygen species as well as lipid peroxidation and aldehydic degradation products are generated and these bring the cell into a so-called pro-oxidant state[551]. These highly reactive secondary radicals can induce DNA damage, although the carcinogens themselves do not react with the DNA.

A major exogenous source of DNA damage is sunlight. The extent of exposure to sunlight is determined by both lifestyle and place of residence. Recreational forms of sun exposure lead to high levels of cyclobutane pyrimidine dimers (CPDs) and pyrimidine 6–4 pyrimidone photoproducts (6–4PPs), the main DNA lesions induced by the UV component of sunlight, in the DNA of skin cells[552].

Also, occupational exposure to genotoxic agents can lead to the induction of DNA damage. High levels of BaP diol epoxide–DNA adducts were detected in lymphocytes of individuals known to be exposed to high levels of BaP, such as coke-oven workers[553], roofers, and foundry workers[554,555]. Generally, the level of adducts in these individuals correlated with the exposure level, but the inter-individual variation was large.

A further external source of DNA damage is formed by the natural background level of ionizing radiation, which is about 2×10^{-3} Gy/year[556]. Gy stands for Grey, which is the unit for the damage done to tissue by ionizing radiation. To put this in perspective, the lowest amount of ionizing radiation which leads to obvious effects, such as vomiting, loss of appetite, and generalized discomfort (mainly known from research on people who survived the atomic-bomb explosions in Japan in 1945) is 1–2 Gy. Our daily background of ionizing radiation seems therefore to be very small and probably induces about 5×10^{-3} SSBs/cell per day. However, it should be noted that there can be large variations in the extent of exposure. For instance, the level of radiation originating from radon (^{222}Rn), which is present in the earth but also in building materials, can vary from 3×10^{-4} to 3×10^{-3} Gy/year[557]. It is also unclear as yet if there is a threshold level below which radiation has no adverse effects. The effect of radon exposure on the rate of chromosomal aberrations in white blood cells has been studied in groups of people chronically exposed to different radon concentrations. Apart from a significant increase of such aberrations with age (see below), high concentrations of radon did not have an effect[558].

In addition, exposure to radiation for medical purposes has been demonstrated to induce DNA damage. Using the γ H2AX assay mentioned in Chapter 4, DSBs have been detected in lymphocytes from individuals undergoing computed tomography examination and found to be repaired within 1 day[559].

6.1.2 DNA DAMAGE AS A FUNCTION OF AGE

Since there are so many different types of DNA lesion forming continuously, either as a consequence of endogenous sources or through environmental exposure, it is important to find those that are most likely to exert adverse effects that may give rise to aging-associated phenotypes. For example, whereas we have seen that the most frequent type of spontaneous damage is probably an AP site, a lesion that is both cytotoxic and mutagenic, the efficient repair of such lesions makes it unlikely that a large number of them persist to cause an effect. It is possible, therefore, that certain slowly accumulating, minor lesions are more important, but those will be difficult to detect.

A major determinant of lesion accumulation is undoubtedly the quality of genome maintenance. As described in Chapter 4, the cell is equipped with a large variety of genome-maintenance systems to prevent the induction of damage, remove it after induction, or eliminate cells that are beyond repair. As discussed, somatic maintenance and repair systems are not maximized, providing a margin of error that could lead to a gradual accumulation of DNA damage with age. An age-related decline in the efficiency of these defense systems would accelerate the rate of such an accumulation during aging. As we have seen from Chapter 4, there is no evidence for drastic declines in genome maintenance with age, which may simply not be tolerated by the organism. Nevertheless, the sole fact that molecular defense systems are imperfect necessitates direct studies on the possibility that DNA damage and other genome alterations accumulate with age. Methodological aspects are of critical importance in such studies. First, methods need to be sensitive enough to detect and quantify low levels of DNA damage. Second, such methods cannot rely on metabolic labeling, but should instead be able to measure DNA lesions in freshly isolated cells or tissue samples. In the following sections I will review the progress that has been made over the last few decades in developing such assays and their application in studying spontaneous DNA damage and its accumulation with age in mammals.

6.1.2.1 DNA breaks/alkali-labile sites

Various attempts have been made to demonstrate an age-related increase in SSBs and DSBs, or damage such as AP sites, that are converted into DNA breaks during treatment with alkali (alkali-labile sites). Traditionally, sucrose-gradient centrifugation has been employed for this purpose under neutral or alkaline conditions. Later, more sensitive assays were developed[560].

By using alkaline sucrose-gradient centrifugation, alkaline elution, and alkaline unwinding, SSBs, DSBs, and alkali-labile sites can be detected. The S1-nuclease assay only detects SSBs, DSBs, or stretches of single-stranded DNA. Using the nucleoid sedimentation assay, in which the sedimentation rate of nucleoids through a neutral sucrose gradient is determined, only SSBs and DSBs can be detected. Nucleoids are what are left after cells are lysed with non-ionic detergent at high salt concentration, which removes membranes, cytoplasm, and nucleoplasm, and retains only the nuclear matrix with the negatively supercoiled DNA (see Chapter 3). Nucleoid sedimentation, however, can easily be influenced by other alterations in the DNA, such as conformational changes. This was shown by Michael Boerrigter, a student in my laboratory[561], and by others[562]. Indeed, the differences between these aspecific methods for detecting DNA damage are often critically important to correctly interpret the results.

More recently, the so-called comet assay, also known as the single cell gel assay, has become one of the standard methods for assessing DNA damage[563]. It is generally used to measure SSBs, DSBs or alkali-labile sites in DNA of individual cells, lysed with a detergent in the presence of high salt and subsequently subjected to electrophoresis in a gel. The migration streaks of such cells, termed comets, are measures of the level of DNA fragmentation under the conditions applied. In its essence the comet assay is a greatly simplified form of nucleoid sedimentation.

Although all these methods essentially lack specificity and are designed for measuring DNA breaks, the use of different conditions (e.g. neutral versus alkaline, the use of lesion-specific endonucleases), make them more versatile and allow detection of, for example, UV-induced pyrimidine dimers, oxidized bases, and alkylation damage.

In the liver, an age-related increase in the level of SSBs and/or alkali-labile sites was found when measured by alkaline sucrose-gradient centrifugation[564]. This was confirmed by results obtained using alkaline elution by Lawson and Stohs[565] and by Erik Mullaart in

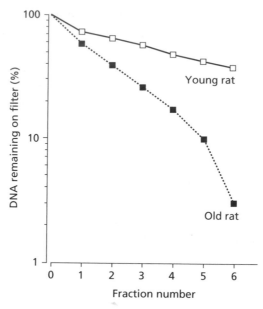

Fig. 6.3 An example of alkaline elution of DNA from rat liver hepatocytes. In these experiments, cells are lysed in an alkaline solution on a membrane filter. The rate of elution of the DNA through the filter is a measure of the number of DNA breaks present in the cells or induced by the alkaline treatment. The more rapid elution of the DNA from cells of the old rats indicates an increase in DNA damage in the liver of these animals with age.

my own laboratory[566] (Fig. 6.3). In addition, we also demonstrated that such an age-related increase in DNA damage in rat liver only takes place in postmitotic parenchymal cells and not in the non-parenchymal liver cells, which still have proliferative activity[566]. By calibrating the alkaline elution profiles to the known amounts of breaks obtained after irradiation of cells with different doses of γ rays, we determined that freshly isolated liver parenchymal cells of young rats contained about 900 alkali-labile sites. The cells from old rats contained almost twice that number, 1600. Presumably, these numbers should reflect the steady-state level of such lesions in cells *in vivo*.

Chetsanga *et al.*[567], found a large age-related increase (5-fold) in the fraction of DNA sensitive to S1-nuclease in mouse liver, indicating an increase in the percentage of single-stranded DNA. However, using the same assay but other mouse strains, no such age-related difference in the level of single-stranded DNA was observed in liver[568,569].

No age-related difference in the level of SSBs and alkali-labile sites could be detected in mouse brain by using alkaline sucrose gradients[564] or in rat brain with alkaline elution[570]. This is in agreement with data from Walker and Bachelard[571], which indicate the absence of an age-related difference in the level of SSBs detected by nucleoid sedimentation in rat brain nuclei. However, using the S1-nuclease assay Chetsanga *et al.*[572] demonstrated an age-related increase in the level of single-stranded DNA in rat brain. In addition, the extent of unwinding of DNA in alkali, a measure for the amount of SSBs and alkali-labile sites, was found to increase with age both in rat brain and in liver cells, but not in intestinal cells[573].

The comet assay was used to study alkali-labile sites in peripheral blood lymphocytes of human individuals of different ages[574,575]. The results did not indicate a different basal level of damage in old compared with young individuals. However, they did point towards increased cell-to-cell variation as well as increased interindividual heterogeneity in the older population. Increased stochasticity is one of the hallmarks of the aging process and can be observed at all levels. This lack of an age-related increase in breaks or alkali-labile sites in human lymphocytes was confirmed by later results obtained by Michael Boerrigter in my laboratory using alkaline elution (M. Boerrigter, unpublished work). More recently, no evidence for an age-related increase in DNA breaks in lymphocytes of rats was obtained by Gedik *et al.*[576].

DSBs were always difficult to detect *in vivo* due to their small numbers. However, with the recent introduction of γ H2AX as a marker for the repair of DSBs, which can be detected by immunohistochemistry using specific antibodies (see also Chapter 4), an important dosimeter has been obtained for this type of damage. Using this assay, it has been demonstrated that γ H2AX foci increase with age in several mouse tissues, including liver and brain, as well as in normal human fibroblasts during senescence, with the number of foci-free cells decreasing[272]. These lesions appeared to be persistent, which was in contrast to similar lesions induced by ionizing radiation. Hence, the DSBs associated with aging or cellular senescence appeared to be unrepairable. Interestingly, others, using this same assay, found that DSBs induced in cultures of nondividing primary human fibroblasts by very low radiation doses remained unrepaired for many days, in strong contrast to efficient DSB repair that was observed at higher doses. The level of DSBs in irradiated cultures was found to decrease to that of unirradiated cell cultures if the cells were allowed to proliferate after irradiation, presumably due to the elimination of the cells carrying unrepairable DSBs[577]. What these findings basically indicate is that highly toxic DSBs can be present in normal cells over extended periods of time, during which they may give rise to chromosomal aberrations.

6.1.2.2 Base damage

I-spots A lack of assays sensitive enough for the detection of specific base damages has precluded the accurate assessment of such lesions in tissues at different ages. One of the first systems suitable for this purpose was the ^{32}P-postlabeling method developed by Gupta *et al.*[578]. The method, which is based on DNA digestion, enrichment of the modified base, radioactive labeling using [γ-^{32}P]ATP, and chromatographic separation of the ^{32}P-labeled modified bases, is used mainly for detecting adducts induced by aromatic carcinogens, such as the aforementioned BaP and 2-AAF. Adduct-containing nucleotides are identified by comparisons with synthetic markers for those adducts one expects to find.

Interestingly, using the ^{32}P-postlabeling assay, the Randerath group found an increased level of several unknown DNA adducts in rat liver, kidney, heart, and lung of 10-month-old rats as compared with 1-month-old animals[579]. They called these adducts I-spots because they arise indigenously. Later, similar adducts were also found in mtDNA at an approx. 2-fold higher level than in nuclear DNA[580]. The highest level of adducts (1000 adducts/cell) was found in liver and kidney DNA, whereas low levels of adducts were present in brain DNA. In these studies the 'old' rats were no older than 10 months. However, using the same assay, Gaubatz observed an approx. 10-fold increase of such aromatic adducts in the heart of old compared with young mice: from about 1200 per cell in 2-month-old mice to about 15 000 per cell in 39-month-old mice[581]. At present the identity of the I-spots is still not known, although they do not appear to co-migrate with any known carcinogen-induced DNA adduct. They may be induced via metabolically activated exogenous carcinogens or stem from metabolically formed, possibly hormone-associated, endogenous electrophiles.

Oxidative lesions   For the detection of small base damages, such as alkylation damage and oxidative damage, in cell culture systems after treatment with mutagens, HPLC has proved to be a method of choice. However, the quantitative determination of base damage in cells and tissues of animals is quite a different matter, due to the relatively low level at which such lesions are present and the multiple potential artifacts that can confound the results.

Bruce Ames and co-workers, because of their interest in ROS as a major cause of aging and cancer, were the first to report quantitative estimates of oxidative DNA lesions, most notably 8-oxoG, in tissues and urine of laboratory rodents[245]. They used HPLC with electrochemical detection, which is about 3000-fold more sensitive than standard UV detection. As mentioned already, their initial estimates of the level of 8-oxoG, probably the main oxidative base lesion, were one or two orders of magnitude too high due to artificial oxidation products during sample preparation. Later, Hamilton *et al.*, using similar methods, reported 8-oxoG levels in rat tissues to vary from 0.01 to 0.03 lesions per 10^5 guanines[503]. Most notably, an age-related increase of about 2–5-fold was observed in all tissues studied. These workers also measured 8-oxoG in mtDNA and found the level of the lesions to be 3–4-fold higher than in nuclear DNA, which confirms the original results obtained by Richter *et al.*[538] (see above), although the total amounts of lesions and the magnitude of the differences found by Hamilton *et al.* were much smaller. Most notably, an age-related increase of about 2–5-fold was observed in all tissues studied. Also, 8-oxoG in mtDNA from liver was found to increase with age by about 2-fold. As reported by the authors, these findings indicate that the nuclear genome of the brain of a young mouse would contain a steady-state level of 180 8-oxoG lesions compared with 640 8-oxoG lesions for an old mouse. Interestingly, this relatively low number of lesions found for the young animals is very similar to an estimate made by others using a quantitative model based on knowledge of endogenous formation rates and repair mechanisms of the lesion[582].

Whereas age-related increases in 8-oxoG have been reported by others as well, and also in other species, such as houseflies[583] and human muscle biopsy specimens[584], others either did not find any age-related increase or reported such an increase for some but not all tissues[585,586]. Hence, the main conclusion from all these studies appears to be that the level of this particular lesion is very low in the first place and that if there is an increase with age then it is not by very much. In fact, dramatically high levels or considerable increases with age of this highly mutagenic lesion would be unexpected in view of the efficient repair mechanisms available, in both the nucleus and mitochondria. As we have seen in Chapter 4, there is also no reason to assume that repair activities for this lesion decline dramatically with age. The same may actually apply to less frequent types of oxidative DNA damage. For example, Le *et al.*[587] reported fewer than 4.3 thymine glycols per 10^9 bases of DNA in cultured cells.

6.1.2.3 Cross-links

As mentioned in Chapter 4, interstrand DNA cross-links are extremely toxic forms of DNA damage, which need to be removed through complicated HR-based pathways in combination with other repair systems, such as NER. Defects in such pathways, as engineered in the mouse, result in a dramatic premature aging phenotype (Chapter 5). This is not surprising because an interstrand DNA cross-link can prevent separation of DNA strands and therefore completely block DNA replication or transcription. The sources of endogenous interstrand DNA cross-links are poorly characterized, and the importance of their repair for normal cellular processes is not clear outside of the clinic, where agents to induce interstrand DNA cross-links, such as mitomycin C and cisplatin, are widely used as antitumor agents. The same is true for DNA–protein cross-links, which can also covalently bind the two strands together and are induced by antitumor chemotherapeutic agents. Potential sources of spontaneous DNA–DNA and DNA–protein cross-links are the aforementioned oxygen radicals and AGEs. In addition, UV and ionizing radiation can induce cross-links under certain conditions.

Whereas a cross-linking theory of aging was postulated in the 1940s[588] and the cross-linking of proteins is still regarded as a major mechanism of aging at the molecular level (see Chapter 7), only few data are available on the age-dependency of the level of interstrand DNA–DNA cross-links and DNA–protein cross-links[403]. DNA–DNA cross-links can be measured in a variety of ways including gel electrophoretic mobility, alkaline elution, relative thermal stability, or template activity. With highly purified liver DNA from rats or mice, no age-related difference in thermal stability or template activity was found[589,590]. These findings suggest the absence of an age-related increase in DNA–DNA cross-linking. However, by using crude preparations from rat and mouse liver Russell *et al.*[589] found an age-related increase in thermal stability and a decrease in template activity, suggesting that there is an age-related increase in the level of DNA–protein cross-linking.

Several authors have reported an age-related increase in protein–DNA cross-links in various tissues of rodents, usually measured by determining the fraction of DNA bound to protein[591–593]. The number of such lesions present at any given time in the DNA of the genome is unclear at present, but unlikely to be extensive because of their toxicity and the availability of sophisticated repair systems. Nevertheless, their role as a major form of aging-related spontaneous DNA damage cannot be ruled out

6.1.3 DNA DAMAGE: A LIKELY CAUSE OF AGING?

From the above we can conclude that although the rate of DNA damage induction is very high, steady-state levels of the most frequently occurring lesions are only modest. Whereas absolute figures of tens of thousands of lesions induced in cells each day suggest a very serious problem, it is difficult to see how steady-state levels of a few lesions per 10^6 undamaged bases can contribute to the adverse effects that develop during the course of the aging process. At such low frequencies, with lesions continuously removed through DNA-repair pathways and re-introduced, it would be unlikely for a lesion to reside in a gene or its regulatory regions long enough to have adverse effects. It seems a reasonable conclusion that DNA damage per se, even in the large absolute numbers of lesions induced in cells each day, is unlikely to play a direct role in causing age-related functional impairment and disease. However, there are at least three scenarios, not mutually exclusive, that may be able to explain how DNA damage could significantly contribute to aging-related cellular degeneration and death.

First, whereas steady-state levels of each of the variety of lesions thus far investigated may be low in normal tissues, such levels can be greatly increased as a consequence of disease. For example, ischemia-induced neuronal injury in animal models is accompanied by increased levels of 8-oxoG and AP sites[594,595]. Increased levels of DNA damage have also been associated with Alzheimer's disease[596,597] and diabetes[598]. According to this scenario DNA damage would not be a primary driver of the aging process but associated with aging-related diseases. However, even in such cases there is no evidence that steady-state levels rise high enough to exert a direct effect. As yet there is no consensus about a possible role of spontaneous DNA damage in the etiology of disease[599].

Second, DNA damage could occur predominantly in certain areas of the genome, in so-called hotspots with a high chance of affecting gene expression. We have already seen how the proteins involved in DNA higher-order structure may also provide protection against genotoxic agents. DNA damage is more likely to be induced in nucleosome-free regions than in the highly compact heterochromatin. Interestingly, in 2004, Lu *et al.*[600] described the age-related reduction in expression of a set of genes in the brain of human individuals, varying in age from 26 to 106 years. To test whether this decreased expression could be due to the accumulation of spontaneous oxidative DNA damage in the promoter regions of these genes, they used formamidopyrimidine DNA glycosylase (Fpg), an enzyme that recognizes 8-oxoG as well as formamidopyrimidines (see the section on BER in Chapter 4). Lu *et al.* used this enzyme to generate SSBs, which in turn render the DNA resistant to PCR amplification. Originally discovered by Ben van Houten (Research Triangle Park, NC, USA), DNA lesions, including oxidative damage such as strand breaks, base modifications, and AP sites, will block the progression of the polymerase in the PCR, resulting in a decrease in amplification of a target sequence. He and his co-workers subsequently used this as a measure of the amount of these lesions[601]. In this case, it would be a measure of the amount of 8-oxoG, with the ratio of intact PCR products in cleaved versus uncleaved DNA, determined using quantitative PCR, as a measure for the amount of damage in a specific DNA sequence. In these experiments Lu *et al.* 600 observed an almost complete loss of intact DNA in 500-bp fragments from the promoter regions of genes that showed reduced expression in brain samples from individuals over the age of 70; the reduction was much less in promoters of non-downregulated genes or in the exon regions of all genes.

In the FPG assay, an almost complete loss of intact DNA would correspond to an average of one lesion in the 500-bp PCR nuclear DNA fragment. If human brain is not significantly more sensitive to oxidative DNA damage than mouse brain, the aforementioned results of Hamilton *et al.* indicate 0.41 8-oxoG lesions per 10^6 dGs in brain of old mice[503]. Assuming a GC content of 40%, this would correspond to approx. 10^{-4} lesions per 500-bp fragment. Hence, the almost complete loss of intact DNA observed by Lu *et al.* would correspond to an overrepresentation of 10 000-fold in promoter regions of genes downregulated with age.

Andrew Collins (Oslo, Norway), who has critically reviewed this field[537], concludes that for human cells the background level of 8-oxoG is between 0.3 and 4.2 8-oxoGs per 10^6 dGs. The data from Hamilton for aged mouse brain cited above are on the low side of this range. Taking the highest value of 4.2 lesions per 10^6 dGs, and again assuming a 40% GC content, this would correspond to fewer than 10^{-3} lesions per 500-bp fragment, still an overrepresentation of 1000-fold. Whereas 8-oxoG may make up only about 5% of all oxidative lesions, it is highly unlikely that FPG would recognize them all. Indeed, according to Collins, FPG mainly recognizes 8-oxoG and the total number of FPG-sensitive sites in normal human cells is not higher than 0.6 per 10^6 dGs, rather lower than the 4.2 lesions per 10^6 dGs used in the above calculation. Hence, also according to this more conservative scenario, the overrepresentation of 8-oxoG observed by Lu *et al.* in promoter sites of upregulated genes in old human brain is remarkably high: at least 1000 times.

The implications of these findings are far-reaching. If spontaneous DNA damage is so selective, then we need to reassess current conclusions that the risk to human health from endogenous oxidative DNA damage has been seriously overstated. Moreover, these results suggest that the accumulation of spontaneous oxidative DNA damage is sufficient to causally explain the aging process in the human brain. For the moment we will need to await further confirmation of these observations.

Finally, DNA damage can have long-term adverse effects through mutagenesis. Whereas extraordinarily efficient in resisting the avalanche of chemical changes threatening the genome's integrity at any given moment, DNA-damage processing has not evolved to a level of perfection that would allow essentially no errors. Errors are being made and while their frequency may be remarkably low compared with the large numbers of lesions induced in our genome every day, mutations may easily accumulate to a level where they might exert adverse effects. In contrast to chemical changes in the DNA double helix, mutations are not recognized by the cell's DNA-damage-processing system and will be just as faithfully replicated as the original, correct sequence. This is why mutations are irreversible and cannot be removed, except through cell-elimination 'repair'. Do mutations accumulate in different organs and tissues of mammals, and, if so, what kind of mutations accumulate and can they have adverse effects?

6.2 DNA-sequence changes

Unlike DNA damage, which can be recognized and repaired through one of the pathways in the network of genome-maintenance and -repair systems, DNA mutations (as well as epigenomic errors) are irreversible. This makes them a true molecular endpoint of the aging process. In Chapter 4 I indicated the various opportunities for DNA mutations to arise. Even in the absence of DNA damage, tautomeric shifts can lead to base-pair substitutions during DNA replication or repair synthesis. In the presence of DNA damage, mutations can arise as a consequence of translesion synthesis. Small deletions or insertions are often associated with defects in mismatch repair and large chromosomal rearrangements can occur after mistakes made during DSB repair. As yet we know very little about possible mechanisms of 'epimutations'.

DNA mutations are expected to occur at a much lower frequency than DNA lesions. However, since in contrast to DNA damage, sequence alterations cannot be repaired, their accumulation can only be prevented through cell elimination, for example, through the apoptotic response. As we have seen, it is likely that such cellular responses come into play predominantly after extensive DNA damage, such as high temporary exposure to genotoxic agents, or the bursts of free-radical damage thought to be associated with phagocytosis. Therefore, whereas there is evidence that tissues from old individuals are characterized by increased spontaneous apoptosis as well as increased numbers of senescent cells (Chapter 7), it is likely that the majority of cells in a tissue will remain intact and may only suffer from a gradually increased mutation load.

Even more than DNA damage, DNA-sequence changes have been found difficult to detect in cells or tissue samples from humans or animals. Hence, whereas the first somatic mutation hypothesis of aging was presented as early as 1958[506], it is only recently that robust methods for measuring genomic mutation loads in cells and tissues from both humans and animals have become available. It should also be kept in mind that the interest in DNA damage and mutations has mainly been driven by the discovery that such events are the cause of cancer. Tumors are clonal derivatives of usually only one cell and characterized by widespread DNA-sequence alterations as compared to the normal tissues from which they arise. This explains why it has been so much easier to detect genome alterations in tumors than in normal tissue. The latter is essentially a mosaic of cells, each with a different spectrum of genomic alterations. Below I will summarize the different types of genome alteration that can occur in cells of animal tissues and the evidence that they accumulate with age *in vivo*, and briefly discuss the likelihood that such changes could contribute to the aging phenotype. Whereas I will focus on mammalian systems, in this case also because most of our knowledge in this field relates to mammals, recent evidence of genomic instability during yeast aging will also be presented.

6.2.1 CYTOGENETIC CHANGES

Cytogenetic changes are genomic alterations that can be visualized by light microscopy in cells, usually during metaphase. They include chromosomal aberrations, abnormalities in the structure or number of chromosomes, and extrachromosomal fragments, such as the extrachromosomal rDNA circles that accumulate in yeast during replicative aging (Chapter 2). A proportion of chromosomal aberrations are unstable and give rise to chromosome fragments without spindle-attachment organelles (kinetochores, centromeres; see Chapter 3). These are termed acentric fragments. When the cell divides, some of these fragments are excluded from the main daughter nuclei and form so-called micronuclei within the cytoplasm, either on their own or in conjunction with other fragments. Stable chromosomal aberrations in germ cells are often responsible for genetic disorders. As we have seen in Chapter 1, chromosomes have been studied for more than a century. However, it was only in 1956 that the number of chromosomes in a human diploid cell was determined to be 46 (22 pairs of autosomes and two sex chromosomes)[602]. Several years later the presence of an extra chromosome 21 was discovered in people with Down's syndrome as well as sex-chromosome anomalies in patients with sexual-development disorders.

Individual chromosomes can be identified in metaphase through staining with different compounds, such as Giemsa or quinacrine, which yield differentially stained regions on chromosomes. The banding patterns appear related to base composition and chromosome loop structure. In this way stains can distinguish between, for example, non-centromeric regions and centromeres. The patterns of bands are found to be specific for individual chromosomes and they allow their distinction as well as the recognition of structural abnormalities or aberrations. Cytogenetic analysis is now an integral diagnostic procedure in prenatal diagnosis. It is also utilized in the evaluation of patients with mental retardation, multiple birth defects, and abnormal sexual development, and in some cases of infertility or multiple miscarriages. Cytogenetic analysis is also useful in the study and treatment of cancer patients since tumor cells generally contain many chromosomal abnormalities with often a particular type of aberration characteristic for a certain type of cancer. An example is the Philadelphia chromosome in chronic myeloid leukemia, the first consistent chromosome abnormality identified in cancer. This is an abnormally short chromosome 22 that results from the reciprocal exchange of DNA between this chromosome and chromosome 9[603]. This translocation takes place in a single bone-marrow cell and, through the process of clonal expansion, gives rise to leukemia. Apart from translocations, the types of chromosomal abnormality that can be detected by cytogenetics are numerical aberrations, duplications, deletions, and inversions.

The late Howard Curtis (1906–1972) and co-workers used cytogenetic analysis to provide the first evidence for an increasing level of genome alterations in a somatic tissue during aging *in vivo*[604]. For this purpose, Curtis and Crowley looked at the main cell type in the mouse liver, parenchymal cells or hepatocytes, which normally do not divide. However, surgical removal of part of the liver results in regeneration through hepatocyte proliferation. When inspecting parenchymal cell metaphase plates after this so-called partial hepatectomy these workers found considerably higher numbers of cells with abnormal chromosomes in old animals than in young animals (from about 10% of the cells in 4–5-month-old mice to 75% in mice older than 12 months).

Later, such large structural changes in DNA—aneuploidy, translocations, and dicentrics—were routinely observed to increase with donor age in white blood cells (which can be stimulated to divide after their transfer into culture) of human individuals; that is, from about 2–4% of the cells carrying chromosome abnormalities in young individuals to about six times more than this in the elderly. It is conceivable that these chromosomal changes reflect changes in the hematopoietic stem cells. The recent use of more advanced methods, such as chromosome painting (see Chapter 3), have amply confirmed the increase in cytogenetic damage with age in lymphocytes from both humans[605] and mice[606]. Also, spontaneous micronuclei were demonstrated to increase with age in human lymphocytes[607].

Hence there is currently universal consensus that the frequency of cytogenetic changes in both human and mouse cells increases with age. In lymphocytes, which is still the only cell type readily accessible for such determinations, the frequency of cells carrying one or more of such abnormalities does not appear to rise much higher than approx. 10%.

However, it should be realized that even with the more recently emerged chromosome-painting methods, the smallest possible type of aberration that can still be detected is 2 million bp. Based on the early work of Howard Curtis, the frequency of cytogenetic alterations may be higher in non-dividing cells, but in the absence of independent confirmation of these results it is difficult to draw conclusions. However, one type of chromosomal aberration—aneuploidy—has been recently found in postmitotic tissues as well.

Aneuploidy is so widespread in tumors that there is probably no tumor that, upon analysis, does not appear to depart from diploidy. Aneuploidy can have major adverse effects through so-called loss of heterozygosity (LOH) due to the loss of one chromosome copy. LOH can have other causes, as we have seen in Chapter 4. However, the gain of a chromosome copy can also adversely affect cellular physiology, as is evident from Down's syndrome, which is caused by trisomy of chromosome 21. Many characteristics of cancer are now explained by aneuploidy, through gene-dose effects, which can greatly alter the networks that underlie cellular function.

In the mouse embryo it has been demonstrated that over 30% of neural progenitor cells display chromosomal aneuploidy, both loss and gain of chromosomes[608]. This high frequency of chromosomally aberrant neurons was also found in normal human brain using interphase fluorescence *in situ* hybridization (or FISH) with probes for chromosome 21[609]. Studies of aneuploidy in postmitotic organs (without the need for metaphases) have become possible through the use of interphase fluorescence *in situ* hybridization. Using specific probes one can now simply count if in a cell there are two or more copies of a given chromosome (loss of a chromosome is more difficult to measure). Chromosome 21-aneuploid cells, including non-neuronal cells and neurons, were shown to constitute about 4% of the estimated 1 trillion cells in the human brain. Since only one autosome was analyzed it is likely that the overall percentage of aneuploidy in the brain is much higher than that. This is in striking contrast with the situation in human lymphocytes in which only 0.6% of the cells were chromosome 21-aneuploid. Routine karyotyping in both human and mouse lymphocytes revealed aneuploidy in only about 3% of such cells. Whereas it is not known whether the frequency of aneuploid cells in the brain increases with age (but see below), the high numbers of aneuploid cells already in the brain of adults indicate that our genome is much less stable than often thought.

A frequent consequence of aneuploidy is cell death. However, the widespread occurrence of aneuploidy in tumor cells already indicates that cell death is by no means inevitable. Indeed, it has recently been demonstrated that aneuploid neurons are functionally active and normally integrated in brain circuitry[610]. This, together with the many other types of random genome alterations that have been demonstrated in somatic cells, leads to genetic mosaicism in the cells and tissues of an individual during development and aging. One may ask whether this genetic mosaicism contributes to individual physiological and behavioral variation, for better or for worse. Thus far this question remains unanswered, but recent studies provide some compelling evidence that large-scale genomic variation is compatible with life and is more common than thought previously. For example, apart from the single-nucleotide polymorphisms, of which there have been over 10 million defined in the human genome, large-scale copy-number polymorphism has been observed among individuals[611,612] and is thought to greatly contribute to genomic diversity and disease susceptibility[613]. However, this is germ-line genetic variation, with its effects subject to selection. Whereas it is certainly possible that some random genomic alterations contribute positively to certain complex physiological functions, such as those encoded in brain circuitry, it is reasonable to expect adverse effects in most cases. This issue, as to how stochastic alterations can have a consistent effect on function, is of major interest for the question that is central in this book: how could random genomic alterations lead to a series of consistent physiological changes that ultimately bring life to a close?

Before I move to the next type of genomic alteration it is important to mention some observations in mice that suggest that chromosomal aberrations in neuronal stem cells accumulate during aging, eventually affecting almost all such cells. It is now generally accepted that vertebrates use stem-cell reservoirs for tissue replacement. For most of the past century it was thought that the adult brain was unable to generate new neurons, but that turned out to be wrong. The adult brain contains a population of multipotent self-renewing cells: stem cells. Stem cells do not usually divide, but when they do the result is always a replacement stem cell and a rapidly dividing progenitor cell. In the previous chapter we saw how genomic integrity of the replacement stem cell can be maintained by template-strand retention, suppression of mitotic recombination, and increased apoptosis.

Neural stem cells can be expanded in culture in the form of cellular aggregates, termed neurospheres, with each neurosphere descending from one neural stem cell. However, some of such neurospheres can be derived from neural progenitor cells, which have limited self-renewal capacity and more restricted lineage potential[614]. Stephen Pruitt (Buffalo, NY, USA) and co-workers analyzed such neurospheres from C57Bl/6×DBA/2 hybrid mice at nine chromosome pairs for genomic areas of structural loss by making use of known single-nucleotide polymorphisms between these two mouse strains[615]. At such polymorphic loci, the loss of one allele indicated LOH. It was found that whereas in 15 neurospheres from young mice no deletions were detected, 16 out of 17 neurospheres from old animals displayed LOH on one or more chromosomes. The results of this elegant study are especially interesting since they indicate the accumulation of chromosomal mutations in stem or progenitor cells, a potential reservoir for cell-based therapies of aging-related degeneration. The impact of aging-related mutation accumulation, even in stem cells, on possibilities for intervention to delay and reverse aging-related decline, will be discussed in Chapter 8.

In summary, cytogenetic studies have demonstrated convincingly that the frequency of chromosomal aberrations increase with age in white blood cells, in both humans and mice. In such cells only a minor portion of the cells, even at old age, is affected. Since alterations affecting fewer than about 2 million base pairs are still undetectable with current methods, it is possible that at old age all cells carry one or more chromosomal alteration. More recent results indicate that chromosomal aberrations, especially aneuploidy, are widespread in brain cells. Little is known about chromosomal aberrations in other postmitotic cells, with exception of the early results from Curtis on aging liver.

6.2.2 GENE MUTATIONS

6.2.2.1 Endogenous selectable marker genes

The frequency of inactivation of specific genes in cells and tissues of animals is difficult to measure in an objective manner since this would require the re-sequencing of thousands of copies of that gene isolated randomly from the DNA of that tissue. With the development of tests based on selectable endogenous target genes, it became possible to assess the mutant frequency at these loci among cells taken from a human or animal and cultured on a medium that selects against the wild-type cells. There are now a number of such tests available, with the hypoxanthine phosphoribosyltransferase (HPRT) locus test the most widely used[616]. HPRT converts the free purine bases, guanine and hypoxanthine, into the corresponding nucleoside phosphate and thus makes them available for the synthesis of nucleic acids. Selection of mutants is based on the differing degrees of toxicity of the synthetic purine base 6-thioguanine for mutant compared with non-mutant cells. In cells with functional HPRT, 6-thioguanine is converted into toxic nucleotides, with the result that the cells die. Loss of HPRT activity results in resistance to 6-thioguanine. The *HPRT* gene is located on the X chromosome in both humans and mice. Hence, when inactivated there is no spare copy of the gene available. Heritable mutations inactivating the *HPRT* gene cause Lesch–Nyhan syndrome, a rare genetic disorder that affects males. Males with this syndrome develop physical handicaps, mental retardation, and kidney problems. Although the *in vivo* function of HPRT, even at the cellular level, cannot be ignored (see below) its inactivation has no apparent consequences for cells growing in culture, because it functions merely in a salvage pathway; with few exceptions, organisms are able to synthesize purines and pyrimidines *de novo*.

Using this test, Richard Albertini (Burlington, VT, USA) and Alec Morley (Adelaide, Australia) were the first to report spontaneous mutations in human lymphocytes[617,618]. Interestingly, a substantial fraction of these spontaneous mutants appeared to be genome rearrangements: large mutations, such as deletions, inversions, and amplifications[619]. Different investigators have now also shown that mutation frequencies at the *HPRT* locus increase with donor age. The results indicate that mutation frequencies in humans (expressed as the number of mutated cells as a function of the total number of cells, corrected for the plating efficiency) increase with age, from about 2×10^{-6} in young individuals to about 1×10^{-5} in middle-aged and old individuals[620]. In mice mutation frequencies have been reported from about 5×10^{-6} in young animals to about 3×10^{-5} in middle-aged mice[621] or old mice[622]. However, in both mice and humans these values could be underestimates. To some extent this is likely due to a certain selection against mutant cells both *in vivo* and *in vitro*. *In vitro*, 6-thioguanine may not only eliminate wild-type cells, but also a number of genuine HPRT mutants. *In vivo*, HPRT mutant frequencies have been found to decrease with time following exposure to mutagenic agents, such as ethyl nitrosourea and radiation[623]. This can be explained by selection against HPRT mutant cells that lack the purine-salvage pathway. Another more general reason why observed mutation frequencies at selectable target loci are underestimates is that large genome rearrangement events could affect not only the target gene, but also neighboring genes essential for survival of the cell.

The HPRT method is also insensitive for certain types of mutation. This has become apparent from studies with other reporters. For example, Grist *et al.*[624] used the *HLA* gene as a marker by selecting cells that had lost a particular HLA antigen due to mutation among human lymphocytes. The results indicated two- or three-times higher mutant frequencies than obtained with the HPRT assay. The additional mutations were ascribed to mitotic recombination events that remain undetected by the HPRT assay, which lacks a homologous chromosome. This important role of mitotic recombination in generating spontaneous mutations was confirmed by work using yet another reporter locus, the adenine phosphoribosyltransferase (*APRT*) gene. The product of this gene catalyzes a similar salvage reaction as HPRT, but then with adenine. Cells lacking APRT are resistant to toxic adenine analogs, such as 2,6-diaminopurine, allowing their selection *in vitro*. However, this gene is on an autosome, both in humans and mice. Using a mouse model in which one copy of the *Aprt* gene was eliminated, work from the laboratories of Peter Stambrook and Jay Tischfield revealed mutation frequencies in lymphocytes and fibroblasts from these animals as high as 1×10^{-4} [342]. Most of the mutations appeared to be LOH due to mitotic recombination events. As mentioned in Chapter 4, these same investigators later showed that the mutation frequency at the *Aprt* locus in mouse ES cells is significantly lower than that in mouse embryonic fibroblasts or in adult cells *in vivo*[410]. Mitotic recombination in ES cells appeared in particular to be suppressed and chromosome loss, which after reduplication leads to uniparental disomy (two copies of the same chromosome from one parent and no copy from the other), represented more than half of the LOH events.

In mice subjected to caloric restriction, the only intervention demonstrated to increase lifespan in mammals (Chapter 2), *Hprt* mutation frequencies were found to increase with age at a significantly slower rate than in the animals fed *ad libitum*[621]. These results were independently confirmed in rats by Aidoo *et al.*[625], and suggest that the level of accumulated somatic mutations reflects biological rather than chronological age. This conclusion was further strengthened by the results of Odagiri *et al.*[435], mentioned in Chapter 5, who demonstrated accelerated accumulation of *Hprt* mutations in peripheral blood lymphocytes of so-called senescence-accelerated mice (SAM). Although the SAM model is not generally accepted as a mouse model of accelerated aging (Chapter 5), these findings nevertheless demonstrate a link between somatic mutation rate and physiological decline.

Interestingly, the HPRT test has been used to determine the frequency of mutant primary tubular epithelial cell clones grown directly from kidney tissue from 2- to 94-year-old human donors. The mutation frequencies found were more than 10-fold higher than in peripheral blood lymphocytes and also increased with age, from about 5×10^{-5} to about 2.5×10^{-4} [626]. Renal tubular epithelial cells perform critical, secretory and resorptive tasks in the kidney and are normally quiescent. It is possible that the observed mutation frequencies are even underestimates, since not all cells in the kidney specimens had the ability to grow into clones, which may well be due to a very high mutation load. The high mutation frequency in the kidney cells could reflect a relatively slow turnover compared with T cells, but it is more likely that there is less stringent selection against mutant cells *in vivo*. In this respect, it is conceivable that the HPRT mutant frequencies in lymphocytes are underestimates because of a high stringency of selection in these cells.

6.2.2.2 Transgenic reporter genes

At the end of the 1980s, when I was finishing my PhD work, the lack of a suitable mammalian model system to measure gene mutations *in vivo* begun to be felt. In spite of its usefulness, the HPRT selective assay is limited to cell types that can still actively proliferate in culture. At this time, the age of transgenic mouse modeling had begun and I designed a mouse model harboring in all its cells the *lacZ* bacterial reporter gene as part of a bacterial DNA virus, a so-called bacteriophage λ vector. The use of a bacteriophage λ vector was inspired by results published at that time by Peter Glazer (New Haven, CT, USA), indicating that such a construct could be recovered from its integrated state in the genomic DNA of mouse L-cell lines using λ *in vitro* packaging extracts[627]. Such extracts contain all the components needed to excise the vectors from the genomic DNA at the so-called cos sites and subsequently package them into viral capsids to produce infectious λ phage. Subsequent infection of *E. coli* is then highly efficient.

The first Mutamouse, as it was later called, was constructed by my then PhD student Jan Gossen (now at Organon, Oss, The Netherlands) and born in 1988. Unfortunately, once we tried to recover the integrated vectors from the mouse genomic DNA in *E. coli*, by using a variety of commercial packaging extracts, the efficiency appeared to be close to zero. To us this was puzzling in view of the results of Peter Glazer. The explanation turned out to be the mechanism of host restriction by which *E. coli* K12 strains manage to get rid of foreign DNA methylated at CpG sites. (The original positive results of Glazer *et al.* can possibly be explained by hypomethylation of their mouse cell lines.) Later, we understood that scientists from Stratagene, the company making a flourishing business out of selling λ packaging extracts which are routinely used in genetic cloning experiments, had taken the same approach. Like us they had been stuck at the recovery stage.

When we looked at the packaging extracts obtained from Stratagene we suspected that these were derived from *E. coli* C, a strain known to be host-restriction-negative[628]. This could explain why these extracts worked so well compared with other commercially available ones. However, the plating cells were *E. coli* K12. Apparently, workers at Stratagene had not realized the importance of the K12 restriction modification, which prevented high-efficiency rescue of bacteriophage λ methylated in the mouse at CpG sites. When testing Stratagene packaging extracts, but now with *E. coli* C as plating cells, one morning our large 25 × 25 cm plates appeared full of beautiful blue plaques with an occasional white one, indicating a *lacZ* gene mutated in the mouse[629]. This first mouse model for mutation detection *in vivo* was described in 1989[630].

A second model, developed by Jay Short and co-workers at Stratagene, was presented in 1991[631]. Their model, named BigBlue, differed from ours in that it harbored not the *lacZ* but the *lacI* gene as a mutation reporter gene. The *lacI* gene is part of the *E. coli lac* operon and encodes the DNA-binding protein called the lac repressor, which inhibits the expression of *lacZ* by binding to the so-called operator, a short region of DNA that lies partially within the promoter. In this system, inactivation of the *lacI* gene results in de-repression of the *lacZ* gene and, hence, the detection of mutants as blue plaques among many colorless ones. This was a significant improvement over our own model since we had to look through thousands of blue plaques to find the occasional colorless ones. Fortunately, quickly thereafter we developed a positive selection system based on the use of *E. coli* C strains rendered deficient in the *galE* gene[632]. Such cells use the β-galactosidase product of the *lacZ* gene to convert lactose or lactose analogs into UDP-galactose, which is toxic and normally further metabolized into UDP-glucose. Since the *galE*-deficient strains can no longer perform this function they will die upon receipt of an intact *lacZ* gene. Hence, in a way that is very similar to the HPRT system we could now positively select for *lacZ* mutants.

Whereas the positive selection system worked well with the bacteriophage λ-based Mutamouse system[633], we later generated another system based on *lacZ* as part of a plasmid rather than a bacteriophage λ system. After our initial studies with the Mutamouse we had begun to realize that the only type of mutation that we detected was base-pair substitutions or very small deletions or insertions. Similar results were obtained by others who either worked with our *lacZ* mouse or with Stratagene's *lacI* mouse. This was strange because results obtained with the aforementioned HPRT assay and other endogenous selectable marker genes clearly indicated that a significant portion of spontaneous mutations were sizable genome rearrangements. In part, this could be explained by the fact that rearrangements, such as large deletions, will go undetected as a consequence of the minimum vector size required for efficient packaging of bacteriophage λ vectors (i.e. between 42 and 52 kb). In addition, genome rearrangements involving the mouse flanking regions

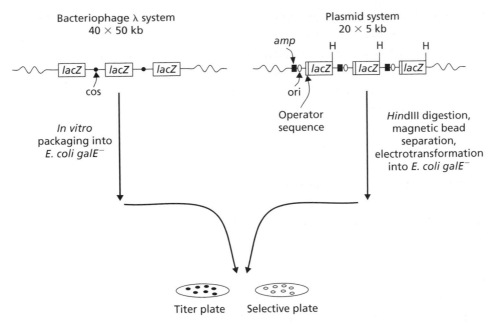

Fig. 6.4 Comparison of bacteriophage λ-based and plasmid-based transgenic mouse reporter models for mutation analysis. In the two most commonly used λ-based models approximately 40 copies of the 50-kb λ vectors are integrated head to tail. They can be excised from genomic DNA and packaged into phage heads using commercially available packaging extracts. Each individual phage results in a plaque on a lawn of *E. coli* cells, but only if the *lacZ* gene is inactive due to a mutation. The *lacZ* plasmids are only about 5 kb in size and present in about 20 copies per haploid mouse genome. Plasmids need to be separated first from mouse genomic DNA before they can be used to transform *E. coli* cells. Each individual plasmid results in an *E. coli* colony, but only if the *lacZ* gene is inactive due to a mutation. *amp*, ampicillin-resistance gene; cos, cohesive ends of λ; *galE*, UDP-galactose-4-epimerase gene; H, *Hind*III site; ori, bacterial origin of replication.

and therefore one of the cos sites cannot be detected (Fig. 6.4). It was not clear, however, why virtually no deletion mutations of moderate size—between 50 and 3000 bp—could be detected with the λ mouse models. It is possible that the large amount of prokaryotic DNA in the λ models inhibits the activity of those DNA-processing systems that in mammals are involved in generating mutations, including DNA rearrangements. This unnatural DNA substrate could also explain the similarity of the mutation spectra between different organs in the phage models, which we observed and later also reported on with the *lacI* system. We did not expect this lack of tissue specificity in view of the profound differences in cellular function and environment and, consequently, in sources of genotoxic stress. The long stretch of prokaryotic DNA comprised by multiple copies of the 50-kb-long bacteriophage λ vector could essentially behave in a prokaryotic way and constrain the generation of a mutation spectrum representative for the mammalian organism. Indeed, both our own *lacZ*-based Mutamouse and Stratagene's *lacI*-based

BigBlue model harbored a stretch of 40 copies (per haploid genome) of the 50-kb-long bacteriophage λ vector. This would amount to 2 million bp of prokaryotic DNA.

The plasmid system subsequently made by Jan Gossen in my laboratory[634] and later more fully validated by Martijn Dollé[635,636], first a student and then a postdoctoral researcher in the laboratory, contained only 20 copies of an approx. 5-kb plasmid, which does not amount to much more than 100 000 bp of prokaryotic DNA. The reason that initially plasmid vectors had not been used as the vector of choice for transgenic mutation models is the notoriously low transformation efficiencies obtained with plasmids excised from their integrated state in the mammalian genome. Efficient intramolecular ligation of excised plasmids requires extensive dilution and the presence of the genomic DNA will also negatively influence the transformation efficiency. By contrast, bacteriophage rescue is a highly efficient process. We greatly improved the efficiency of plasmid rescue by first separating the excised plasmids from the mouse genomic DNA. For this purpose, we first collect the plasmids with magnetic beads coupled to the LacI repressor protein, which binds very tightly to the operator region in the *lacZ* gene. After washing away the mouse genomic DNA, the pure *lacZ* plasmids are transferred into *E. coli* by electroporation. Figure 6.4 compares the λ and plasmid rescue procedures.

A distinct advantage of the *lacZ*-plasmid system over the bacteriophage λ systems is its capability to detect genome rearrangements with one breakpoint in a *lacZ* gene and the other elsewhere in the mouse genome (Fig. 6.5). Excision with the restriction enzyme *Hin*dIII (just outside the *lacZ* gene) would then result in a truncated plasmid attached to a piece of mouse DNA from the breakpoint to the first *Hin*dIII site in the mouse genome. This can be on the same chromosome (in case of a deletion or inversion) or on another chromosome (in case of a translocation). The mouse DNA fragment cloned in the *lacZ* mutant plasmid is generally small because *Hin*dIII cuts rather frequently in the mouse genome. It is possible to physically characterize such mutants by sequencing the mouse DNA fragment. Searching the mouse genome sequence (which is completely known) for this fragment quickly yields the identity of the breakpoint. Hence, the *lacZ* plasmid system is able to detect a broad spectrum of mutations, including large genome rearrangements. However, it cannot detect them all. Homologous (mitotic) recombination, for example, leading to deletion of entire reporter-gene copies (LOH) is a frequent mutational event, as indicated by the results obtained with the *Aprt* mice, but goes undetected in all transgenic assays.

By now, a number of mice equipped with transgenic reporter genes have been made. In most of these models the location of the reporter gene does not appear to influence the mutation frequencies, with some exceptions. For example, we made one *lacZ* transgenic line with the bacteriophage λ vectors integrated near the pseudoautosomal region of the X chromosome and found spontaneous somatic and germ-line mutation frequencies that were up to a hundred times higher than in all other lines tested[637]. Somewhat later Peter Glazer's group also identified a mouse line, with the integrated λ vectors on

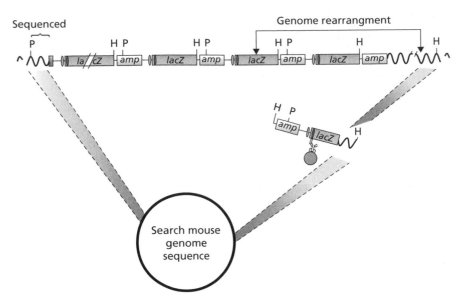

Fig. 6.5 Genome rearrangements with one breakpoint in a *lacZ* gene and the other elsewhere in the mouse genome can be recovered with the plasmid-based mouse mutation reporter model. Both *Hind*III and *Pst*I can be used to excise individual plasmid copies. *Hind*III digestion will result in the recovery of internal copies plus the 3′ mouse flanking copy. The 5′ copy cannot be recovered because it lacks the ampicillin-resistance gene (*amp*). *Pst*I digestion will only yield internal plasmids because it disrupts *amp*, which cannot be regenerated by ligation in case of a mouse flanking copy. H, *Hind*III site; P, *Pst*I site.

chromosome 7, in which the frequency of spontaneous mutations was unusually high: 20-fold higher than in other transgenic mice carrying a similar number of vectors at other chromosomal locations[638].

With the development of transgenic mouse models harboring chromosomally integrated reporter genes, it became possible to directly test the hypothesis that somatic mutations in a neutral gene accumulate with age in different organs and tissues. Using the *lacI* mouse model, Lee *et al.*[639] were the first to demonstrate an age-related increase in mutant frequency (expressed as the number of mutant reporter genes as a function of the total number of recovered reporter genes) in spleen, from about 3×10^{-5} in mice of a few weeks old to $(1–2) \times 10^{-4}$ in 24-month-old animals. Shortly thereafter, in my own laboratory, Martijn Dollé demonstrated age-related increases in the mutation frequency in some, but not all, organs in the mouse. For example, Dollé demonstrated that mutation frequencies at a *lacZ* transgene increase with age in the liver, from about 4×10^{-5} in the young adults to about 15×10^{-5} in old animals (about 30 months), whereas such an increase was virtually absent in the brain[640]. Also spleen, heart, and small intestine showed increased mutation frequencies with age[641,642], but not the testes[643]. These organ-specific patterns of mutation accumulation at the *lacZ* transgene locus[488] are shown in Fig. 6.6.

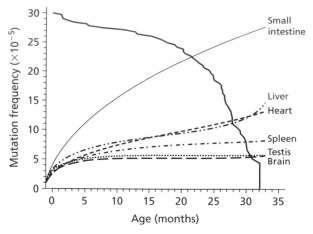

Fig. 6.6 Spontaneous *lacZ* mutant frequencies with age in brain, testis, spleen, liver, heart, and small intestine of the *lacZ*-plasmid transgenic mice. The lines represent hand-drawn curves matching mean mutant frequencies at different age groups. The survival curve of these mice, set to 100% at birth, indicates 50% survival at 26.5 months of age. This was not different from non-transgenic control animals of this strain. Re-drawn with permission from ref. 488.

The organ-specific mutation accumulation observed in the *lacZ*-plasmid mice appeared to be very similar to that obtained by others. For example, Tetsuyo Ono (Sendai, Japan) and co-workers used the original bacteriophage λ mouse model made by Jan Gossen in my laboratory. They also observed an age-related increase of *lacZ* mutation frequency in liver, heart, and spleen, but virtually no increase in brain and testes[644]. Also Stuart *et al.*, analyzing the mutation frequency in liver, brain, and bladder of *lacI* mice, observed a tissue-specific aging-related pattern, with mutations accumulating in liver, more rapidly in bladder, but not in brain[645]. Also using the *lacI* reporter mice, Hill *et al.* analyzed spontaneous mutation frequency in middle to late adulthood of the mouse, at 10, 14, 17, 23, 25, and 30 months of age in adipose tissue, liver, cerebellum, and male germ cells[646]. These authors observed elevation of the mutation frequency with age in adipose tissue and liver, but not in cerebellum or male germ cells, thereby extending and confirming the tissue specificity mentioned above. Mutation accumulation in male germ cells at the *lacI* locus in the mouse was studied in the laboratory of Christi Walter (San Antonio, TX, USA). Interestingly, an initial decline in mutation frequency was observed during spermatogenesis, with a subsequent increase in spermatogenic cells obtained from old mice[647]. The mutation frequencies eventually obtained in old mice were similar to the ones obtained in the aforementioned studies (at either young or old age). Whereas several laboratories reported a generally lower spontaneous mutation frequency in both brain and male germ cells[646], there seems to be consensus that in these tissues there is no increase with age (or a very small one). Hence, it is possible that these investigators were able to accurately measure *lacI* mutations at much lower levels than others, in which case the aging-related increase escaped detection by other investigators.

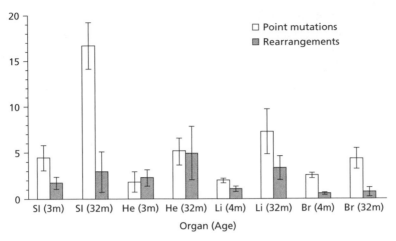

Fig. 6.7 A comparison of young and old mice for the type of mutations found to accumulate at the *lacZ* transgene locus. Note that in the small intestine (SI) virtually all such mutations are point mutations, but not in liver (Li) or heart (He). Also note the virtual lack of genome rearrangements in brain (Br). m, months.

As mentioned above, bacteriophage λ models do not indicate a profound organ or tissue specificity with respect to the spontaneous mutation spectra. By contrast, in the *lacZ*-plasmid model, with its approx. 20-fold smaller concatamer, Martijn Dollé in my laboratory observed striking organ specificity with respect to the mutational spectra of the old animals. Mutations in the *lacZ* reporter mouse model can be characterized by restriction digestion and nucleotide sequencing of the positively selected *lacZ*-mutant plasmids recovered from a mouse tissue. Mutations that do not alter the restriction pattern are point mutations—single base changes or very small deletions. Mutations that cause changes in the restriction pattern are deletions and other types of rearrangement. Only a few of these were deletions within the *lacZ* gene. Most of these mutations appeared to involve genome rearrangements between the *lacZ*-plasmid cluster and the mouse chromosomal DNA. As mentioned, such events can be physically characterized based on the small mouse fragment that indicates the breakpoint in the mouse genome. Already, by simply separating mutations on the basis of whether or not they changed the restriction-enzyme pattern in these two classes, striking organ specificities were observed. For example, as shown in Fig. 6.7, whereas in heart and liver both point mutations and genome rearrangements were found to accumulate, in small intestine the age-related increase was entirely due to point mutations[641].

Further characterization of the point mutations by nucleotide sequencing of entire mutant *lacZ* genes obtained from different organs at young and old age indicated that initially, at young age, the mutation spectra were more or less the same for all organs. However, during aging they started to diverge significantly, resulting at old age in drastic differences between organs, with a high frequency of GC to AT base-pair substitutions at CpG sites in postmitotic organs, such as brain and heart, and a more varied pattern in the more proliferative tissues[275]. Especially in brain and heart of old animals, GC to AT transition mutations at CpG sites were dominant. As mentioned in Chapter 4, such mutations are a typical consequence of the spontaneous deamination of 5-methylcytosine.

Further characterization of the genome rearrangements also required nucleotide sequencing. Whereas some conclusions could be drawn about the nature of such mutations based on the restriction patterns, characterization of the breakpoint in the mouse genome allowed their identification as deletions, inversions, or translocations. Such physical mapping of 49 genome-rearrangement mutations, mainly from heart and liver of young and old mice, indicated intrachromosomal deletions or inversions, varying from smaller than 100 kb to 66 Mb, as well as translocation events[648]. A significant increase in the frequency of these genome rearrangements with age was found in heart and liver, but not in brain. Assuming that the *lacZ* plasmid reporter locus is representative for the overall genome, Martijn Dollé in my lab extrapolated the genome rearrangement frequencies from the 3-kb *lacZ* transgene to the entire 6×10^9-bp diploid genome. The outcome was subsequently divided by a factor of two, because only one of the two breakpoints needs to occur in a *lacZ* reporter gene. The results indicate that up to 37 (in the aged heart) genome rearrangements can accumulate in an aged cell[648].

There are multiple explanations for organ specificity in the rate of mutation accumulation and the type of mutations found during the course of the aging process. A most logical one is that the mutation frequency is simply a function of organ-specific cell-proliferative activity. Indeed, it is reasonable to assume that with higher proliferative activity there is a higher chance of replication error, one of the major mechanisms of mutation generation (see above). This could explain the relatively high mutation frequency in small intestine, which increases about 5-fold during aging[641]. It does not explain, however, the relatively low age-related increase in spleen, and the absence of an increase in testes; both these organs should contain plenty of actively proliferating cells. If proliferative activity alone is important, then it is also not clear why mutations accumulate in heart, a postmitotic organ par excellence.

To resolve at least some of these issues, Ana Maria Garcia, a postdoctoral researcher in my laboratory, investigated the possibility of intra-organ variation in the two tissues that were most dissimilar in terms of mutation accumulation: the highly proliferative small intestine and the postmitotic brain. By separating the inner part or mucosa of the intestine, which mainly contains the rapidly dividing epithelial cells, from the outer part or serosa, a layer of postmitotic, smooth muscle cells, she was able to directly compare the *lacZ* mutation frequency between proliferating and non-dividing cells in this organ from 18-month-old mice. The results indicate significantly higher mutation frequencies in the highly proliferative mucosa than in the serosa (Fig. 6.8). Most of the mutations in the mucosa were point mutations, suggesting replication errors as their source. However, Ana Maria also observed a significant increase in *lacZ* mutation frequency in the serosa, underscoring the fact that DNA replication is not strictly necessary for mutations to accumulate with age.

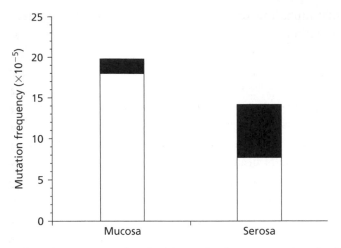

Fig. 6.8 More mutations have accumulated with age in the mucosa than in the serosa of the small intestine from 18-month-old mice. The excess mutations found to accumulate in the mucosa were all point mutations (white). A. Garcia, submitted for publication.

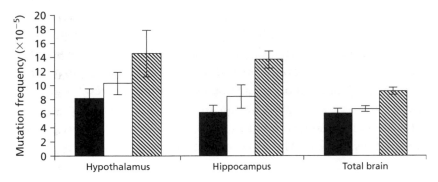

Fig. 6.9 Mutation frequencies are higher in the hypothalamus and hippocampus than in the whole brain. Also the rate of age-related mutation accumulation is significantly higher in these structures. Black bars, young mice; white bars, 18-month-old mice; gray bars, 30-month-old mice. A. Garcia, submitted for publication.

With respect to the brain, Garcia was interested in the possibility that while in the overall organ mutations were not (or only slowly) accumulating, certain sub-structures could nevertheless show increased mutagenesis with age. Indeed, when subdividing different brain parts, she observed that in the hypothalamus and hippocampus mutations were accumulating with age with the mutation frequency higher than in the brain overall at all ages (Fig. 6.9). Therefore, factors other than DNA replication errors, possibly closely related to the function of a particular organ or sub-structure, are likely to play a role in determining the patterns of genomic instability as they arise during aging. As we will see in Chapter 7, work from another member of my laboratory, Rita Busuttil, convincingy demonstrates that mutations do not need replication for their induction. It all depends on the type of mutation.

6.2.2.3 Summary

In summary, using both endogenous selectable marker genes and bacterial reporter transgenes, strong evidence has been provided for the age-related accumulation of spontaneous mutations in mammals. The results indicate organ and tissue specificity of the mutation frequencies and the results obtained with the *lacZ*-plasmid model also indicate organ specificities of the types of mutation that were accumulating. In interpreting these data it should be noted that the observed increases were modest (varying from less than 2-fold, to little more than 4-fold) and appeared to level off at middle age (Fig. 6.6). This leveling off was found by multiple investigators using different mutation target genes and can be explained to a large extent by a rapid early increase in mutations[649]. This pattern contrasts with the previously mentioned age-related increase in chromosomal anomalies, which is exponential[606]. However, the variation between individuals, even in inbred mice, is large, which constrains any definite conclusions regarding the kinetics of mutation accumulation. It should also be noted that none of the currently available systems are sensitive for all types of mutation and it is therefore likely that the real mutation frequency is significantly higher. Most of the transgenic models also do not account for mutational hotspots and such important functional endpoints as cell death are missed. Indeed, to put the results on mutant frequencies of different organs and tissues at various age levels into context, it will be necessary to also assess cell proliferation and cell death. Most importantly, it will be necessary to determine at some point the critical level of cellular mutation loads in terms of physiological consequences. This will be further discussed in Chapter 7.

6.2.3 POTENTIAL HOTSPOTS OF DNA MUTATIONS

Thus far, I have described the results obtained with assays measuring genomic instability at sequences thought to be representative for the genome overall. However, as discussed in Chapter 3, the genome consists of different types of sequence, some of which have particular characteristics that could make them either unusually prone to mutation or lend a mutational event an extraordinarily high functional relevance. Below I will discuss such potential genomic hotspots of mutation.

6.2.3.1 Telomeric instability

A genetic instability hotspot that has received a lot of attention is the telomere. Telomere shortening, due to the loss of minisatellite repeat elements, has been observed to occur during repeated passaging of human cells in culture. As we have seen, this causes replicative senescence and in the absence of p53 leads to crisis. This latter stage is characterized by chromosomal aberrations, dominated by end-to-end fusions. The importance of telomerase in preventing this is underscored by the observation in the laboratories of Woody Wright and Jerry Shay (both in Dallas, TX, USA) that its reconstitution prevents replicative senescence and essentially immortalizes normal human cells[650]. Also in view of the important role of replicative senescence as an *in vitro* model of normal aging, this is generally interpreted as strong evidence for telomerase as a major and perhaps the most important genome-stability system in actively proliferating human cells. Its potential importance for aging *in vivo* is underscored by the observation of premature aging of mice with an inactivated telomerase RNA component (see Chapter 5) and a number of functional studies, which will be discussed in the next chapter.

As mentioned in Chapter 4, in the genetic disorder dyskeratosis congenita, telomere shortening is accelerated, and patients have premature onset of many age-related diseases and early death[651]. Is there any evidence for telomere attrition *in vivo* to the extent that it can be considered as a major cause of aging, for example, through increased replicative senescence? Telomere length can be measured as the mean length of terminal restriction fragments, which involves a standard genomic Southern blot with terminal restriction fragments detected by hybridization to a probe containing telomeric repeats. Whereas alternative methods allowing the measurement of single telomeres were later developed[652,653], most data on telomere length during aging were obtained with analysis of terminal restriction fragments. Average telomere length in white blood cells and other tissues of humans have been measured by a number of groups and found to decline from approx. 12.2 kb in a fetus to 7.2 kb in a 72-year-old[654–656]. The greatest loss was found to occur at relatively young age. No individuals have been found with a mean telomere length of less than 5 kb, even among 100-year olds[655,656].

More recently, using a quantitative PCR method in a study of 143 normal unrelated individuals over the age of 60 years, an association was found between telomere length in blood and mortality[657]. However, this was not confirmed in a later study involving 598 normal individuals over 85 years[658]. Whereas it is therefore questionable whether telomere attrition by itself has phenotypic consequences *in vivo*, it should be kept in mind that average telomere length in blood is not the same as the average length of telomeres in an individual cell or the shortest telomere of an individual chromosome. Hence, even if average telomere length is not a predictor of survival, telomere loss could nevertheless be one of many different forms of genomic instability contributing to normal aging, in this specific case, for example, by resulting in an increased number of senescent cells.

6.2.3.2 mtDNA mutations

The 16.5-kb mitochondrial genome has been implicated as a major target for somatic mutations during aging[659]. According to the mitochondrial theory of aging somatic mutations in mtDNA accumulate with age as a consequence of the free radicals generated so close to their target. The resulting degenerative changes in this important cell organelle are then expected to result in impaired function of the respiratory chain, leading to increased free radical production and more mutations. The logic of this vicious cycle has appealed greatly to gerontologists and explains the importance assigned to the mtDNA as a target for age-related mutations. However, as we shall see, recently obtained results shed some doubt on vicious-cycle ideas.

The main reasons to consider the mitochondrial genome as an important target for aging-related spontaneous DNA mutations are its physical proximity to a major source of free radicals, the lack of protective histone proteins, and the presumed absence of mtDNA-repair systems. As we have seen in Chapter 4, this last point turned out not to be entirely correct; several repair pathways have been described for mtDNA, most notably an efficient BER system, and possibly also mismatch repair and recombinational repair[660]. Nevertheless, as noted earlier in this chapter there is evidence for a relatively high susceptibility of mtDNA to oxidative DNA damage. Apart from their increased vulnerability to DNA damage, mitochondria are the key organelles for respiration through electron transfer and oxidative phosphorylation, arguably the most important cellular function in eukaryotes.

Encoded by the mitochondrial genome are two rRNAs, 22 tRNAs, and genes encoding 13 mitochondrial protein subunits. They are necessary to produce essential enzyme complexes involved in oxidative phosphorylation. However, most mitochondrial components are encoded in the nuclear genome. What is the evidence that mutations in the mitochondrial genome accumulate to a level that starts contributing to aging-related cellular degeneration and death?

Within human populations, the mitochondrial genome has a high rate of sequence divergence with multiple polymorphic variants. This has led to its major role as a tool for examining human evolutionary history. mtDNA is inherited exclusively through the mother and a number of human heritable diseases are caused by specific mutations in the mitochondrial genome. These mutations typically impair oxidative energy metabolism, causing such aging-related diseases as cardiomyopathy and neurodegeneration. Somatic cells contain hundreds (epithelial cells) to hundreds of thousands (cardiomyocytes) of copies of the mitochondrial genome per cell and disease-causing mutations are often present in only a fraction of the total. This situation is called heteroplasmy. A mutation is called homoplasmic when it is carried by all copies of the mitochondrial genome. Usually the clinical severity of the disease is correlated with the proportion of the mutated mtDNA in the target tissues. There is a threshold above which a disease phenotype becomes apparent[661]. This threshold may be 60% for deletions and as high as 90% for point mutations. Similar to the situation with nuclear DNA mutations, deletions have a higher impact because they affect a larger part of the mitochondrial genome. However, the possibility cannot be excluded that physiological effects manifest at frequencies of deleted mitochondrial genome copies as low as 20%.

For somatic mutations arising with age the situation is very similar to the mitochondriopathies. Somatic mtDNA deletions increase with age in a tissue-specific manner[487]. In particular, muscle, heart, and brain accumulate relatively high levels of deleted mtDNA. Intra-organ variation has been demonstrated for the brain, with an accumulation in the cortex, but not in the cerebellum[662]. Hence, this situation is highly reminiscent of what has been observed for nuclear DNA with the *lacZ* reporter mice, in which similar striking inter- and intra-organ variation in mutation accumulation has been observed (see above). Also in this case, the question is whether or not the observed mutation frequencies can lead to physiological consequences. For the nuclear genome this question cannot really be addressed at this stage, but the mitochondrial genome is much smaller and relatively easily accessible due to its high copy number per cell. Nevertheless, measuring mutations in the mitochondrial genome remains technically challenging.

Before a mutation can have a phenotypic consequence, it must first clonally expand. There is now ample evidence that mitochondrial mutations not only accumulate with age, but can also clonally expand to reach levels that may be physiologically significant. Like so many studies aiming to detect and quantitate low-frequency events, in this field one has to be extremely cautious in interpreting the work that has been done. Most assays are based on PCR, which is notoriously prone to artifact. Some of the best work in analyzing mtDNA mutations in aging is from the laboratory of Konstantin Khrapko (Boston, MA, USA), who developed a PCR-based method to scan significant parts of mitochondrial genomes from single cells[663]. Using this assay he demonstrated that in human cardiomyocytes from old but not from young donors, mtDNA deletions occur at a frequency of about one in 15 cells[664]. In such cells deleted mtDNA copies represented up to 65% of the total number of copies, although in most cells this fraction was much less.

Excellent work in this field has also been done by the groups of Doug Turnbull (Newcastle upon Tyne, UK) and Judd Aiken (Madison, WI, USA). Turnbull's group was the first to demonstrate mtDNA deletion mutations in muscle fibers from normal elderly subjects with very low activity of cytochrome c oxidase (a marker of respiratory-chain deficiency), suggestive of a mtDNA defect[665]. Aiken's group observed clonality of deletion mutations in the mitochondrial genome in defective muscle fibers in aged rats[666] and rhesus monkeys[667]. Hence, in the case of muscle there is concrete evidence that mtDNA deletions may actually cause defects in oxidative phosphorylation. Recent data from Khrapko's laboratory, using a novel single-molecule PCR approach, indicate high fractions of deleted mtDNA in cytochrome c oxidase-deficient neurons in the substantia nigra of elderly humans[668].

For point mutations the situation is more complicated because such mutations are more difficult to detect. Using a similar single-cell PCR approach, Khrapko's group sequenced about 5% of the mitochondrial genome in a number of cells from tissues as diverse as buccal epithelium and heart muscle obtained from donors of different ages[669]. They discovered that many cells contained high proportions (varying from about

30 to 100%) of clonal mutant mtDNA expanded from single initial mutant mtDNA molecules. Virtually all expanded mutations were observed in tissues from elderly. They concluded that in an aged human body, there is on average approximately one expanded mtDNA point mutation per cell.

Rather than postmitotic cells, such as cardiomyocytes or muscle fibers, Turnbull's group analyzed intestinal colonic crypts, clonal populations of cells derived from one or two stem cells, for mtDNA mutations[670]. They sequenced the entire mitochondrial genome from individual colonic crypts obtained by laser microdissection from elderly individuals. Pathogenic point mutations were detected in most but not all of the cytochrome c oxidase-deficient crypts. Since the frequency of such crypts increased with age these findings support a more general role of mtDNA mutations, including point mutations, in aging.

In summary, the accumulation of mutations in the mitochondrial genome, both deletions and point mutations, in aging human tissues has been demonstrated conclusively. It also seems highly likely that through the expansion process at least some of these mutations reach physiological significance. The question remaining is whether enough of these highly penetrant mutations accumulate with age to causally contribute to aging in a substantial fraction of cells in a tissue. While this still remains to be demonstrated, it does seem clear that the original vicious cycle of mtDNA mutations leading to respiratory defects, which in turn increase ROS production, thereby causing increased mutagenesis, is unlikely to be true. Indeed, when more free radicals would generate new mutations one would not expect the clonality described above and observed by multiple investigators. Moreover, as already described in Chapter 5, the POLG-defective mitochondrial mutator mice display greatly increased mtDNA mutation rates but no increase in oxidative stress[490]. Nevertheless, although a direct link between mtDNA mutations and physiological consequences in aging tissue has not been established, the association—by multiple research groups—of such mutations with respiratory defects implicates the mitochondrial genome as a major functional target of the aging process.

6.2.3.3 Repeat-element loci

The families of repeat elements that are such a major component of the mammalian genome are also recognized mutational hot spots. Before I discuss the nature of the various forms of instability and the evidence that mutations at these loci accumulate with age, it is important to mention the potential role of some of these repeat families in the regulation of gene-expression patterns. As early as 1969, Roy Britten (Corona del Mar, CA, USA) and Eric Davidson (Pasadena, CA, USA) proposed that networks of dispersed DNA repeats, thereby focusing on transposons, coordinate the expression of batteries of genes involved in development[671]. As we have seen, there is evidence that both the mobile genetic elements (mainly retrotransposons) and simple, tandem repeat loci are unstable.

Such instability would allow rapid evolutionary change at the level of gene regulation rather than through protein structural alterations. Whereas, as yet, evidence has emerged for networks of miRNAs that act as additional layers of gene regulation there are no consistent data pointing towards a role of the main repeat families in the genome; that is, retrotransposons and tandem repeats. Nevertheless, it is conceivable that age-related instability of these sequences has adverse effects on normal patterns of gene regulation.

As described in Chapter 3, there are two major families of retrotransposons, termed LINES and SINES, with L1 and Alu, respectively, as their main representatives in humans. L1 repeats, the only active human mobile element, encode the machinery to move both itself and Alu, which is much shorter and does not contain an open reading frame. The capability of such elements to retrotranspose has been demonstrated in human cell lines[672]. The recently obtained evidence for both expression and retrotransposition of L1 elements in mouse neural progenitor cells in the brain as well as in such cells *in vitro* has also been discussed in Chapter 3[192]. There was some evidence that L1 insertion occurred preferentially in or near genes with an effect on differentiation of the neural progenitor cells. It has not been possible to accurately determine the frequency of the insertion events, which may vary from one in 10 to one in 1000 cells. Hence, there is no unambiguous evidence yet that they contribute significantly to somatic instability and nothing is known about a possible increase in transposition activity with age. The authors suggested that the observed L1 transposition could positively contribute to brain plasticity, but in my opinion it is more likely that such events would contribute to the random accumulation of mutations in the genome to possibly exert adverse effects. Indeed, overexpression of L1 proteins leads to toxicity[673] and L1 is able to induce a p53-dependent apoptotic response[674]. Of note, L1 is preferentially expressed in the germ line. Its lower overall expression level in somatic cells may be a consequence of methylation[673]. We will later see that there is evidence for a general demethylation of genomic DNA sequences during aging in a variety of tissues, which may contribute to increased instability of L1 elements with age, but this is unknown at present. Indeed, to my knowledge there are no studies that have been focused on instability of this type of repeat family during aging.

The second group of potentially unstable repeat elements comprises tandemly repeated DNA loci, which include the well-known minisatellites and microsatellites. In general mutation frequencies at these loci are about 1000-fold higher than the aforementioned mutational reporter genes, including *HPRT* and the transgenic reporter loci[675]. Classic hypervariable minisatellites, as originally characterized by Alec Jeffreys (Leicester, UK)[676], have germ-line mutation rates varying from 0.1 to 15% per transmission in humans[677]. These values were obtained by comparing minisatellites in blood DNA between parents and offspring or by direct analysis of these loci in sperm. Whereas it is probable that a high proportion of minisatellite mutations occur during meiotic recombination, they could actually occur at any point after fertilization, during development and aging. The situation is somewhat similar for microsatellite germ-line mutation rates, but it should be noted that mutation rates are highly locus-specific, depend greatly on the length of the allele, and may differ between humans and mice. Germ-line mutation rates at a single locus can vary by as much as three orders of magnitude for alleles within the normal range, but on average remain much higher than rates of mutation in the genome overall. Indeed, minisatellites and microsatellites are the most unstable loci in the human genome. Mutations in these sequences can arise during meiosis in the germ line, but they are not restricted to meiosis and also occur during development and (as we will see) aging. As already indicated in Chapters 3 and 4, microsatellite instability is probably a result of slippage replication errors, but minisatellite mutations may arise as a consequence of recombination processes, such as unequal crossing-over.

To study the effect of radiation on the tandem repeat loci, germ-line minisatellite mutation rates have been evaluated in four irradiated groups of human individuals from Japan and the former USSR[675]. The results were somewhat variable, but in general suggest that minisatellite instability is increased in people exposed to radiation. This is similar to the results obtained when mice were exposed to radiation[678].

Studies of microsatellite instability have been restricted mainly to tumor DNA, in which mutation rates at these loci can be so high (for example, in case of defective mismatch repair; see Chapter 4) that mutant fragments can simply be detected as a different PCR product after amplification of the locus from total tumor DNA. Age-related microsatellite instability has also been reported in normally aged tissues, most notably human T lymphocytes[679]. However, the results of these studies, in which individuals were scored qualitatively as to the presence or absence of a microsatellite-instability phenotype, are not very reliable due to the difficulties in analyzing bulk DNA from normal tissues[680]. This is complicated because in spite of the high mutation rates at microsatellite loci as compared with the overall genome, they are still too low to be detected directly in bulk DNA from normal cells. Analysis in such cases is somewhat complicated by the tendency of the *Taq* polymerase used in the PCR to undergo slippage, which leads to a characteristic stutter band pattern. It is not possible to dissect such complex banding patterns to derive truly quantitative mutation frequencies. Alternatively, so-called small-pool PCR, in which multiple aliquots of a dilution of the DNA sample (from one to 100 genome equivalents) are subjected to PCR amplification, with the products separately analyzed to discover mutant fragments, dramatically increased the sensitivity of detection. Using this approach it has been possible to analyze microsatellite instability in normal constitutive tissues and detect mutant fragments. A high degree of automation has further facilitated the routine application of this method to study microsatellite instability during aging.

Results from a recent study in which blood DNA from 17 normal individuals of a variety of ages was analyzed by small-pool PCR at six microsatellite loci indicated an age-related increase from 1% in 20–30-year-old individuals to 4% in 60–70-year-olds[681]. While of great interest as a symptom of aging, it is difficult to judge at this stage the potential functional consequences of microsatellite instability. Some microsatellite loci are contained within coding regions and others may affect gene regulation[682], but most are probably not functionally relevant.

Another type of repeat family for which age-related changes has been reported is the tandemly repeated rRNA genes. Already as early as 1972 work from the late Bernard Strehler (1925–2001) and co-workers indicated instability of these genes[683]. Analyzing DNA from different tissues of beagle dogs of different ages, as well as human heart-biopsy samples, these investigators reported a loss of hybridizable rRNA gene copies with age. In the beagle dogs such a gene copy loss was only found for postmitotic tissues: muscle, brain, and heart[684]. These results were not confirmed in a later study by Peterson *et al.*, studying mouse heart[685]. This will also remind the reader of work in Leonard Guarante's laboratory on rRNA gene instability in aging yeast, giving rise to extrachromosomal rDNA circles (Chapter 2). Brad Johnson (Philadelphia, PA, USA), then in Guarente's laboratory, investigated the possible occurrence of rDNA circles in various human tissues (heart, kidney, liver, spleen, skeletal muscle) from donors aged 4 days to 84 years. Evidence for such signs of rDNA instability was not obtained. The same was true for human and mouse colon crypts. There was also no evidence for changes in rDNA copy number with age in the same human tissues, although the number of individuals was small and individual variation large (B. Johnson, personal communication). Interestingly, in later studies from several different laboratories evidence was obtained for an age-related decline in rRNA gene activity in fibroblasts and lymphocytes from human individuals of different ages[686,687].

6.3.4 MUTATIONS IN GERM CELLS

The level of genomic instability in germs cells is of special interest for somatic mutation theories of aging. As discussed multiple times, mutations in the germ line are a necessary evil to guarantee evolvability: the generation of individual variation that permits the perpetuation of life when circumstances change. A question that has been raised in the past is how mutation accumulation in somatic cells can cause aging while the germ line is apparently immortal. The answer is 2-fold. First, germ cells may have superior genome-maintenance systems to keep mutation loads relatively low. This is illustrated by results from Olsen *et al.*[688], indicating more efficient BER activities in human and rat male germ cells than in white blood cells. As we have seen, stem cells are also likely to have superior genome-maintenance systems (Chapter 4). Second, and more importantly, germ cells (and to some extent also stem cells) have the opportunity to get rid of genetically damaged cells through selection.

Even superior genome-maintenance systems are unable to completely prevent mutation accumulation and it is generally assumed that an age-related increase in chromosomal abnormalities and other types of mutations in parental germ cells is responsible for increased risk of offspring with a genetic defect, such as Down's syndrome[689]. Chromosomal abnormalities, including aneuploidy and chromosome structural rearrangements, occur in 10–20% of spermatozoa and oocytes[690]. At least 15% of all recognized human pregnancies abort, most likely as a consequence of chromosomal mutations[691]. In fact, 50% of all spontaneous abortions contain a gross chromosomal mutation. An age-related increase in the number of sperm with chromosomal breaks, translocations, and aneuploidy has also been observed in mice[692].

The germ-line mutation rate in males is generally much higher than in females, which was first established by J.B.S. Haldane, who was also the one to first estimate the human mutation rate[693]. This higher paternal mutation rate is now ascribed to the much larger number of germ-cell divisions[173]. As discussed, replication errors are an important factor in the generation of mutations, although not the only one. It is therefore not surprising that most mutations causing genetic disease are derived from the father. This can be determined for sporadic cases of the disease using polymorphic markers linked to the disease gene. Disease incidence increases dramatically with age, most likely because mutations accumulate in the germ cells as they do in somatic cells. Interestingly, in all such cases the mutation is a point mutation, possibly because they are almost all small genes and rearrangements, such as deletions, could remove the entire gene, leading to early lethality. However, no such paternal effect is observed for heritable diseases caused by mutations in very large genes. In such genes the disease-causing mutation is often a large deletion[173]. As we shall see in Chapter 7, Rita Busuttil in my laboratory, working with somatic cells from our *lacZ*-plasmid reporter mice, observed a striking replication dependency for point mutations, but not for large deletions and genome rearrangements in general.

The picture that emerges of the germ line is that of an effective sieve that allows most mutations to be lost by negative selection; that is, through the death of germ cells or early embryos. Apoptosis may play an important role in eliminating defective cells from the germ-cell pool. In somatic tissues such a sieve as a mechanism to select against cells rendered dysfunctional by mutations could also be present, but is destined to be less effective because cell death itself would lead to functional decline. Indeed, aging of somatic tissues can be considered as a continuous battle between maintaining genomic integrity and ensuring sufficient cell-functional mass. In a recent publication with my colleague, Yousin Suh (San Antonio, TX, USA), we called this the zero sum game of aging[694]. Altering the balance between cell survival and cell elimination can cause an increase in either cancer or non-cancer degenerative symptoms of aging, as we have seen in the previous chapter. Only by improving the processes that prevent genome deterioration without the need for cell elimination can we expect to alter the zero sum game and significantly extend lifespan (Chapter 8).

With the recently emerged possibility of cloning mammals by transferring nuclei from adult somatic cells into enucleated oocytes, it should be possible to directly investigate the functional consequences of genome deterioration with age. Indeed, the success rate of cloning animals from somatic cells of old individuals should be markedly lower. Although the results thus far obtained with cloned animals suggest that cloning is less efficient from cells of adult animals as compared with fetal or embryonal cells, it is difficult to draw conclusions from the relatively small number of cloned animals obtained from somatic nuclei of adult animals. There are no systematic studies in which different age groups were compared side-by-side with respect to the efficiency of cloning from somatic nuclei. Whereas fetal cells may have much less genetic damage than adult somatic cells, the success rate of cloning using any cell type is very low. However, this may largely be due to technical problems or problems of genomic reprogramming by oocytes rather than the accumulation of mutational damage in adult somatic cells[695–698].

Some anecdotal results are notable. In one study, somatic-cell cloning technology was used to produce eight newborn calves from a 16-year-old, infertile bull[699]. Whereas bulls have a maximum lifespan of about 30 years, this result would certainly not suggest that at middle age somatic mutation accumulation with age is so high that it leaves not a single cell with an intact genome. In reality it merely indicates that one can still find a cell with a genome sufficiently intact to successfully direct development of a new calf. On the other hand, there is some evidence for signs of premature aging in the first cloned sheep, including premature telomere shortening[700]. Finally, in an interesting series of experiments mice have been cloned from cumulus cells (small cells that develop in the ovary and support the growth of oocytes) from several successive generations of cloned mice[701]. The sequentially cloned mice showed no signs of premature aging, at least up to 1.5 years, when they were studied. However, to put this in perspective, only about 2% of enucleated oocytes receiving a cumulus cell generally develop to live-born pups, a cloning efficiency which appeared to drop in successively cloned generations[701]. Hence, selection for mutation-free cumulus nuclei (which are extremely young anyway) cannot be excluded.

Apart from the well documented higher incidence of abnormal reproductive outcome among older parents, as discussed above, there is one report describing an effect on the lifespan of their offspring. Gavrilov *et al.*[702] studied the effect of paternal age on human longevity in European aristocratic families with well known geneology. Interestingly, these workers found a significant adverse effect of having an older father (50–59 years) on daughters, who lived on average 4.4 years less than daughters of young fathers (20–29 years). No significant effect was found for sons.

6.3 Changes in DNA modification and conformation

Thus far, most studies of genome alterations in aging have focused on DNA damage (chemical changes) or mutations (sequence changes). However, more and more attention is now focused on another source of phenotypic variation that is not based on variations in the base sequence of DNA. Such so-called epigenetic variation is heritable information encoded by modifications of genome and chromatin components that affect gene expression. As we have seen in Chapter 3, the structural basis for epigenetic variation involves DNA methylation or histone-modification patterns. Epigenetic information is only partially stable and destined to change during development and in carrying out essential somatic functions. Hence, it would not be surprising if stochastic changes in the epigenome turned out to be more relevant as a potential cause of aging than DNA sequence changes. Epigenetic patterns of transcription control have now been conclusively shown to be essential in the regulation of gene expression (Chapter 3). After embryogenesis, when epigenetic patterns are erased and reset, development and differentiation are largely facilitated by epigenetic programs[382]. Random changes in these patterns are genuine mutations and likely to have adverse effects, similar to changes in DNA sequence. Called epimutations by Robin Holliday (Sydney, Australia)[703], they have now been demonstrated to be critically important as causal factors in cancer[704]. Attention has thus far been focused on changes in methylation at cytosines, primarily in CpG dinucleotides of promoter regions. Of special interest in this respect are so-called CpG islands, which are usually unmethylated in normal cells. Hypermethylation can silence tumor-suppressor genes, thereby facilitating tumor formation. However, hypomethylation can also contribute to cancer, for example, by facilitating genomic instability[705]. Epigenomic alterations are now also increasingly recognized as part of aging and its associated pathologic phenotypes[706,707]. Early evidence indicated a general demethylation in most vertebrate tissues with aging[708]. In mitotically active tissue, it is possible that this reflects some deficiency in maintenance remethylation as this normally occurs after DNA replication. As discussed in the previous section, such a general demethylation during aging may desuppress silenced retrotransposons, such as L1, to increase genome instability.

Genome-wide demethylation has been suggested to be a step in carcinogenesis, based on the results of a study suggesting that defects in DNA methylation might contribute to the genomic instability of some colorectal tumour cell lines[709]. Evidence that supports this notion comes from another study in which elevated mutation rates were found at both the endogenous *Hprt* gene and an integrated viral thymidine kinase (*tk*) transgene in mouse embryonic stem cells nullizygous for the major DNA methyltransferase (*Dnmt1*) gene. Gene deletions were the predominant mutations at both loci. The major cause of the observed *tk* deletions was either mitotic recombination or chromosomal loss accompanied by duplication of the remaining chromosome[710]. These results suggest that the role of methylation may include the suppression of genomic instability. This may be especially important in repeat-rich genomic regions, to prevent illegitimate recombinations. Whether the observed hypomethylation in aged tissues also promotes genomic instability is unknown at present. A similar approach cannot easily be taken to study hypomethylation *in vivo* since *Dnmt1* nullizygosity is embryonically lethal. For this purpose, it will be necessary to construct hypomorphic alleles for this gene.

Most studies on the role of methylation in aging have focused on specific genes or their promoter regions. In early approaches, including some studies by my own laboratory[711], pairs of so-called isoschizomeric restriction enzymes were used. Such enzymes recognize the same site except when this is methylated, in which case one of the pair is no longer able to digest the sample. The best example is the restriction enzyme pair *Hpa*II and *Msp*I, which both recognize CCGG. The former is sensitive to CpG methylation, but the latter is not. Southern-blot analysis using specific probes then allows the assessment of these sites in or around the gene of interest by comparing hybridization patterns. More recently, sequence-specific methods have been developed utilizing the sequence differences between methylated alleles and unmethylated alleles that occur after sodium bisulfite treatment[712]. DNA is modified by sodium bisulfite treatment converting unmethylated, but not methylated, cytosines to uracil. Comparison of modified with unmodified sample by sequencing, or using a PCR test with specifically designed primers, will distinguish methylated from unmethylated DNA in bisulfite-modified DNA.

There are now numerous reports in the literature on changes in methylation status of individual genes during aging[713]. However, the functional relevance of these changes is largely unknown. This situation is different for cancer, in which hypermethylation has been associated with the silencing of genes involved in cell-cycle regulation, tumor-cell invasion, DNA repair, apoptosis, and cell signaling[704]. Similar to mutations that inactivate these genes through DNA-sequence alterations, hypermethylation provides tumor cells with a growth advantage, allows them to metastasize, and increases their genetic instability, which will allow them to acquire further advantageous genetic changes. The target for such hypermethylation events are the CpG islands located in gene promoter regions. Methylation of CpG islands is strongly correlated with transcriptional suppression. As mentioned in Chapter 3, CpG islands are generally unmethylated in normal somatic cells. As discussed in that chapter, the changes in chromatin structure resulting from hypermethylation would effectively silence transcription. By now, the methylation status of CpG islands of a large number of genes in many tumors has been assessed using a method called restriction landmark genomic scanning (RLGS)[714]. RLGS is a genome-wide method based on two-dimensional separation of DNA fragments obtained by *Not*I digestion, which is methylation-sensitive.

RLGS has been used in the laboratory of Bruce Richardson (Ann Arbor, MI, USA) to investigate whether CpG islands change methylation status during aging[715]. After comparing more than 2000 loci in T lymphocytes isolated from newborn, middle-aged, and elderly people, the conclusion was that 29 loci (approx. 1%) changed methylation status with age: 23 cases of hypermethylation and six cases of hypomethylation. Most of these methylation changes happened before middle age. Interestingly, the same CpG islands affected in T cells also changed with age in other tissues, suggesting that methylation instability is locus-specific. Interestingly, in T cells and various tissues from aging mice no methylation changes were observed.

When interpreting these results it has to be realized that not all CpG islands were investigated in this study (there are probably more than 20 000 CpG islands in the human genome) and that there are many genes that lack CpG islands. Nevertheless, the results of this study basically confirm on a global scale the results of the many previous, isolated studies in which certain genes were found to change methylation status with age, sometimes an increase and sometimes a decrease, but most were not. It should be noted that stochastic changes in methylation pattern cannot be analyzed because the techniques to select for genes or genome sequences that have undergone an isolated methylation change in one or few cells are currently lacking. At any rate, if it is true that about 1% of all genes suffer from changes in methylation status with age in a fraction of cells that is sufficiently large to be detectable, then the conclusion should be that the epigenome is much less stable with age than the genome.

There is other evidence pointing in the same direction. In a recent study, global and locus-specific epigenetic differences between identical twins were reported to arise with age[716]. In this study, involving 80 twins varying in age from 3 to 74 years, a variety of epigenomic markers, including X-chromosome inactivation, total genomic 5-methylcytosine content, sequence-specific DNA methylation (by bisulfite genomic sequencing), and global methylation status (by RLGS), were assessed in blood in comparison with measures of gene expression (using microarrays and quantitative real-time PCR). The results indicate that while these monozygotic twins are epigenetically indistinguishable early in life, at older age they start to exhibit remarkable differences in overall content and genomic distribution of 5-methylcytosine DNA and histone acetylation, affecting their gene-expression profiles.

Evidence for a relaxation of epigenetic control of gene expression during aging has been reported by multiple authors, with the stability of X inactivation and genomic imprinting the obvious targets of study. For studying the reactivation of inactive X-linked genes mice are often used harboring an X-autosomal translocation, so that the normal (i.e. non-translocated) X chromosome is always the one that is inactivated. Hence, when the activity of a gene can be assessed specifically on the normal X chromosome (because the other one is defective) it is possible to verify whether during aging a supressed gene is re-activated. For example, in the mouse an approx. 50-fold age-related increase in liver cells that had re-activated the gene ornithine transcarbamoylase on the inactive X chromosome has been observed. This was ascribed to an age-related reduction in DNA methylation[717].

Age-associated loss of epigenetic repression of X-linked genes has been analyzed quantitatively for the gene *Atp7a* in various tissues of the same X–autosomal translocation mouse model[718]. The results indicated a mean organ-specific relaxation of gene repression in the oldest group of mice of up to 2.2% (this is the fraction of transcripts from the inactive X relative to its active sister chromosome). Using the same quantitative methodology, the *Igf2* gene, imprinted in one mouse strain but not in another, was shown to increase relaxation of its epigenomic repression at old age by up to 6.7%.

In humans, the situation may be different. Indeed, in a study of the X-linked *HPRT* locus in 41 women heterozygous for mutations at this locus (causing Lesch–Nyhan syndrome), an increase with age in $HPRT^+$ skin fibroblast clones over the expected 50% (as a consequence of random X inactivation) was not found[719]. Further studies showed that the silent locus does not detectably re-activate spontaneously in culture, but only in response to treatment with 5-aza-2-deoxycytidine, a potent inhibitor of methylation. Whereas this can be interpreted in terms of locus-specific determinants of the stability of epigenomic control (as is also suggested by the aforementioned RLGS results), Robin Holliday proposed that age-related reactivation of a silent X-linked gene may be many times higher in mouse cells than in human cells as part of the much tighter genetic control systems in long-lived versus short-lived animals[720]. Hence, this could be another example of increased genome maintenance capacity of humans as compared to mice. On the other hand, it would be in conflict with the aforementioned RLGS results indicating substantially more alterations in global methylation status in humans compared with mice. Suffice it to say that also in this case more, extended data-sets are needed to draw definite conclusions.

Decreased epigenomic control during aging would be in keeping with the so-called dysdifferentiation theory of aging proposed by Richard Cutler (Phoenix, AZ, USA)[721]. According to this hypothesis the primary process of aging is the time-dependent drifting away of cells from their proper state of differentiation. Some experimental support for the hypothesis had already been obtained earlier in the laboratory of Cutler himself in the form of increased hybridization of globin gene probes to RNA extracted from brain and liver of old versus young mice[722]. This was well before the emergence of PCR, and even before the age of Northern-blot analysis (this method, with the name playfully coined after Southern-blot analysis, is still a standard technique for identifying specific sequences of RNA, in which RNA molecules are separated by electrophoresis, transferred to nitro-cellulose, and identified with a suitable probe).

Methylation patterns are intricately related to changes in histone-modification patterns. As we have seen, together these epigenomic control systems are two sides of the same coin: chromatin remodeling. DNA methylation acts synergistically with histone deacetylation to repress transcription. DNMT1, for example, can associate with HDAC2 and DNMT-associated protein 1, a co-repressor, to form a complex that is recruited to replication foci through interaction with PCNA. In view of the relationship between chromatin structure and gene expression, it is important to know the possible conformational changes that may occur in the DNA of the genome during the aging process. For example, changes in the stability of nucleosomes, the fundamental unit of chromatin structure, could lead to repression or derepression of gene expression.

An interesting example of a possible role of DNA higher-order structure in controlling aging is the progressive destabilization of repeated DNA in yeast, which has been demonstrated to be the cause of aging in this species[149] (see also Chapter 2). The packaging of DNA and histones into heterochromatin in yeast occurs at telomeres, mating-type loci, and rDNA and is mediated by the Sir2/3/4 complex. As described, extrachromosomal RNA copies are generated by HR between the repeats of the rDNA array, possibly as a result of erroneous HR repair of DSBs (Chapter 4). The resulting ERCs are segregated to the mother cell, accumulate, and ultimately cause senescence. An important function of heterochromatin therefore seems to be the suppression of recombination. This would suggest an age-related loss of heterochromatin, which is in keeping with reported changes in nucleosome spacing with age and an altered sensitivity to nucleases[723]. Such a heterochromatin-loss model of aging has indeed been proposed[724], but suffers from a chronic lack of data confirming a general relaxation of transcription control with age and its possible structural relationship with a loss of heterochromatin and increased genomic instability. Interestingly, David Sinclair (Boston, MA, USA) and co-workers recently demonstrated that in mice SIRT1 is bound to repetitive DNA and to promoters of a number of genes. In the brain, such binding was found to decrease with the age of the mouse. In mouse ES cells SIRT1 was released from repetitive DNA in response to genotoxic stress in the form of treatment with hydrogen peroxide (D. Sinclair, personal communication).

What are the possible mechanisms that drive epigenomic changes with age, such as hypo- or hypermethylation? First of all, metabolism and environmental factors, such as diet, may play very important roles by influencing the supply of methyl groups and/or by affecting the activities of DNMTs[725]. A diet deficient in methyl donors can promote hypomethylation. Dietary factors are likely to be critically important very early on, when methylation patterns are established. As mentioned in Chapter 3, genome-wide demethylation occurs before embryo implantation, which is followed by *de novo* global remethylation before organs develop. This allows the establishment of gene-specific methylation patterns, which subsequently determine tissue-specific transcription. Imprinting—the repression of either the maternal or paternal allele—is another example of DNA methylation early during development. Insufficient development of DNA-methylation patterns can lead to embryonic lethality and developmental malformations. There is even evidence that dietary factors *in utero* modulate disease risk in later life by affecting methylation patterns.

An example of a nutrient important for methylation is folate, a water-soluble vitamin B. Folate acts in the methionine cycle and is critically important in the synthesis of S-adenosylmethionine, the common methyl donor for DNA methylation. Folate deficiency leads to a decrease in S-adenosylmethionine and an increase in homocysteine. Dietary deficiencies in folate can decrease genome-wide methylation, both in humans and animals, and methylation is corrected by folate supplementation. The mechanisms contributing to the age-dependent changes in global DNA methylation may include a decrease in expression of DNMT1. Interestingly, micronutrient deficiency can mimic radiation or chemicals in damaging DNA by causing SSBs, DSBs, and oxidative lesions[726].

While undoubtedly important for optimizing methylation and histone-modification patterns during aging, nutritional and metabolic factors cannot be the only causes of a general loss in transcription control through epimutation. As discussed in Chapter 4, epigenomic status needs maintenance during and after DNA transactions. Errors during this process may be inevitable, similar to the DNA-sequence mutations that occur after the repair of chemical damage in DNA. However, it is also possible that epimutations are consequences of DNA-sequence changes. Since the local structure of DNA is nucleotide-sequence-dependent, mutations can be expected to influence the chromatin, thereby causing gene-expressional changes. The same is true for DNA damage, which has been demonstrated to cause changes in nucleosome positioning[727]. DNA methylation and DNA–histone and DNA–non-histone protein interactions together are thought to organize the genome into transcriptionally active and inactive zones. DNA damage and mutations can disturb this pattern of regulation, thereby causing normally silenced alleles to be expressed and vice versa. It is therefore possible that age-related changes in DNA-methylation patterns and histone protein changes are secondary rather than primary events in the aging process.

6.4 Summary and conclusions: a DNA damage report of aging

Of the three categories of DNA alteration defined at the beginning of this chapter it has by now been fairly well established that all of them occur with age. Their relative contribution and possible causal relationship with aging is unknown, but in spite of the scarcity of high-quality data on a broad range of human and animal tissues and the large variation in technical approaches in studying a given genome alteration we can at least begin the issuing of a genome damage report. I have summarized some of the data presented in this chapter in the form of a list of genome alterations for those categories for which we have reliable quantitative information at old and young ages (Table 6.2). This is limited to data obtained with mutation reporter loci, such as *HPRT* and the *lacZ* and *lacI* transgenes. Martijn Dollé in my laboratory extrapolated the mutation frequencies from the various reporter genes to the entire 6×10^9-bp diploid genome. This gave us absolute values for the average mutation load in a diploid cell. Needless to say, this is based on the assumption that these reporter genes are representative for the genome overall. However, since there are no major differences between results obtained with different reporters, apart from those that can be expected on the basis of the type of gene and/or assay used (e.g. X-linked or autosomal, neutral or functional), this seems a reasonable assumption. I have also limited this report to the nuclear genome, because we already know that mtDNA mutations in several tissues accumulate to a level that is close to exerting adverse effects. Table 6.2 is also restricted to mammals (humans and mice) in view of the virtual lack of data for other organisms, with exception of yeast (see Chapter 7).

Table 6.2 Estimated mutation loads per diploid genome in young and old animals

Target sequence	Organ/ cell type	Mutation frequency ($\times 10^{-5}$)		Target size (bp)	Mutation load per genome		Reference
		Intra	GR		Intra	GR	
HPRT	Mouse or human T cells	0.3/2.6	0.0/0.5	1323/40 000	12/116	0.0/0.3	620, 621
HPRT	Human kidney cells	4.3/21.3	0.8/3.8	1323/40 000	193/964	0.6/2.8	626
lacZ transgene	Mouse brain	2.6/4.4	0.5/0.6	3000	52/89	4.7/5.9	640, 644
	Mouse liver	2.2/7.8	0.9/2.7	3000	43/156	9.1/27.3	640, 644
	Mouse heart	2.1/6.4	1.9/3.7	3000	42/128	19.4/37.0	641, 644
	Mouse small intestine	4.6/17.1	1.5/2.4	3000	92/343	15.4/23.9	641, 644
lacI transgene	Mouse brain	3.0/4.0		1083	166/222		645, 646
	Mouse liver	3.0/8.0		1083	166/443		645, 646

Columns for mutation frequency and load show young followed by old animals. Intra refers to *lacZ* internal mutations. GR means genome rearrangements. For *HPRT* the target sizes shown are for the coding region and the total gene.

The estimate that can be derived from Table 6.2 suggests that the total mutational load for most organs or cell types increases with age, some times quite substantially. The increase, however, as well as the total mutation load, is highly cell-type- or tissue-dependent. At old age, cells suffer from a total mutation load of at least several hundred alterations, including a number of genome rearrangements. This is in reasonable agreement with estimates of the human mutation rate in somatic cells, made by Alec Morley on the basis of *HPRT* and *HLA* mutational reporter genes[728]. Expressed per unit of time rather than per cell division, these results suggest a mutation rate of 73×10^{-4}, or almost one in 1000, over a 100-year lifespan. Extrapolated to the entire diploid genome this would amount to about 400 mutations.

The question whether such mutation loads are high enough to cause the pathophysiological changes associated with aging is the subject of the next chapter. However, we should first realize that the numbers provided in Table 6.2 are underestimates, for at least two reasons. First of all, they only include mutations that affect the reporter gene to a sufficiently large degree to inactivate its function almost entirely. We already discussed that in the case of *HPRT* this is likely to lead to false-negative results. In general, one would expect that there are many mutations with an adverse effect that is not strong enough to be selected in a reporter-based assay, but does contribute to aging-related deficiencies. In this respect, the results obtained with different reporter assays are not directly comparable in view of the differences between genes in sensitivity to mutations. For example, the *lacI* gene is about three times smaller than the *lacZ* gene and one would therefore expect to find lower mutation frequencies at this locus. Instead, as shown in Table 6.2, the mutation frequencies for this reporter are only slightly lower. This is even more surprising in view of the fact that with the *lacI* gene, which is based on bacteriophage λ vectors, all the larger mutations, such as genome rearrangements, go undetected (see above). The explanation for the relatively high *lacI* mutation frequency is a larger number of sites that when mutated give rise to phenotypic effects. We never systematically analyzed the saturation of the *lacZ* gene in terms of the number of amino acids mutated, but this has been done for the *lacI* gene and another reporter, the λ *cII* gene. The *cII* gene is much smaller than the *lacI* gene, but generally yields higher mutation frequencies, due to its increased sensitivity to mutations[729].

The second reason that the results in Table 6.2 are probably underestimates is of course the demonstrated instability of large regions in the genome, sometimes two to three orders of magnitude more unstable than what has been found with the aforementioned reporter genes. For example, based on an estimated number of about 30 000 microsatellite loci in our genome, the observed increase in mutant microsatellites at several loci during aging of 1–4%[681] corresponds to 300–1200 mutants per diploid genome. Many other classes of genomic sequence remain uninvestigated in relation to aging, but it is very likely that, for example, transposition events will eventually turn out to be major contributors to the level of genomic change with age. Most notably, I have not been able to include any epigenomic changes since these cannot be estimated on a per-cell basis due to the lack of suitable methods of detection.

In conclusion, the data thus far available on genome alterations indicate that in spite of the large amounts of chemical damage to the genome induced continuously throughout the lifespan, this type of alteration is unlikely to contribute much to the aging phenotype. If so, and I should make it clear that the evidence for this is not yet fully conclusive, then this lack of a major effect is almost entirely due to the extraordinary capacity of our genome-maintenance systems to quickly repair virtually all such lesions. The importance of these systems for the cell clearly warrants the enormous investments in them (Chapter 4). The outstanding performance of our genome-maintenance systems is also responsible for the extremely low rate of errors that prevent the two other types of genome alteration, genetic and epigenetic changes, from accumulating much faster. In most cases this guarantees that an organism can mature and reproduce, often multiple times. Eventually, however, even extremely rare errors add up, especially in a genome that is robust; that is, designed to tolerate many random alterations. As we have seen, tolerance is an efficient way to maintain the function of large genomes, which are impossible to fully protect from random alteration by genome-maintenance systems alone. However, even slight increases above the tolerable level of genome alterations should then give rise to a decline in fitness. Such a gradual loss of genome integrity is in keeping with the nature of aging, which is a gradual process of fitness decline and does not involve dramatic changes over short periods of time. Based on the results thus far obtained it is likely that the loss of genome integrity as a possible cause of aging is based on instability at multiple types of loci acting in concert. As we have seen, even manipulation of the frequency of mtDNA mutations, often considered as a cause of aging on its own, is insufficient to bring forward in time all the known senescent phenotypes. Together, however, the various types of genome alteration observed could very well destabilize the genome to the extent that this leads to the adverse phenotypic effects observed during aging. It is to the nature of these phenotypic effects and their possible causal relationship with genome destabilization that I now turn.

7 From genome to phenome

Greatly increased knowledge about the genome, including the complete DNA sequence for a multitude of species, has brought us much closer to understanding how adult mammalian physiology and pathology are embedded in the structural characteristics of DNA, chromatin, and cellular nuclei. The main challenge in our postgenomic era is to unravel the genomic basis of variations in proteins, cells, and systems and identify the causative networks that determine the development of form and functional differentiation in multicellular organisms. This work is far from complete and essentially differs from studying how structural alteration in somatic genomes adversely affects organismal structure and function as part of the aging process, which presents its own unique challenges.

In Chapter 1, I introduced the term phenome: the sum total of biochemical, physiological, and morphological characteristics of an organism. The phenome relates to the phenotype as the genome relates to the genotype, as its manifest physical properties. Although the semantics of the -omics world are sometimes confusing[730], in my definition a phenome is not restricted to phenotypic traits with genotypic origins, but includes environmental and lifestyle influences. To identify the contribution of irreversible, random alterations in genome structural information to the aging process, it is necessary to reconstruct the sequence that may lead from somatic mutational events to the phenotypic features that describe the aging phenome. In addition, since on the basis of both evolutionary and phenomenological considerations aging is a multifactorial process affecting most if not all bodily functions, genome deterioration may be a universal, conserved cause of senescent deterioration, but does not necessarily underlie each and every aging-related phenotype. Other primary causes of aging not residing in the somatic genome may exist and it is important to distinguish those from genome-dictated causes.

In this chapter I will discuss a possible causal relationship between stochastic accumulation of genome alterations as described in the previous chapter and the aging phenome. As described previously, genome alterations are those changes that irreversibly alter the information content of the genome. This includes chemical changes in DNA primary structure, alterations in DNA sequence or DNA sequence organization, as well as epigenomic alterations, such as changes in methylation patterns or histone modification. For the purpose of this discussion, it is necessary to define the aging process, not on the basis of the demonstrated mortality increase in aging cohorts, but as a series of degenerative processes occurring in individual organisms over their lifetime. Whereas the former, population-based definition of aging has been very useful for understanding aging as a biological process, it is less informative for an investigation into what happens in the individual.

Many if not all organs, tissues, and cell types show an age-related decline in function, often accompanied by an increase in disease incidence. For example, muscle strength declines with age, atherosclerotic plaques build up in the arteries, immune functions decline, bones often weaken, and—the most visible aging phenotype—skin wrinkles and loses underlying fat. These changes often interact, making it difficult to distinguish between primary and secondary age changes. For example, wasting of skeletal muscle may not be caused primarily by changes in muscle cells, but by a decrease in circulating IGF-1, produced by the liver primarily in response to growth-hormone stimulation. The fact that growth hormone itself is produced by the pituitary gland further complicates the situation.

Whereas cell autonomous age changes may be difficult to find, some have been described, most notably in stem cells. For example, highly purified hematopoietic stem cells were found to undergo significant changes in their capacity to generate lymphoid and myeloid progenitors, with reduced B lymphocyte production. This was reflected in the up- or downregulation of genes involved in specifying these lineages, as assessed by microarray analysis[731]. It was suggested that gene-expression changes in these stem cells would ultimately be responsible for the well documented age-related loss of immune function and increased incidence of various leukemias. Interestingly, transplantation of old stem cells into young bone marrow and vice versa indicated that the adverse effects of aging in these cells were not due to the microenvironment of the aging bone marrow. The study did not make it clear what ultimately caused the autonomous effects of aging on hematopoietic stem cells, other than that they could be due to age-dependent changes in gene expression. Whatever the cause of the problem, there is no evidence for a loss of hematopoietic stem cells with age. In fact, the number of such cells has been found to increase with age, but they become less functional[732].

Aging-related cell-autonomous loss of function is not a general characteristic of stem cells, as indicated by work with muscle stem cells. Skeletal-muscle satellite cells are normally quiescent in adult muscle, but act as a reserve population of cells, able to proliferate in response to injury and give rise to regenerated muscle and to more satellite cells. Tom Rando (Stanford, CA, USA) and co-workers demonstrated that a decline in skeletal-muscle stem-cell activity, resulting in impaired regeneration of aged muscle, could be reversed through parabiotic pairings (connecting the shared circulatory systems) between young and old mice, thereby exposing old mice to young serum[733]. This suggests that the age-related decline of progenitor cell activity in muscle is not autonomous since it can be modulated by systemic factors that change with age. Also in this case the age-related impairment of muscle regeneration is not due to depletion of stem cells. Whereas the number of such cells may decrease with age by 20–30%, more than enough of them remain to mount a regenerative response.

A question that may come up at this point is whether a process that is as complex, multi-factorial, and interactive as the aging process can be captured for a given individual in some sort of global index. This has been attempted in the form of programs to establish biomarkers. Biomarkers are usually associated with disease. For example, cardiac biomarkers are enzymes, proteins, and hormones that are associated with heart function, damage, or failure[734]. Some of the tests are specific for the heart whereas others are also elevated with skeletal-muscle damage. Cardiac biomarkers are used for diagnostic and prognostic purposes, along with other laboratory and non-laboratory tests, to detect heart failure as well as to help determine prognosis for people who have had a heart attack. There are also biomarkers for susceptibility to a disease, as well as biomarkers for drug responses.

Biomarkers for aging are supposed to be age-related changes that measure biological rather than chronological age and predict the onset of age-related diseases more accurately than by counting years. This is significantly different from the disease-based concept of biomarkers mentioned above, which is practically oriented. Indeed, whereas disease biomarkers recognize both individual variation and the variation inherent in the complexity of the disease process, biomarkers for aging are based on the assumption that, in spite of the complexity of its phenotype, aging has an underlying common cause distinct from disease that can be captured in some rate measure to predict individual mortality and morbidity. It is certainly possible that aging reflects some basic mechanism common to all somatic cells. Indeed, it is the premise of this book that such a basic mechanism involves instability of the somatic genome. However, multicellular organisms contain many specialized cell types, which alone or in combination create multiple physiological systems. There is no evidence for markers that presage aging rate in every one of these systems, and it is also highly unlikely that they all degenerate at the same pace. Indeed, different organ systems age at various rates within the body, which essentially constrains the possibility of capturing biological age and mortality risk using a limited number of biomarkers[735]. Nevertheless, biomarkers for aging-related disease phenotypes can be used in combination to obtain measures of age or functional profile relative to an individual's chronological peers. Studies of this kind have given us an intriguing glimpse of a possible relationship between aging and disease.

An example is the approach taken by Ravi Duggirala (San Antonio, TX, USA), who studied the heritability of biological aging in a Mennonite population[736]. Such relatively isolated, homogenous populations lend themselves well to genetic investigations of complex phenotypes such as aging. To measure biological aging, the phenotypic data available for this population were used as the predictor variables of chronological (actual) age with the predicted age relative to chronological peers estimated using regression analysis. The residuals (i.e. the difference between chronological and predicted ages) were used as a measure of biological age; individuals with negative residual values are considered biologically older and individuals with positive residual values biologically younger compared with their chronological ages. The phenotypic data involved as many as 41 predictor variables, including such blood chemistry variables as glucose, high-density lipoprotein/cholesterol ratio, and total iron, and other variables such as blood pressure, body mass index, hand steadiness, and grip strength. Hence, these predictors are really biomarkers for complex, aging-related diseases, such as diabetes, coronary heart disease, hypertension, renal function, and general functional deterioration in the elderly.

The total phenotypic variance of the residuals (i.e. biological age) was partitioned into genetic and random environmental components. Heritability could be determined on the basis of family relationships among the study population. As it appeared, these residuals were significantly heritable, but independent of the genetic and environmental effects associated with the predictors. Hence, the factors that influence the genetic variance in biological age may be unrelated to the genetic factors that underlie the age-dependent phenotypic expression of the predictors, the quantitative markers of various physiological or functional processes. Therefore, these results suggest that there is a general, intrinsic process of aging with a genetic component of its own, predisposing to aging-related disease phenotypes, each of which can also have independent genetic and environmental risk factors (Fig. 7.1). This is not contradictory; it simply reflects the notion of a process of intrinsic aging, possibly universal to all organisms, which can be modified by factors that may be specific for the species or the individual. The recent dramatic improvements in high-throughput methodology for identifying genes controlling complex phenotypes in population-based studies provides the most likely way to test this hypothesis, as will be outlined at the end of this chapter.

Attempts to capture biological age on the basis of hypothetical biomarkers, or more realistically on the basis of functional status and disease, raise the question of what is

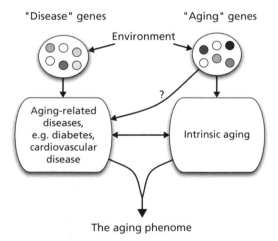

Fig. 7.1 The hypothetical relationship between intrinsic aging and aging-related disease. It is likely that during evolution, late-onset, disease gene variants accumulated in the germ line as dictated by the evolutionary theory of aging for which there is now universal consensus. Intrinsic aging, on the other hand, may involve universal mechanisms, such as genome instability, occurring in individuals of all species. It is possible that intrinsic aging processes contribute to aging-related disease.

and what is not part of the aging phenome. To some extent, I already addressed this question in Chapter 5 by defining an aging phenotype as an observable characteristic of an organism that falls outside the realm of natural selection because it has no negative impact on reproductive success. This definition would clearly include diseases, as long as they arise after the reproductive period. Some have suggested that aging is not a disease and that it is necessary to carefully separate late-life disease phenotypes from the intrinsic process of irreversible change predisposing to disease. Whereas it is obviously possible and also necessary to distinguish between aged tissue that does and aged tissue that does not contain pathological lesions such as tumors, I do not believe that age-related disease can be separated from aging. While attempts to do this are discussed below, it should be kept in mind that we often lack information about the natural cause of death of individual organisms at extreme old age. For the best-studied (mostly Western) populations of *H. sapiens*, by far the most meticulously researched species in terms of aging-related pathology, arteriosclerosis (especially atherosclerosis and its sequelae, such as coronary and cerebrovascular disease), diabetes, hypertension, senile dementia, and senile amyloidosis are the particularly important pathologies and, together with cancer, account for the bulk of mortality at old age. Infectious disease, most notably pneumonia, is a common final cause of death in old people. Other diseases, such as osteoarthritis and osteoporosis, are also prevalent among the aged but usually not lethal. However, especially among the oldest old, for those who manage to escape the most common diseases of aging, the cause of death is often not known. In such cases it is not possible to conclude that those who escaped disease will die of intrinsic aging. Because of the complex interactions between aging phenotypes, natural death may ultimately always be traced back to disease processes. For example, subtle aberrations in cell structure, including tissue atrophies, neuropathies, and microvascular disease, may underlie the sudden deaths of old people when subjected to extreme temperatures.

Our relative ignorance about the phenotypic associates of natural death at old age is even more obvious for other species. Even in mice, the species with the second best-characterized aging phenome, patterns of age-related pathology are incompletely described and we often do not know the cause of death. Little is known about aging-related pathology in invertebrate species (see further below).

Different causal factors can lead to very similar aging phenotypes. It is therefore possible that aging has one common denominator, for example, genome deterioration, and a multitude of additional, private causes determined by both genotype and environment. Some such factors may reinforce existing aging-related disease phenotypes. For example, the genetic factors causing familial forms of Alzheimer's disease accelerate the age of onset and increase the severity of the disease. Indeed, neurofibrillary tangles and senile plaques, the neuropathological hallmarks of this disease, are merely less abundant and widespread but not absent in the brains of intellectually intact elderly people[737].

An aging phenome, therefore, is the sum total of multiple proliferative and degenerative processes which can manifest as disease (e.g. atherogenesis, diabetes, cancer), relatively benign non-disease symptoms (e.g. baldness), or symptoms generally recognized as biomarkers of aging, such as collagen cross-linking, changes in T-cell subsets, and cataracts. It is questionable whether any grouping of aging phenotypes into categories of disease-related or non-disease-related phenotypes is biologically meaningful. They are presumably all a result of diminishing selection pressure after the reproductive period and the interactive nature of aging phenotypes would make it very difficult to distinguish between a non-disease-related, degenerative change and a disease-related one. Systematic approaches to better define aging phenomes of multiple species through phenotypic ontologies are imperative (see also Chapters 1 and 5). Such standardized vocabularies of terms would provide the means to digitize species-specific phenomes, thereby making them computationally accessible. Interactive databases of such digitized aging phenomes in different species and genotypes, seamlessly integrated with molecular data-sets at the various -omics levels, in combination with the dramatic advances in gene network theory as outlined in Chapter 1, would open up the possibility of tracing back individual aging phenotypes to alterations in the action of specific functional pathways. In this way we may ultimately succeed in reducing aging in a particular species to a relatively limited number of primary events (Fig. 7.2). The question that is central to this chapter and, indeed, for the book as a whole, is whether such primary events include random genome alteration and, if so, what the mechanisms are that lead from these stochastic changes to consistent functional endpoints common to different individuals, populations, and even species. Clues to this end can be obtained from the etiology of cancer, the only aging-related phenotype that is generally recognized as likely to be directly due to increased genome alterations, both genetic and epigenetic.

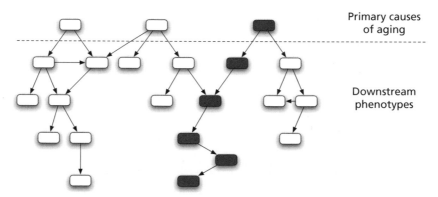

Fig. 7.2 Delineation of an aging cascade (gray shading) from its phenotypic manifestation to its primary cause.

7.1 **The causes of cancer**

Cancer stands out as a major aging-related pathology in mammals and in the histopathological analysis of animals it is common to distinguish between neoplastic and non-neoplastic lesions. In mammals a large fraction of all pathological lesions at the end of life are neoplastic lesions, with age as the single most important risk factor of the commonest forms of cancer (Fig. 7.3)[738]. Because genome alterations are generally assumed to be the main driver of carcinogenesis, it is relatively easy to make the case that mutation accumulation is responsible for the age-related increase in cancer incidence and mortality. However, as discussed below, other factors that are subject to changes with age may contribute as well.

In Chapter 5 cancer was considered as the opposite of aging, with the former a result of increased, uncontrolled cell division and the latter frequently associated with cell loss or loss of proliferative capacity. This antagonism was only fully realized when scrutinizing the results of the action of the two main cellular systems that defend our tissues against cancer: apoptosis and cellular senescence. While anti-cancer, these processes are generally pro-aging and we have already seen how syndromes of accelerated aging could be a result of a steep increase in such cellular responses to DNA damage (Chapter 5). Before turning to genome instability and how it causes cancer, we will first take a closer look at the systems that are supposed to defend us against cancer and the possibility that their actions contribute to some aspects of the aging phenome.

7.1.1 SENESCENCE AND APOPTOSIS

While not without logic, the perceived antagonism between cancer and aging is somewhat misleading. Indeed, it is entirely based on model systems in which cellular or organismal

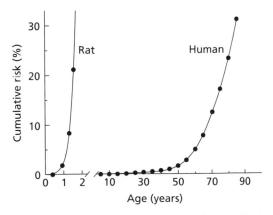

Fig. 7.3 The exponential rise in cancer incidence with age is related to biological, not chronological, age. Taken from reference 536.

aging is greatly exaggerated. For example, whereas cellular senescence *in vitro* (literally, aging under glass) appears as a dramatic example of what may happen with cells during the aging of an organism (Fig. 7.4), the situation of such cells is really incomparable with the situation in their natural environment *in vivo*. As discovered by Leonard Hayflick (San Francisco, CA, USA), human fibroblasts derived from embryonic or adult tissues, including skin, invariably undergo cessation of proliferation after a fixed number of population doublings in culture[418]. This has often been interpreted as mirroring the aging process *in vivo*. To some extent this reflects August Weismann's original idea that through natural selection the division potential of somatic cells becomes finite, thus limiting the regeneration of the soma and the lifespan of the organism[739]. It is now clear that replicative senescence is a major tool in the extensive arsenal of mammalian cells to control cancer; however, as such it likely contributes to some of the degenerative aspects of aging.

Replicative senescence in normal human fibroblasts in culture is due to progressive telomere shortening and can be prevented by ectopic expression of telomerase in these cells[650] (see also Chapters 4 and 6). Telomere-directed replicative senescence is a characteristic of primates in general and also for ungulates (hoofed mammals), but not for rodents[740]. In contrast to humans, rodents express telomerase in most of their adult tissues[741]. Rodent fibroblasts reach senescence very early after only a few population doublings and generally acquire the capacity for unlimited proliferation. As we have seen,

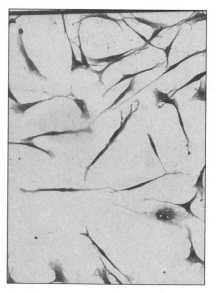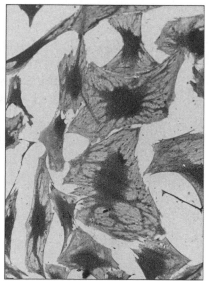

Fig. 7.4 The dramatic morphological changes that normal fibroblasts undergo as a consequence of the senescence response are difficult to miss. Left: early-passage rat skin fibroblasts are spindle-shaped and divide rapidly. Right: after several population doublings virtually all proliferating cells have stopped dividing and are now greatly enlarged, with a number of structural characteristics typical for such cells.

results obtained by Judith Campisi and co-workers have now made it clear that mouse cells become senescent primarily because of the high oxygen tension under normal culture conditions[353]. When mouse embryonic fibroblasts are cultured under conditions of low oxygen, senescence was not observed. It is unclear whether this apparent immortality is due to immortalization of cells which then gradually take over the population or to the lack of telomere-directed senescence which would make these cells immortal from the beginning. Nevertheless, since after repeated passaging at 3% oxygen mouse embryo fibroblasts no longer undergo senescence, even at 20% oxygen[353], the cells must have at least adapted to culture conditions.

When interpreting these results it should be kept in mind that virtually all studies with rodents have been done with laboratory mice (and sometimes rats). Wild mice have much shorter telomeres[742] and to my knowledge telomerase expression levels have never been measured in adult tissues of such animals. Therefore, the possibility cannot formally be excluded that cells from these animals undergo senescence like human cells. Nevertheless, it is conceivable that, as discussed extensively in Chapter 4, rodent cells miss some important defense system(s) which human cells do have that protects them against the effects of high oxygen. This does not mean that human cells do not suffer the adverse effects of oxidative stress. Indeed, increasing oxygen tension also accelerates replicative senescence in normal human fibroblasts, to some extent by accelerating telomere attrition[743]. Apart from their increased capacity to survive and replicate under adverse conditions, human cells have acquired the telomere-counting mechanism that, in the absence of the enzyme telomerase, so effectively induces senescence as a protection against cell immortalization and transformation.

Replicative senescence is primarily a DNA-damage response and can be induced in cultured cells (in the presence of intact telomeres) by various mutagenic agents, most notably agents that cause genome-rearrangement mutations, such as ionizing radiation or hydrogen peroxide. Senescence is generally distinguished from a phenomenon called crisis. If senescence fails, for example, due to the inactivation of the p53 DNA-damage-response pathway, cells will continue to divide until they reach a stage of mitotic catastrophe, which is characterized by a high frequency of chromosomal aberrations, apoptosis, and cellular senescence (see also Chapter 4).

It is not clear how relevant cellular senescence is for aging *in vivo*. While there is evidence for an accumulation of senescent cells in the skin of aging humans[501], a lack of reliable markers has essentially constrained an accurate quantification of their exact numbers, which even at very old age may be only modest. Nevertheless, especially in proliferative tissues a loss of functional cells may exert an adverse effect at old age. For example, it has been demonstrated that areas of the endothelial lining of the blood vessels exposed to high hemodynamic stress undergo more rapid telomere loss with age[744]. This suggests some importance of focal replicative senescence *in vivo* in areas of high cell turnover, with a possible role in cardiovascular disease (see also below).

Also using telomere dysfunction as a marker, John Sedivy (Providence, RI, USA) and co-workers demonstrated an exponential increase of senescent cells in the skin of aged baboons, up to about 15%[745]. Finally, important evidence that replicative senescence could be important *in vivo* was provided by work from Rita Effros (Los Angeles, CA, USA). Cytotoxic T cells, the immune cells responsible for control of viral infection (see below), undergo replicative senescence in culture after repeated rounds of proliferation similar to fibroblasts. Effros and co-workers demonstrated that in both elderly people and in people chronically infected with HIV there are high proportions of senescent cytotoxic T cells[746]. This indicates that the replicative capacity of such cells is exhausted, which may explain the compromised immune response in both elderly and AIDS patients.

Even when senescent cells would only comprise a very small fraction of the total, the possibility cannot be ruled out that they will exert an adverse effect. For example, accumulation of non-dividing senescent cells may compromise tissue renewal or repair. In addition, senescent cells encode secreted proteins that can alter the tissue microenvironment, stimulating the proliferation of cells that harbor pre-neoplastic mutations[747]. Nevertheless, under normal conditions, the impact of cellular senescence *in vivo* is likely to be small and it certainly never rises to the level observed in the cell-culture model.

The other major anti-cancer mechanism with potentially pro-aging side effects is apoptosis (also called programmed cell death), which is as clean in its action as senescence is messy[748]. Cells that undergo apoptosis disappear extremely rapidly. They are often phagocytosed when they are still alive. Hence, they do not induce an inflammatory response and there is therefore no side effect of their action other than loss of cells. (Another form of cell death, necrosis, is associated with pathological states and does cause inflammation by leaking the contents of the cells.) Apoptosis is different from senescence, which as we have seen not only represents a loss of functional cells, but also leaves them around to exert adverse effects. Apoptosis is an essential component of animal development[749]. It is important for the establishment and maintenance of tissue architecture based upon formation and then removal of specific structures. This enables greater flexibility, as primordial structures can be adapted to different functions at various stages of life or in different sexes.

Of more relevance to aging is the role of apoptosis as a defensive strategy to remove infected, mutated, or otherwise damaged cells. Indeed, it has been suggested that one mechanism through which caloric restriction would extend lifespan was by promoting the death of senescent cells[750]. Defects in apoptosis can certainly promote cancer and other disorders. On the other hand, increased spontaneous apoptosis is likely to be one of the mechanisms underlying premature aging in mice with defective DNA repair (Chapter 5). However, slightly increased rates of apoptosis (which are extremely low under normal conditions, even at very old age) have been found associated with long-lived, calorically restricted mice[751]. This would be in agreement with the hypothesis mentioned above that apoptosis benefits health span.

Whereas there can be no doubt that both senescence and apoptosis occur in tissues during normal aging, their low spontaneous rates make it unlikely that we age because we run out of cells. In one of my own research projects, extensive macroscopical and microscopical examination of various tissues of aging mice, including measurements of organ weights, do not indicate dramatic changes in mass or cellularity (D. Beems, unpublished work). Nevertheless, tissue atrophy is a component of normative aging, which may involve not only cell loss but also fibrosis, decreases in size, and functional degeneration. The possibility cannot be excluded that cellular senescence and cell death in key tissues, possibly of small subpopulations, are a major determinant of aging. Of note, the view that cell loss per se may not be a major contributor to human or animal aging does not invalidate the antagonism between cancer and aging, as this is observed in the DNA-repair-defective mouse models discussed in Chapter 5. There is no reason to doubt the validity of the premature aging symptoms observed in these animals, which almost certainly result from impairment of processes involved in cellular proliferation. However, under normal conditions these same symptoms may be caused not so much by a loss of cells but by a loss of functional cells, which may be due to genome instability. This would explain why aging is associated with both an increase in the incidence of cancer and progressive degeneration of organ and tissue function.

7.1.2 CANCER AS A RESULT OF GENOME ALTERATIONS

While numerous endogenous and environmental factors influence cancer risk, there is abundant evidence that cancer is primarily a disease of genomic instability. A lifetime of accumulating mutations is the most likely, albeit not the only, explanation for why cancer strikes older people so much more frequently than young people. In typical full-blown cancer, cells are rife with genetic abnormalities. This conclusion goes back to at least Theodor Boveri (1862–1915), who over 100 years ago proposed that cancer is based on aberrant chromosome combinations (now known as aneuploidy)[752]. As we have seen in Chapter 1, this was the time when the general theory of chromosomal inheritance came to fruition. Boveri's proposal that malignant tumors could be the result of an abnormal chromosome constitution has now generally been accepted, albeit with a much broader repertoire of genomic alterations than aneuploidy alone.

Early on, it was realized that cells achieve tumorigenic potential through specific phenotypic changes, allowing them to escape growth control and progress into a metastatic tumor. Initially, the amount of genome instability observed, as well as its variation from tumor to tumor, was difficult to reconcile with the relatively small number of specific changes that were thought to distinguish a normal from a tumor cell. Later, patterns were discerned, most notably the now generally adopted scheme first proposed by Bert Vogelstein (Baltimore, MD, USA) and co-workers, which consists of different phases from benign growth through a pre-cancerous stage and ultimately to metastasis[753]. More and more genes were subsequently discovered that if altered were responsible for helping the cell through these different phases. The wide variation from tumor to tumor in the type of genetic variation was soon ascribed to the multiple ways a particular pathway can be affected. As described in Chapter 1, genes almost never act alone, but in pathways and ultimately networks. Hence, a particular attribute, such as the capacity to invade a tissue or provide for blood vascularization, can be obtained through different genetic alterations affecting the same pathway.

Because cancer is clonal, with a chance process being responsible for transforming a single cell, a simple model of tumorigenesis through genomic alterations is as follows. Cells accumulate random alterations in their genome as a consequence of imperfections in genome maintenance and the continuous onslaught of DNA damage, as discussed in Chapters 4 and 6. Most alterations would not contribute at all in setting the cell on its pathway to cancer. They would have either no effect at all, adversely affect one or several cellular functions, or may occasionally be lethal. By chance, random mutations would release some of the bonds that restrain the cell from uncontrolled growth. This would result in clonal outgrowth. A second set of genomic alterations, again occurring by chance in one or more of the cells in the clonal outgrowth, would provide an additional growth advantage for that cell and its progeny. This scenario would be repeated to guide tumor development from clonal lineage to clonal lineage, eventually resulting in a full-blown malignant tumor.

What this model does not address is the rate at which mutations accumulate or the number and type of mutations necessary to provide for all the necessary attributes. While in cell-culture models manipulation of only a few genes is sufficient for tumorigenic conversion of normal epithelial and fibroblast cells[754], many more specific changes are necessary to provide a cell with all the characteristics it needs to survive the many cellular safeguards that prevent uncontrolled cellular growth. Therefore, cancer needs many more mutations than the normal background mutation frequency can provide. This was first realized by Larry Loeb, who subsequently proposed his mutator phenotype hypothesis[755,756]. According to this hypothesis, mutations that increase mutation rates are essential for tumors to evolve. Target genes for such mutations are genes that control the fidelity of DNA replication, the efficacy of DNA repair, chromosome segregation, damage surveillance (e.g. checkpoint control), and cellular responses (e.g. apoptosis). Mutations in genes involved in genome maintenance can greatly increase the rate of genome instability, including point mutations, microsatellite instability, and LOH. Whereas this increases the chance of creating new mechanisms to escape normal cellular restraints, it would also greatly increase mortality. Hence, once a cell has acquired its necessary attributes, mutants will be selected in which survival is maximized by genome restabilization.

Cancer is a problem that is typically associated with renewable tissues. In nematodes or fruit flies virtually all cells of the body are postmitotic and cancer is not a part of the pattern of aging-related pathology in these animals. It is possible that the evolution of renewable somatic tissues provided an organism with increased longevity, since worn-out cells can be replaced. However, such tissues are also highly vulnerable to unwanted proliferation, which may lead to cancer according to the scenario described above[421,757]. It is now generally assumed that only certain cells from a tumor—cancer stem cells—have the capacity to initiate new, full-fledged tumors. Since tumor-forming stem cells have not been identified for all tumors it is difficult to say whether this is true for cancer in general. It is equally difficult to judge whether cancer stem cells arise directly from mutations in the normal stem cells that sustain various tissues, or only when these cells start dividing to replenish a tissue.

It is often argued that cell proliferation is critically important to generate the mutations that initiate the process of carcinogenesis. As we have seen in Chapter 4, errors during DNA replication are a major source of the somatic mutations that drive malignant tumorigenesis. However, mutations can occur equally well in non-dividing cells, although not all types of mutations. Figure 7.5 shows the results of an experiment in my laboratory by a student, Rita Busuttil, who treated embryonic fibroblasts from the *lacZ* plasmid reporter mice described in the previous chapter with two different DNA-damaging agents, ultraviolet light or hydrogen peroxide. To test for the influence of cell proliferation on the efficiency of mutation induction in these cells, she allowed one population to divide in an unlimited manner, providing the cells with medium containing serum. The other population was starved of serum by giving them only 0.5% serum rather than the usual 10%. This results in so-called quiescent cells. (In the complete absence of serum the cells would die.) The results show that while UV-induced mutagenesis is highly dependent on cell proliferation, hydrogen peroxide treatment resulted in slightly more mutations in the quiescent

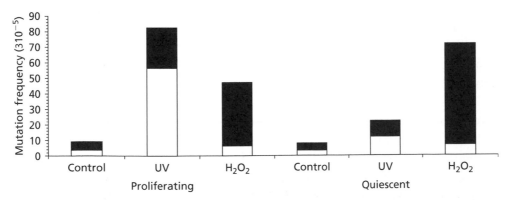

Fig. 7.5 Mutations in the *lacZ* transgene in proliferating and quiescent mouse embryonic fibroblasts after UV irradiation or treatment with hydrogen peroxide. UV irradiation mainly induces point mutations (white bars), most of which are replication-dependent. Hydrogen peroxide on the other hand mainly induces large genome rearrangements (black bars), which are independent of replication. R. Busuttil, submitted for publication.

cells than in the actively proliferating ones. Interestingly, while the replication-dependent, UV-induced mutations were mainly point mutations, hydrogen peroxide treatment resulted almost exclusively in the genome-rearrangement mutations discussed in the previous chapter. Hence, these results suggest that point mutations are highly dependent on replication errors, while genome rearrangements might be a result of mistakes during annealing of double-strand breaks known to be induced by hydrogen peroxide. This latter type of event apparently does not require DNA replication for its fixation.

There is also evidence from other research groups that mutation accumulation in cells during aging is less dependent on proliferation rate—the number of generations—than on chronological time[728]. This means that tumor induction does not necessarily require cell division and that non-dividing cells can accumulate a substantial number of mutations during aging. Nevertheless, it is obvious that cells already destined to divide regularly, such as epithelial cells, are more likely to develop into tumors than postmitotic cells, such as neurons and cardiomyocytes. This is, of course, exactly what is generally observed.

Based on the above, it is logical to assume that a gradual accumulation of mutations with age drives the increased incidence of cancer. However, other factors are likely to play a role as well[437]. For example, it is highly conceivable that the aging process is accompanied by gradual alterations creating an intra- and extracellular environment that is increasingly permissive of the formation and progression of cancer[758]. This could involve gradual declines in immune response, DNA repair, and apoptosis. Interestingly, another possibility, proposed by Judith Campisi and briefly mentioned above, brings us back to replicative senescence. Campisi was the first to report the accumulation of senescent cells during human aging, mainly in the skin. The changes in these cells include the secretion of biologically active molecules, such as matrix metalloproteinases, inflammatory cytokines, and growth factors, which can promote the proliferation and neoplastic transformation of preneoplastic epithelial cells in stromal/epithelial co-cultures and in mice[747]. Hence, cellular senescence *in vivo* can actually promote cancer, a somewhat surprising conclusion with regard to a mechanism that is essentially anti-cancer. The logic behind this potential pro-cancer effect of senescence *in vivo* is that the number of senescent cells only begins to have such an effect long after the period of first reproduction. In this concept, the adverse effect of cellular senescence is a pleiotropic effect of its beneficial effect of keeping cancer at bay for a long enough time to reproduce[421].

Whatever the importance of other host factors in explaining the increased cancer incidence with age, there is no dispute about the role of genome alterations in tumor initiation and progression. Genome alterations may play a role in cellular outgrowths in general, including hyperplastic nodules and atherosclerotic plaques (see below). However, other aging-related phenotypes, not based on clonal expansion, are generally not ascribed to mutations. In the past, random mutations were thought to be so rare that they were considered unlikely candidates to cause phenotypes such as aging-related organ dysfunction. This situation has changed to some extent with the realization that genomes are highly dynamic, even in somatic cells. Indeed, as outlined in the previous chapter, a wide variety of genome alterations have been demonstrated to occur in mammalian organs and tissues, with the resulting mutations in some cases shown to accumulate with age. Indeed, as we have seen, even in the absence of cell proliferation, aging will continue to promote genotypic divergence of tissues and organs by making errors during the processing of spontaneous DNA damage. Below, I will briefly review aging phenotypes other than cancer and discuss the possibility that they are also caused by random genome alterations.

7.2 Genome instability and tissue dysfunction

It is now generally accepted that genomes, including somatic genomes, are more plastic than initially thought, essentially turning tissues of adult organisms into genotypic mosaics by randomly altering the genome of their cells. However, there are no accurate estimates of average genomic mutation loads of different cell types during aging, not even for the small mitochondrial genome (see Chapter 6). Even if available, such data alone would not be sufficient to establish causal relationships. Before discussing how to test for such causality, I will first review some common aging-related, non-cancer phenotypes, mainly in humans, to see if they possibly could be due to cellular degeneration as a consequence of random genome alterations. I will also discuss possible alternative causal factors, such as protein alterations.

7.2.1 HUMAN AGE-RELATED DEGENERATION

As has been stressed many times, aging has an extremely complex phenotype of which cancer is a major part and in a category of its own due to its clonal nature and the known general mechanism of its initiation and progression. Aging phenotypes other than cancer are varied and usually become noticeable between the ages of 40 and 50.

7.2.1.1 Hearing

As people age, they may hear less well, and balance may become slightly harder to maintain. These changes occur because some structures in the cochlea (the part of the inner ear that converts vibrations into nerve impulses sent to the brain) or auditory nerve (connecting the inner ear to the brain) deteriorate slightly. Presbycusis, as this is called, appears to be based on the age-related degeneration of the stria vascularis, which maintains the ionic gradients necessary for sound signal transduction and the complex interaction between the inner and outer hair cells, possibly in combination with hair-cell loss or ganglion cell loss[759]. Both the stria vascularis and the hair cells, which are all postmitotic, are highly metabolically active and would be compromised by a deficiency of intracellular ATP due to mitochondrial dysfunction[760], which might ultimately lead to cell degeneration and death. Interestingly, in mice lacking the antioxidant enzyme SOD1 (also known as CuZn-SOD), age-related cochlear hair cell loss was enhanced[761]. Hence, in this case alterations in the mitochondrial and/or nuclear genome could evidently underlie the degenerative effects.

7.2.1.2 Vision

There are a number of pathological causes of vision loss among the elderly and the most common are age-related macular degeneration, glaucoma, cataract, and diabetic retinopathy. An important pathological change in the lens (and often used as a biomarker for aging) is cataract, likely to be caused by damage, cross-linking, and precipitation of crystallins, contributing to a loss of lens clarity. Age-related cataracts are the most common cause of blindness worldwide.

Around the age 40, most people notice that seeing objects closer than 1 m becomes difficult. This change in vision, called presbyopia[762], occurs because the lens in the eye stiffens. Normally, the lens changes its shape to help the eye focus. When the lens stiffens, the eye cannot easily focus on objects that are close. The lens is enclosed by a thick capsular basement membrane under which lies a layer of cuboidal epithelial cells. These cells differentiate into new lens fibers, progressively overlaying the existing lens fibers. While the younger, outer lens fibers are metabolically active, the older fibers degrade their nuclei and organelles, enabling the lens to maintain clarity. The lens fibers are packed with transparent proteins called crystallins. This structure is unique in that there appears to be no protein turnover in the center throughout one's lifespan[763]. Presbyopia, therefore could be due entirely to age-related, post-translational protein modifications, such as aspartic acid racemization (loss of optical purity), oxidation, glycation, and deamidation. Since this would involve no cells, it essentially rules out random genome alteration as a causal factor. However, changes in the ciliary body, a ring of smooth-muscle fibers arranged concentrically around the opening in which the lens is suspended, have also been implicated in the phenotype and such changes could, in theory, be due to genome alterations.

7.2.1.3 Taste and smell

Generally, when people are in their 50s, the abilities to taste and smell start to gradually diminish. Originally, taste decrements in the elderly were ascribed to a loss in the number of papillae and taste buds. However, as evident from later studies, there may not be such age-related losses and little is known about degenerative changes in gustatory neural pathways[764]. Age-related decrements in odor perception could be due to a variety of structural and physiological changes throughout the olfactory system, from olfactory epithelium to the olfactory bulb and the olfactory cortex. There is no evidence for or against a causal relationship with genome instability.

7.2.1.4 Skin

As people age, the skin tends to become thinner, less elastic, drier, and finely wrinkled[765]. The layer of fat under the skin thins and is replaced by more fibrous tissue. The number of nerve endings in the skin decreases, as do the number of sweat glands and blood vessels. Blood flow in the deep layers of the skin decreases. While all aging phenotypes strongly depend on environmental, lifestyle, and genetic factors, skin aging is of course heavily dominated by the degree of sun exposure and skin pigment. Chronological skin aging, which depends on the passage of time and not on UV irradiation, is often distinguished from photoaging, with which it shares some important molecular features. From this point of view, photoaging is basically the superposition of UV irradiation from the sun on intrinsic aging. While photoaging could work directly through the induction of damage to DNA and other macromolecular structures, UV irradiation also activates cell-surface receptors, triggering downstream signal-transduction pathways, which leads to photochemical generation of ROS. While the mechanism through which this happens is unclear, it is conceivable that chronological skin aging also works through ROS, perhaps in a similar way to other tissues.

It is tempting to speculate that accumulation of genome alterations in skin cells, resulting from DNA damage induced by UV irradiation or ROS, causes the loss of fat cells and the reduced activity of sebaceous and sweat glands that are such an important component of skin aging. However, protein cross-linking, for example due to glycation and the formation of advanced glycation endproducts (AGEs; see Chapter 6) is often implicated as a major cause of the loss of flexibility of collagen fibers just below the dermis and the stiffening of elastin, the protein that gives skin its flexibility. Protein changes should therefore be considered as a potential primary cause of skin aging.

Naturally, mutation accumulation has been implicated in the striking aging-related vulnerability to skin cancer in humans, with suggestions of a possible role of an age-related decline in DNA repair[766]. Also, telomere erosion has been implicated in skin aging, especially with regard to its degenerative, non-cancer features[767,768]. Of note, in the keratinocytes of the continuously renewed epidermis telomerase is expressed and therefore able to counteract telomere erosion.

7.2.1.5 Musculoskeletal system

Apart from thinner, wrinkled, leathery skin, the most visible evidence of aging is the strikingly altered musculoskeletal system[769,770]: the bones, joints, and skeletal muscles (striated muscles). In the course of the aging process, bones tend to become less dense and therefore weaker and more likely to break. Bones become less dense partly because the amount of calcium they contain decreases. Part of the reason for this is that less calcium is absorbed in the digestive tract and levels of vitamin D (which helps the body use calcium) decrease slightly. Osteoporosis is a major health problem, especially for women after the menopause. Older people often appear to have lost height due to kyphosis, which has already been described as a frequent aging phenotype appearing prematurely in mice with defects in genome maintenance (Chapter 5). Also, the amount of bone marrow decreases, with fewer blood cells produced.

Bone mass is maintained through a delicate balance between bone formation by osteoblasts and bone resorption by osteoclasts[771]. This process of bone remodeling is possibly a consequence of the continuous stress exerted on bone and may lead to the replacement of about 10% of all bone each year in humans. Hence, dissolving older bone mass by osteoclasts and new bone formation by osteoblasts is the mechanism for the repair and continuing strength of bone. It is the excess activity of osteoclasts (common after menopause in women) that produces osteoporosis. Bone homeostasis depends upon the intimate coupling of bone formation and bone resorption, which is subject to a complex circuit of cytokines, growth factors, and hormones[772].

The major influence on age-associated bone deterioration is generally considered to be a deficiency of progesterone. However, the disease can also be caused by mineral and vitamin deficiencies, drugs, poor eating habits, lack of exercise, too much cortisol, and too little testosterone. This multitude of possible causes does not rule out the possibility that ultimately, when all other factors are optimal, osteoporosis may still occur, but then as a consequence of genome instability in osteoblasts and osteoclasts. Indeed, it is possible that genome instability in the stem-cell pools that give rise to these cells impairs new bone formation and we have already seen that the increased DNA-damage responses in some mice with accelerated aging adversely affect the bone cells, leading to kyphosis and osteoporosis in these animals (Chapter 5).

Specialized populations of bone cells form, maintain, and remodel the cartilage that lines the joints. The cartilage tends to become thinner with age and, consequently, the surfaces of a joint may not slide over each other as well as they used to. This makes the joint more susceptible to injury with repeated injury often leading to osteoarthritis, which is a common disorder in later life. Ligaments, which bind joints together, tend to become less elastic as people age, making joints feel tight or stiff. Like the aforementioned changes in collagen fibers and elastin in the skin, and the changes observed in the eye lens, these changes in ligaments are often ascribed to cross-linking of collagen molecules as a result of oxidation or glycation. Currently there is no evidence that the activity of the cells responsible for laying down and maintaining the ligaments undergo functional decline as a consequence of genome instability.

As people age, the amount of muscle tissue (muscle mass) and muscle strength tend to decrease. This process is called sarcopenia. Loss of muscle mass begins around age 30 and continues throughout life. Muscle mass decreases because the number and size of of muscle fibers decreases. This change may occur because the levels of growth hormone and testosterone, which stimulate muscle development, decrease with aging. Also, muscles cannot contract as quickly in old age.

From the previous chapter we also know that mutation accumulation, especially in the mitochondrial genome, in muscle fibers is often considered as a most likely explanation for age-related muscle wasting. However, systemic factors (which, in turn, can also have a basis in genome alterations) play a major role in modulating adverse effects of age on muscle regeneration. We have already seen that the failure of muscle satellite cells at old age to regenerate lost muscle could be corrected by young serum[733]. Likewise, the age-related decline in circulating IGF-1, which is well documented[773], could underlie loss of muscle mass and strength. Interestingly, this latter systemic defect could be reversed locally through overexpression of IGF-1 in muscle cells[774].

It is difficult to investigate whether genome instability could ultimately be the primary cause of systemic deficits. Indeed, IGF-1 decline may be due to a reduction in GH, secreted by the pituitary gland into the bloodstream, thereby reducing the production of IGF-1 by the liver. GH and IGF-1 levels (as well as other related hormones, such as somatostatin) in the body are tightly regulated by each other and by sleep, exercise, stress, food intake, and blood-sugar levels. To make it even more complex, it is the hypothalamus that produces hormones that regulate the pituitary gland. This serves to illustrate the difficulties in deciding where the ultimate cause of muscle dysfunction may reside.

Muscles, of course, have to function almost everywhere. We have already seen that muscle weakening may contribute to the aging of the lens, and elsewhere, for example the muscles used in breathing, such as the diaphragm, tend to weaken as people age. Apart from the skeletal, striated muscle there is smooth (or visceral) muscle, forming the muscle layers in the walls of the digestive tract, bladder, various ducts, arteries, and veins, and cardiac muscle, a cross between the smooth and striated muscles, comprising the heart tissue.

7.2.1.6 Brain

In contrast to what was originally thought, as people age, the number of nerve cells in the brain decreases only slightly; this loss is focal, not global[775]. Several factors help compensate for this decline. As cells are lost, new connections are made between the remaining nerve cells. New nerve cells may form in some areas of the brain, even during old age. In addition, the brain is redundant in the sense that it has more cells than it needs to perform most activities. Some mental functions may be subtly reduced. They include vocabulary, short-term memory, the ability to learn new material, and the ability to recall words. After about age 60, the number of cells in the spinal cord begins to decrease. As a result, older people may notice a decrease in sensation. At old age, nerves may conduct signals more slowly. Also, the nervous system's response to injury is reduced. Nerves may repair themselves more slowly and incompletely in older people than in younger people.

In humans the most dramatic changes in the brain are associated with disease, most notably Parkinson's disease[776] and Alzheimer's disease[777]. Parkinson's disease is a movement disorder with a resting tremor and problems in initiating or stopping movement. Although there are many other neuronal groups affected in different brain regions in Parkinson's disease, the loss of dopaminergic neurons in the substantia nigra is the most consistent feature of the disease. In addition, the disease is characterized by the presence of so-called Lewy bodies and Lewy neurites in surviving nigral neurons, intracellular aggregates of lipids and proteins, including ubiquitin and α-synuclein. As mentioned earlier in this chapter, it is not always easy to distinguish pathology associated with normal aging from the situation in a patient. This is especially true for Alzheimer's disease. The hallmark phenotypes of this disease are the so-called amyloid neuritic (or senile) plaques, found outside the cell and consisting of amyloid protein knotted up with tendrils of malformed brain cells, and the neurofibrillary tangles, made of a protein called tau clustered inside neurons. Although generally much less widespread, both these lesions also occur in the brain of normally aged individuals[737].

Parkinson's disease is usually ascribed to loss of function or dysfunction of certain candidate genes, including α-synuclein and parkin, leading to a decrease in dopamine levels[776]. Likewise, faulty processing of the amyloid precursor protein (APP) that increases levels of the toxic amyloid β peptide is usually considered to underlie the pathogenesis of Alzheimer's disease[778]. However, there are several possible alternative causal factors to explain the age-related nature of neurodegenerative diseases such as Parkinson's and Alzheimer's diseases. These explanations are not mutually exclusive and also do not rule out the more proximal candidate causes underlying these diseases. Oxidative stress has been implicated in both diseases and in neurodegeneration and aging in general[779]. Oxidative stress could, for example, underlie protein aggregation, a hallmark of both diseases. Protein aggregates are complexes of misfolded or unfolded proteins that are essentially insoluble and metabolically stable under normal physiological conditions[780]. Their frequency of occurrence in tissues and organs (not only in brain) is increased with age, with the two best known forms being amyloid and lipofuscin. Neither the mechanisms of their formation, nor their possible role as a causal factor in aging and aging-related disease, even for Alzheimer's disease or Parkinson's disease, are known with certainty. Protein damage is a likely causal factor, but stochastic alterations in the stoichiometry of different proteins and their polypeptide chains may play a role as well. The latter could be caused by increased genome instability, which is likely to lead to aberrant patterns of transcription (see below).

DNA damage and genome instability have long since been implicated in Alzheimer's and Parkinson's diseases. Evidence has been obtained, including data from my own laboratory, that defects in DNA repair underlie familial forms of both Alzheimer's disease and Parkinson's disease[781,782] and increased DNA damage in the form of DNA breaks and alkali-labile sites was found in the cortex of patients with Alzheimer's disease[596]. More recently, high levels of mtDNA deletions were found in the substantia nigra neurons in both aged controls and individuals with Parkinson's disease[668,783]. Based on the surprisingly high frequency of chromosomal aneuploidy in the brains of both humans and mice and the dramatic increase in LOH events in neuronal stem cells of mice during aging (Chapter 6), it is conceivable that increased instability of the nuclear genome will also eventually be found in selective cell populations of the elderly or patients with neurodegenerative disease. However, such analyses are more difficult to carry out than searching for mtDNA deletions, and data are still lacking.

7.2.1.7 Heart and blood vessels

There are a number of different types of change in the heart and blood vessels that occur with age. There is some cell loss and an increase in size of cardiac myocytes. Some cells are replaced with fibrous tissue. The walls of the heart become stiffer, and the heart fills with blood more slowly. There are deposits of lipofuscin and amyloid, but amyloid deposition occurs only in half of people over 70 years. Overall, the cardiac physiology of physically fit elderly people is very similar to that of younger individuals, but they often require a compensatory mechanism called the Starling mechanism[784].

We are of course all familiar with the effects of aging on the blood vessels. The walls of the large arteries thicken and become stiffer, due to alterations in elastin and calcium deposition. This process of arteriosclerosis[785] increases blood pressure and could be a reason for the observed thickening of the heart muscle (hypertrophy). It can involve the arteries of the cardiovascular system, the brain, the kidneys, and the upper and lower extremities. Possible mechanisms underlying arteriosclerosis include the now familiar cross-linking of collagen and other protein changes. For example, AGEs have been implicated in increased vascular and myocardial stiffness, as well as in atherosclerosis. This is suggested by the observation that patients with diabetes, who suffer from an accelerated rate of advanced glycation of proteins, manifest arterial stiffening at a younger age.

Arteriosclerosis differs from atherosclerosis, which involves the build-up of fatty deposits in the innermost lining of large and medium-sized arteries. Such lesions, termed plaques, often lead to coronary heart disease, strokes, and other disorders because of the occurrence of blood clots which form in the narrowed arteries or from ruptured plaques. However, they also affect the medium-sized and large arteries of the brain, heart, kidneys, and legs. Atherosclerosis develops over a period of decades and has a complex pathogenesis. It probably begins with damage to the arterial wall, possibly due to transport of oxidized low-density lipoprotein across the endothelium. This results in inflammation and the formation of regions of foamy lipid-laden macrophages and smooth muscle cells, termed fatty streaks, just beneath the intimal surface of large arteries. Continuous migration of smooth muscle cells into the intima converts fatty streaks into the fibrous plaques that protrude into the arterial lumen. Later, calcification can occur and fibrosis continues yielding a fibrous cap that surrounds a lipid-rich core. This formation may also contain dead or dying smooth muscle cells. In myocardial infarction such fibrous plaques rupture and the subsequent formation and release of thrombi may ultimately occlude vessels.

Ultimately, therefore, atherosclerosis is defective wound repair and the result of a failure in endogenous inhibitory systems that normally limit such repair. One such inhibitory pathway is the transforming growth factor-β signaling pathway, which inhibits clonal expansion of the vascular smooth muscle cells in atherosclerotic plaques. A somatic mutation within a microsatellite repeat tract in the coding regions of the transforming growth factor-β receptor gene in human atherosclerotic lesions has been reported and proposed to account for the frequent clonal expansion of the vascular smooth-muscle cells[786]. However, subsequent studies did not find such mutations or only at low frequency[787]. Nevertheless, the accumulation of both nuclear and mitochondrial DNA mutations, including genomic rearrangements, in vascular endothelium and smooth-muscle cells may be important for the development of atherosclerosis[788].

The involvement of mutations was originally inferred from the apparent monoclonality of the smooth-muscle cells, the focal proliferation of which is characteristic of plaques and likely to be an early step in the pathogenesis of atherosclerosis[789]. There are a number of reports describing chromosomal alterations in human atherosclerotic plaques as compared with adjacent normal tissue. Most of these studies were based on the use of microsatellite markers to study LOH (similar to the LOH found to increase with age in mouse neuronal stem or progenitor cells mentioned in the previous chapter). For example, in a recent study DNA samples from 78 aortic plaques were compared with corresponding DNA from blood taken from the same individuals[790]. Using 63 microsatellite markers, LOH was found in 37 out of the 78 atherosclerotic plaques. Interestingly, 33 histopathologically normal aortic tissue samples were also compared with their corresponding blood samples and LOH was detected in 6 out of 33 specimens. While this indicates a significant association of allelic imbalance with aortic atherosclerosis, it also illustrates how widespread genomic instability is, even in seemingly normal tissue. As in most other complex, aging-related disease phenotypes, many different cell types interact to give rise to atherosclerosis. Because not all these cell types have been analyzed for changes in DNA or chromatin it is as yet too early to conclude that genome instability is likely to play a role in the etiology of this disease.

Finally, relatively little is known about microvascular changes with age. However, there is increasing evidence for its importance and microvascular alterations with aging can greatly contribute to the development of stroke, skin and kidney aging, and vascular dementia. Changes in vascular endothelial cells may contribute to microvascular degeneration and the possible role of telomere-directed cellular senescence in aging-related degeneration of such cells has already been mentioned.

7.2.1.8 Alimentary system

Changes in the digestive system during aging are relatively minor. The muscles of the esophagus contract less forcefully, but movement of food through the esophagus is not affected. Food is emptied from the stomach more slowly, and the stomach cannot hold as much food because it is less elastic. But in most people, these changes are too slight to be noticed. The digestive tract may produce less lactase, an enzyme the body needs to digest milk. In the large intestine, materials move through a little more slowly. In some people, this slowing may contribute to constipation. A significant fraction of elderly suffers from atrophic gastritis, a process of chronic inflammation of the stomach mucosa (probably autoimmune in origin) leading to loss of gastric glandular cells and their eventual replacement by intestinal and fibrous tissues. As a result, the stomach's secretion of essential substances such as hydrochloric acid and pepsin is impaired, leading to digestive problems, vitamin B_{12} deficiency, and pernicious anemia. The age-related increase in both mtDNA and nuclear DNA mutations in the intestine has been discussed extensively in the previous chapter.

The organ that has probably been studied most extensively in relation to aging is the liver. Multiple age-related changes in structure and function have been described for this organ in both humans and rodents, but many of the observations are contradictory[791]. There is some reduction in liver mass and total blood flow. There is some evidence that the liver tends to become smaller because the number of cells decreases. However, declines in cell number have been reported for all organs at some point in the past, with later contradictory reports after some more careful studies, brain being the best example. Liver is the best example of an organ undergoing polyploidization. However, this happens early in life and is not associated with functional decline or disease. As we have seen, liver has the capacity to regenerate after hepatectomy; for example, after surgical removal of a tumor. The rate of liver regeneration after hepatectomy declines in old animals, but the regenerative capacity remains unchanged, given sufficient time.

The best described functional change that occurs with age in the liver is the reduced clearance of certain drugs and a marked increase in the frequency of adverse drug reactions. This is illustrated by the results of an experiment conducted in my laboratory in which young and old mice of two different strains were compared for their capacities to repair DNA adducts induced by BaP. As described in the previous chapter, BaP needs to undergo metabolic activation to yield more reactive forms that can then interact with nucleophilic centers in DNA. Using the ^{32}P-postlabeling assay we determined the levels of the main adduct in different organs of these mice. The results[394] indicated that the maximum formation of this adduct was more than 3-fold higher in young as compared to old mice. This is likely to be due to decreased activation of BaP by enzymes of the cytochrome P450 family in the old animals. Indeed, others have demonstrated that the total microsomal cytochrome P450 content is significantly reduced in rodent liver at old age[792]. The liver is therefore less able to help rid the body of drugs and other substances and the effects of drugs, intended or unintended, last longer. It is unclear whether the age-related accumulation of mutations or the observed epigenomic changes are causally related to liver functional decline.

7.2.1.9 Kidney and urinary tract

As people age, the kidneys tend to become smaller. Similar to liver, this may be because the number of cells decreases and there is less blood flow. Beginning at about age 30, the kidneys begin to filter blood less well[793]. This is associated with glomerulosclerosis, or scarring of the tiny blood vessels called glomeruli, interstitial fibrosis, tubular atrophy, vascular wall thickening, and declining numbers of glomeruli. This results in declining glomerular filtration rates and creatinine clearance. The kidneys may also excrete too much water, making dehydration more likely. Moderate to severe renal failure occurs in one-fifth of the elderly population. However, to a large extent renal failure is facilitated by complications resulting from other diseases, such as diabetes and hypertension. As we have seen, such interaction among different aging-related pathologies is the norm rather than an exception.

As discussed in Chapter 6, mutations at the *HPRT* locus were found to accumulate with age in renal tubular epithelial cells to a very high frequency[626]. These cells are metabolically active, consume large amounts of oxygen and perform a majority of the secretory and resorptive work needed to process glomerular filtrate. It is possible that this points towards a role of somatic mutation accumulation in the loss of renal function and elevated risk of malignancy during aging.

The urinary tract changes with age in several ways that may make controlling urination more difficult. Indeed, among people over 60 years old, 15–30% suffer from leakage of urine. At old age, the bladder can hold less urine. As women age, the urethra (which carries urine out of the body) shortens and its lining becomes thinner. The muscle that controls the passage of urine through the urethra (the urinary sphincter) is less able to close tightly and prevent leakage. These changes may result from the decrease in the estrogen level that occurs with menopause and from muscle damage associated with childbirth. As men age, the prostate gland tends to enlarge. In many men, it enlarges enough to partly block the passage of urine. This often leads to hydronephrosis, accumulation of urine in the kidney with the ultimate loss of renal function.

Hence, whereas in men lower-urinary-tract symptoms are most often attributed to prostatic obstruction, in women hormonal abnormalities are often the cause. However, there could also be a similar pathophysiology based on defects in the detrusor muscle, the smooth muscle in the wall of the bladder that contracts the bladder and expels the urine. Coordinated contraction and relaxation of the detrusor myocytes is required for normal organ function and alterations in myocyte structure and function have been implicated in the etiology of various diseases of the lower urinary tract. However, other factors that can play a role in the voiding dysfunctions observed in the elderly include fibrosis of the detrusor muscle and consequent impairment of contractility, cross-linking of collagen and elastin, and loss of acetylcholinesterase-positive nerves. Since urinary function is controlled by the coordinated activity of the urinary tract and the brain, non-cell-autonomous age changes as a potential cause of the dysfunctions cannot be ruled out.

7.2.1.10 The reproductive system

Adverse effects on the reproductive system are probably among the earliest components of the aging phenome and mainly driven by changes in the neuroendocrine axis. The effects of aging on the reproductive system are more obvious in women than in men. In women, most of these effects are related to menopause, when the levels of circulating female hormones (particularly estrogen) decrease, menstrual periods end permanently, and pregnancy is no longer possible. The decrease in female hormone levels causes the ovaries and uterus to shrink. There is often hyperplasia of thecal cells, specialized stromal cells, in atrophic ovaries. The tissues of the vagina become thinner, drier, and less elastic. The breasts become less firm and more fibrous, and they tend to sag. Most men remain fertile until death, even though testosterone levels decrease, resulting in fewer sperm and a decreased sex drive. Most men can continue to have erections and reach orgasm throughout life. However, erections may not last as long or may be slightly less rigid. Erectile dysfunction (impotence) becomes more common as men age.

7.2.1.11 Neuroendocrine system

Already in Chapter 2 I discussed extensively how the neuroendocrine system, from nematodes to mice, exerts its effects on lifespan. Indeed, one can almost say that the aging process is driven, at least in part, by neuroendocrine activity rather than by its decline with age. Nevertheless, we have already noticed that hormonal decline is ubiquitous during aging and can be a source of many degenerative cell and tissue changes. As people age, the levels and activity of some hormones (in addition to sex hormones) decrease. For example, the levels of circulating GH and IGF-I decrease, which may contribute to the reduction of muscle mass with age. The level of aldosterone, a hormone produced by the adrenal glands, also decreases. This decrease may contribute to the tendency of older people to become dehydrated more easily.

7.2.1.12 Immune system

The immune system is probably the most extensively researched host-defense mechanism in aging[794]. In his now classic book *The Immunological Theory of Aging*[795], Roy Walford (1924–2004), a pioneer in aging research, proposed that the normal process of aging is caused by faulty immune processes. In general, older adults experience more infections and a greater severity of infections than younger adults. Whereas other responses to infection, such as the capacity to mount a fever, are also reduced in the elderly, it is reasonable to suspect first and foremost a decline in immune response. This is especially true because vaccines do not work as well in older adults as they do in younger individuals[796]. The immune system is a complex combination of organs, cells, and signaling molecules that together make up our defense against invasion by bacteria, viruses, fungi, parasites, toxins, and foreign bodies in general, such as tumor cells. It carries out this protective action by distinguishing between self and non-self. However, increased numbers of autoantibodies at old age suggest that this fundamental component of the immune system is also compromised by the aging process.

There are two complementary forms of immunity: natural or innate immunity and adaptive or acquired immunity. Natural immunity includes dendritic cells, which take up, process, and present antigens to the $CD4^+$ T-helper lymphocytes in the form of small peptides complexed with their own class II major histocompatibility complex (MHC) proteins. Whereas older people may have fewer dendritic cells, they generally retain their capacity for presenting antigen. Another cell type active in natural immunity is the macrophage. This cell type is specialized in phagocytosis and intracellular killing of microorganisms, is cytotoxic to tumors, and is also an efficient antigen-presenting cell. On top of that, these cells produce signaling molecules—cytokines—which initiate the acute-phase response, a form of defense during acute illnesses involving the increased production of so-called acute-phase proteins. In elderly individuals, the rate of antigen clearance by macrophages is decreased and the toxicity of these cells against tumor cells is low.

A third cell type of the natural immune response is the natural killer (NK) cell. NK cells can kill target cells spontaneously and they are involved in host resistance to various tumors and infectious diseases. The number of NK cells appears to increase with age but their activity decreases, making them less capable of destroying tumor cells[797]. The main soluble effector of natural immunity is the complex enzyme system in blood termed the complement system. The system consists of interacting plasma proteins, which defend against microorganisms through cytolysis and activation of inflammation. Complement factors can also attach antigens to phagocytes, increasing the efficiency of the engulfment of microbes. Like many components of the immune system, the complement system can be responsible for unwanted physiological effects that increase during aging, such as ongoing or chronic inflammation. This causes damage to the injured tissues and has long been linked to rheumatoid arthritis and osteoarthritis, but more recently also to other aging-related diseases, including neurodegenerative diseases and atherosclerosis. In this respect, it has been suggested that the reduction in lifetime exposure to infectious agents has contributed to the decline in old-age mortality in the last two centuries[798].

Adaptive immunity resides in the T and B lymphocytes, which are capable of specifically attacking everything that is foreign, based on their recognition of all possible specific antigens presented to them. As first proposed by Frank Macfarlane Burnet[799] (1899–1985) and Niels Jerne (1911–1994)[800], they can do that through a mechanism called clonal selection, which is eerily reminiscent of Darwinian selection. To summarize the situation, B and T cells circulate around the body in great numbers of great diversity, distinguished by various combinations of the DNA segments that make up their immunoglobulin or T-cell-receptor genes, respectively. As already mentioned in Chapter 4, this variation is generated through the process of V(D)J recombination, in which the DSB repair pathway of NHEJ plays such an important role. The variation is such that all possible antigens can be recognized. Indeed, after the initial clonal selection, subsequent somatic hypermutation serves as a fine-tuning mechanism to further optimize antibody–antigen binding[801] (see also Chapter 4). In species such as mice and humans, after selection B cells migrate to lymphoid follicles. There, together with antigen-specific T cells and follicular dendritic cells, they form germinal centers and start to replicate at considerable rates. Mutations are introduced in the V regions of their receptors at a rate 10^5–10^6 times higher than the background DNA mutation frequency. The resulting variants are selected on the basis of their affinity for the antigen presented by follicular dendritic cells and destined for the so-called memory compartment. Such cells have a much higher affinity for antigen than do germ-line cells. The similarity of the entire process of selection and optimization of an immune response to adaptive biological evolution is striking.

T cells are the most important cells in an immune response. Responsible for cellular immunity, they originate in the thymus where they undergo selection to weed out those T cells expressing receptors that can recognize self antigens. The molecules responsible for self recognition are the so-called MHC antigens, a highly polymorphic system responsible for rejection of organ transplants. Only T cells that can recognize MHC molecules complexed with foreign peptides are allowed to pass out of the thymus. Different T-cell populations are distinguished on the basis of their expression of different types of CD (cluster of differentiation) molecules.

Cytotoxic or killer T cells ($CD8^+$) do their work by releasing lymphotoxins, which cause cell lysis. Regulatory T cells (Tregs) serve as managers, directing the immune response. They secrete the lymphokines, signaling molecules that can stimulate cytotoxic T cells and B cells to grow and divide and enhance the ability of macrophages to destroy microbes. The $CD25^+$ $CD4^+$ Tregs represent a unique population of lymphocytes capable of powerfully suppressing immune responses. These T-cell populations actively control the properties of other immune cells by suppressing their functional activity to prevent autoimmunity and transplant rejection but also influence the immune response to allergens as well as against tumor cells and pathogens. Also $CD4^+$ helper T cells are necessary for effective cell-mediated immunity, possibly mainly in the form of memory T cells, programmed to recognize and respond to a specific pathogen once it has invaded and been repelled. Thus, multiple, sometimes rare, cell types must communicate during cell-mediated immune responses and it is far from clear how such interactions are orchestrated within organized lymphoid tissues[802].

B cells control the humoral immunity. They are the antibody-producing cells and are only stimulated to maturity when an antigen binds to its surface receptors in the presence of a T-helper cell. Most of them then become plasma cells producing highly specific antibodies. The other B cells become long-lived memory cells as described above.

Also for the adaptive immune response the overall number of lymphocytes is not greatly affected in the elderly. There is some evidence that older human individuals have significantly lower absolute numbers of lymphocytes per volume of blood, but the decline is not great and the individual variation very high[803]. Also in this case it is unlikely that the age-related decline in immune response is caused by a lack of cells. Instead, it is more the ratios of the different cell types that are affected and the way they react to an infection. Perhaps the most important aging-related changes in adaptive immunity are found in the T-cell compartments. The most visible sign of this is involution of the thymus, the primary lymphoid organ in which bone-marrow-derived T-lymphocyte precursors mature and differentiate into functional T lymphocytes[804,805]. Regression of the human thymus starts after birth and continues at a constant rate until middle age. Because there is very little effect of thymectomy in old age, thymic involution must be an evolutionary adaptation, with the peripheral lymphoid organs taking over the role of the thymus. However, alterations in T cells do appear to underlie aging-related functional decline of the immune response. Both T-helper and cytotoxic functions decline with age. Aging also leads to a dramatic increase in the proportion of antigen-experienced memory T lymphocytes with a concomitant decrease in naive T lymphocytes, thus lowering the ability to respond to new antigens.

Aging-related changes in the B-cell compartments have also been observed. The amount of antibody produced in response to foreign antigens decreases with age, possibly due to a decrease in the number of circulating and antigen-responsive B lymphocytes. At old age, B lymphocytes produce antibodies that bind antigens less well. In view of the important role of T lymphocytes in regulating B lymphocytes it is possible that defects in humoral immunity are really caused by changes in cellular immunity.

All together, there is sufficient evidence to believe that adaptive immunity is seriously affected by the aging process, probably more so than innate immunity. Naturally, it would be extremely important to test the hypothesis that such a functional decline can ultimately be traced back to a basic molecular cause, such as genome instability, in some or all of its constituent cells. The enormous complexity of the different cellular interactions that control the immune response, especially adaptive immunity, make the unraveling of such a cause-and-effect relationship a daunting task. In 1989 Claudio Franceschi (Bologna, Italy) proposed a general theory suggesting that aging is indirectly controlled by a network of cellular and molecular defense mechanisms (the network theory of aging)[806]. He proposed that alterations in this network of cellular defense mechanisms could seriously disturb cell physiology, with the neuroendocrine and the immune systems particularly affected since they are themselves networks of interconnected cells, dependent on each other as well as other factors. He argued that the two major challenges in testing this hypothesis would first be to understand how each component process affects function and viability of the network as a whole and then to disentangle the interactions between different processes and levels of functioning. I would add a third challenge and that is to systematically organize the different activities of the components of the network and their changes with age in a way that would allow us to identify the primary cause(s) of their decay and collapse with age. Whereas we are currently far from knowing the molecular basis of aging-related degenerative changes in this network of interactions, it certainly is not inconceivable that they all ultimately derive from mutation accumulation, adversely affecting the signaling cascades in these highly differentiated cells.

7.2.1.13 Summary and implications

The best way to summarize this brief overview of aging phenotypes other than cancer is to say that age-related functional degeneration is evident at all levels. It is conceivable that in most cases cellular changes are the ultimate cause of the degenerative phenotypes. Hence, random genomic mutation could, directly or indirectly, be a primary cause of many aging phenotypes. As we have seen, this possibility has been explored in only few cases. Most work in this respect has been focused on studying mutation accumulation in the mitochondrial genome, to a large extent because it is better experimentally accessible than the nuclear genome. This is especially true when studying individual cells or small cell populations. Hence, there is very little we can say at this point in time about genome alterations in select cell subpopulations.

A major alternative causal factor appears to be protein cross-linking, which is implicated in a number of commonly observed aging phenotypes, most notably aging of the lens or collagen cross-linking of ligaments in the joints. Proteins are subject to a variety of spontaneous degradation processes, including oxidation, glycation, deamidation, isomerization, and racemization. These nonenzymatic modifications can produce functionally damaged species that may reflect the action of aging at the molecular level. However, there is no evidence that defects in protein repair, for example, the conversion of L-asparaginyl and L-aspartyl degradation products by methyltransferases, cause accelerated aging[807]. Likewise, mice lacking methionine sulfoxide reductase (which repairs oxidized methionine residues) have a shorter lifespan and exhibit enhanced sensitivity to oxidative stress[808], but do not show a pattern of premature aging symptoms so characteristic for the many mouse models described in Chapter 5 with defects in genome maintenance.

There is some evidence in flies that overexpression of the repair methyltransferases leads to lifespan extension, but only at higher temperatures[137]. Similarly, overexpression of the enzyme peptide methionine sulfoxide reductase A, which catalyzes the repair of oxidized methionine in proteins by reducing methionine sulfoxide back to methionine, also extends the lifespan of fruit flies[138]. These transgenic animals were more resistant to oxidative stress and the onset of both their age-related decline and their reproductive capacity were markedly delayed.

Taken together, there is no strong evidence that protein damage is a major primary cause of aging. However, current data are inconclusive and additional experiments are needed to test causality. Meanwhile, it is impossible to rule out cellular and non-cellular changes other than DNA or protein changes. Indeed, the accumulation in the germ line through genetic drift of mutations with adverse effects at old age, as described in Chapter 2, would predict a sheer infinite number of autonomous forms of degeneration. It is obvious that there are many forms of age-related pathological lesion that are unlikely to be caused directly by genome alterations. However, they can certainly be an indirect result of such events, based on the aforementioned extensive networks of communication between different cell types in the body. This makes it absolutely critical to further define the aging phenome into a hierarchy of age changes, from primary to secondary and so on. Genome-wide genetic-association studies of aging human populations are likely to play an important role in this endeavor (see below).

As we have seen, in most tissues and organs cell loss does not appear as a major cause of aging-related functional decline, although atrophy in key tissues cannot be ruled out as a causal factor. Interestingly, hematopoietic stem cells in mice can outlive their original, aged donor, as demonstrated by repeated serial transplantation in lethally irradiated recipients[809]. What is changing in stem cells and most certainly in their differentiated derivatives, is the quality of the function they provide. Indeed, depending on the mouse strain, there is a limit to serial transplantations and functional decline in hematopoietic stem cells has been documented[732]. Nothing is known about the possible intrinsic causes of hematopoietic stem-cell aging or stem-cell aging in general. However, we have already seen that neuronal stem cells accumulate genomic mutations during aging and there is no reason why other stem-cell types would not display loss of genome integrity. Indeed, evidence has been obtained for chromatin dysregulation during hematopoietic stem-cell aging in the mouse (M. Goodell, personal communication) and for stabilization of the chromatin structure through overexpression of a polycomb group protein (involved in histone methylation and deacetylation; Chapter 3) preserving the replicative potential of these cells (G. de Haan, personal communication). Could random genomic or epigenomic alterations be a likely mechanism to explain aging-related deterioration of stem-cell compartments?

As discussed in Chapter 4, stem cells appear to have generally lower spontaneous mutation loads than non-stem cells, possibly due to increased sensitivity to apoptosis and/or retention of the parental DNA strands in the stem cells after asymmetric cell division. Apart from such aids in maintaining a pristine genome, stem cells have much better opportunities to rid themselves from mutations (including epimutations) through selection than differentiated cells. There is evidence from plants that stem cells with high mutation loads can be purged from the population[810], which may explain why such organisms (similar to *Hydra*, the primitive metazoan mentioned in Chapter 2) generally do not show signs of aging. Random mutations in stem cells would also not cause any problem for the organism, as long as they do not affect the future functions the stem-cell descendants need to carry out. However, adult stem cells are much closer to the differentiated state and their plasticity is not entirely clear. This may explain why the stem cells that occur in many tissues appear to show symptoms of aging-related degeneration. While unknown at present, it is conceivable that at least some of this degeneration is due to random genome alteration.

Finally, it is important to notice that in all cases the magnitude of the adverse effects of aging appears to be surprisingly modest, at least in the absence of complications. The most frequently occurring complications are those caused by disease, often a result of unusual vulnerability due to environmental, lifestyle, or genetic factors. As discussed above, it is very difficult to distinguish disease from normal aging, except when the cause of the disease is clearly unrelated to aging, as in certain heritable or environmental diseases. An obvious example is lung cancer, which is rare during normal aging but very frequent in smokers. The enormously increased risk of disease at older ages explains why in early studies the effects of aging were often overestimated. Indeed, in the absence of disease, there is often no good case for significant adverse effects. This was the rationale behind the SENIEUR protocol[811]. Originally this represented a set of inclusion criteria for immunogerontological studies in humans to exclude as many endogenous and exogenous influences on the immune system as possible and standardize the population under study. Later, the protocol was adopted as a more general approach to separate pure aging from those phenotypes related to exposure to environmental and disease factors. As later realized, whereas it is useful to include the presence of disease in the interpretation of data derived from an aging population, it is not possible to superimpose aging-related diseases on a normal aging pattern that is nominally disease-free[812]. Rigorously applied, the SENIEUR protocol would eventually exclude all subjects from population-based studies on aging, because it is conceivable that natural death is always the result of disease-related complications. We may not always be aware of it, because the cause of death in the oldest old is often not known. Rather than adopting a model of disease-free elderly subjects, representing the normal pattern that all old people would exhibit were it not for their illnesses, it is more realistic to assume that different cell types and/or functional sub-populations of cells age at different rates, with those aging the fastest and causing problems ('disease') that ultimately lead to death. Genetic, environmental, and lifestyle factors would then determine the aging phenotype and the cause of death. It is possible, but far from proven, that the rate and type of genome alteration in these different cell populations ultimately leads to the functional deterioration causing the problems and the ultimate demise of the individual.

7.2.2 AGING IN OTHER SPECIES

As mentioned above, our knowledge of aging-related phenotypes in species other than humans is limited. However, what is known about other mammals generally confirms the observations made in humans. This is not surprising because all mammals share immune, endocrine, nervous, cardiovascular, skeletal, and other complex physiological systems. In principle, they can all also develop the common diseases associated with such shared physiology, including cancer, atherosclerosis, hypertension, diabetes, osteoporosis, cataract, and many others. However, there are some notable differences between humans and other mammals, including non-human primates. For example, whereas there are similarities in the spectrum of vascular changes, such as increases in thickness of the intima media in the carotid wall, rodents or non-human primates are highly resistant to atherosclerosis[813]. In addition, whereas all mammals, to some extent, display age-related cognitive decline, brain histopathological changes can differ substantially. Whereas mice (apart from transgenic mice overexpressing mutant genes associated in humans with neurodegenerative diseases) never show brain lesions such as senile plaques or neurofibrillary tangles, aged dogs and non-human primates do[814]. In these cases, however, the histopathology is not exactly the same as in humans[815]. Aged dogs display diffuse amyloid in their brain, not the neuritic plaques seen in humans. The situation for non-human primates is controversial and it is as yet unclear whether such animals develop Alzheimer's disease. Indeed, it is really only the human species in which a broad scala of age-related neuropsychiatric illnesses is displayed. The reason for this is unknown, but it may involve the long lifespan of humans permitting these phenotypes to manifest. Finally, cancer is a major aging-related disease in all mammals, with some distinct differences in tumor spectra. A curious exception seems to be the naked mole rat, a long-lived mammal in which tumors have not as yet been observed[816]. Overall, however, the aging phenome of all mammals is very similar. For invertebrates this situation is distinctly different.

Two of the most popular animal model systems for aging are undoubtedly the nematode worm *C. elegans* and the fruit fly *D. melanogaster*. Nevertheless, we do not know a lot about the aging phenome in these invertebrate animals. For nematodes, the most remote from mammals of the two model species, some limited information is now available[817,818]. *C. elegans*, which has a mean lifespan of 9–23 days (depending on the temperature[819]), is a self-fertile hermaphrodite organism with a total of 959 somatic cells at maturity, all of known ancestry. Aging of this organism generally results in substantially slower movement, a lower pharyngeal pumping rate, accumulating lipofuscin, a flaccid appearance, and a decline in progeny production. Most of this can be readily observed through the dissecting microscope. Using other microscopic techniques, including electron microscopy and fluorescence microscopy to detect green fluorescent protein as a transgene under control of tissue-specific promoters, it has been possible to look at specific cell types such as neurons and muscle cells in aging nematodes. The results indicate muscle deterioration resembling human sarcopenia[817]. This may easily explain the locomotory impairment mentioned above. Surprisingly, the nervous system appears to be remarkably preserved, at least in the N2 strain, the only one examined thus far with such methods. Interestingly, these studies revealed extensive variability in age-related degeneration, both among different, genetically identical animals and among cells of the same type within individuals. This emphasizes the potentially important role of stochastic effects in aging and I will come back to that below in a model that could explain how accumulating informational randomization of the aging somatic genome might be responsible for the aging phenome in different organisms.

D. melanogaster is closer to mammals than *C. elegans*. It was and still is the model system of choice in genetics and has recently found a wealth of applications in studying complex physiology. Several human disease models have been developed in *Drosophila*, particularly for neurological diseases[820]. *Drosophila* is a popular model organism for studying the genetics of aging for the same reason as *C. elegans*, because it is genetically tractable and has a short lifespan. Genetic screens have also in the fly allowed the identification of single genes that control lifespan (Chapter 2). Nevertheless, we know even less about aging phenotypes in flies than in nematodes, a notable exception being the heart. *Drosophila* has a circulation and a tubular structure that contracts spontaneously to transport hemolymph, carrying energy substrates from the abdomen to the thorax and head. Investigators found that maximal heart rate is significantly and reproducibly reduced with aging in *Drosophila*, analogous to observations in elderly humans[821]. In spite of such progress, the causes of spontaneous death in flies are as unclear as in nematodes.

What do we know about the possible molecular basis of aging in flies and nematodes? Is there any evidence that instability of the somatic genome plays a role in the cellular degeneration and death that is also likely to underlie aging in invertebrates? One major difference with vertebrate organisms is of course the lack of proliferative cell populations in adult flies and worms, in which cell division is confined to the germ line. However, while this may influence the spectrum of genome alterations, it is unlikely to suppress loss of genome integrity. As we have shown, certain types of mutation can be efficiently induced in non-dividing cells (see above and Fig. 7.5) and it is simply incorrect to assume that mutation always depends on DNA replication.

In both *C. elegans* and *D. melanogaster* there is evidence for the accumulation of mtDNA mutations, analogous to the situation in humans, non-human primates, and rodents[822,823]. In *Drosophila*, the accumulation of deleted mtDNA appears to be tissue-specific: occurring mainly in the thorax. Similar to the situation in mammals, it is unclear whether such mtDNA mutations are a cause or an effect of aging in these animals. Virtually nothing is known about nuclear-genome changes in either nematodes or flies. Results from Klass *et al.*, in the early 1980s[824], suggest that for nematodes there is some evidence for an altered DNA template function at old age. More recently, progressive and stochastic changes in nuclear architecture (e.g. nuclear lamina, chromatin) were observed in most non-neuronal cell types of aging nematodes[220], somewhat similar to the recently observed age-related nuclear defects in human fibroblasts[219]. Interestingly, as yet unpublished work by Simon Melov (Novato, CA, USA) and colleagues provided the first evidence that large-scale instability of the nuclear genome is associated with increasing age in the worm. Using a combination of real-time PCR (see below), in combination with morphological methods for visualizing DNA, these investigators found stochastic variability in genome copy number with age within and among individual nematodes. Their results indicate that genome copy number becomes dramatically deregulated with age, similar to that seen with neoplasms arising with age in human beings. This molecular variability appears to be tissue-specific, with most of it probably restricted to the reproductive tissues. The changes observed appeared to involve mainly large-scale genome amplification, although at very old age deletions were also detected (S. Melov, personal communication).

Hence, it is unclear as yet whether aging in animals, vertebrates, or invertebrates is caused by random genome alteration. This is different for another highly popular model organism for studying aging: *S. cerevisiae* or baker's yeast. This organism has proved to be a great model for studying a variety of processes in higher organisms. While aging phenotypes differ a lot more between yeast and mammals than between flies and mammals or even nematodes and mammals, there are a great many genes in these species that encode very similar proteins[825]. Probably the best examples are the two close homologs (Ras1 and Ras2) of the mammalian *ras* proto-oncogene, which are essential for yeast viability. Expression of the mammalian H-ras sequence in a yeast strain lacking both *ras* genes restores this viability, indicating conservation not only of sequence, but also of biological function. Humans share 79, 39, 31, and 11% gene homologs with mice, fruit flies, nematodes, and yeast, respectively (Table 3.1). As we have seen, this homology extends into the realm of aging, with for example the sequence of the human *WRN* gene being highly similar to that of the yeast *SGS1* gene, which also encodes a DNA helicase. *SGS1*-mutant yeast cells have a markedly reduced lifespan and share other cellular phenotypes with cells from individuals with Werner's syndrome[826]. There is a fair amount of evidence that the molecular pathways that regulate longevity are partially conserved throughout evolution, ranging from yeast to humans (see Chapter 2).

Aging in yeast can be studied as replicative aging or chronological aging. The former can be studied by counting the number of daughter cells produced by individual mother cells, whereas chronological aging is measured in non-dividing populations in water or minimal medium. It is not clear if genome instability is the only cause of replicative or chronological yeast aging, but there is certainly good evidence that it is at least one major cause. First, replicative or mother-cell aging can be caused by the previously mentioned accumulation of rDNA circles, which is a form of genome instability. As we have seen, this is controlled by Sir2, the absence of which accelerates aging, while its overexpression extends lifespan in yeast. (Overexpression of the Sir2 homolog in *C. elegans* also extends lifespan of this animal[153]; see Chapter 2.) Independent of the rDNA circles, evidence has been provided for increased LOH in yeast contributing to replicative aging in this species[151]. Expression profiling of middle-aged mother cells revealed an apparent DNA-damage response[827]. For yeast undergoing chronological aging recent results indicate mutation accumulation in the reporter gene *CAN1*[828]. These mutations were decreased in long-lived *sch9* mutants, which is consistent with the idea that the accumulation of random mutations drives aging.

These results with yeast strongly suggest that genome instability is a major cause of aging in this organism. Two factors were critically important for this conclusion to be reached. First, its small genome greatly facilitates the detection of genomic instability, either directly (the rDNA circles) or indirectly, using ready available marker systems, such as, for example, marker genes affecting colony color or selectable reporter genes. Second, yeast lends itself extremely well to genetic studies, allowing the comparison of lifespan and genomic instability in a great many different genetic backgrounds in a very short time. Before I discuss how we may be able to take conceptually very similar approaches in mammals, including humans, it is necessary to address the question of how we should envisage the translation of genomic instability, a stochastic process, into aging-related cellular degeneration, neoplastic transformation, and death.

7.2.3 GENOME INSTABILITY: STOCHASTIC EFFECTS ON TRANSCRIPTION

For genome instability to underlie the relatively subtle changes in cellular function likely to be causally related to aging-related degenerative phenotypes, an explanation must be sought that obviates the argument that consistent functional decline cannot be due to random genome alterations, such as DNA mutations, because the frequency of such events is too low. As we have seen, the assumption that genome alterations are too rare to explain aging is now essentially refuted by the abundant evidence for genome instability at all levels at sometimes surprisingly high frequency. Nevertheless, it still remains obscure how such random changes could adversely affect cellular function yielding aging phenotypes that are consistent from individual to individual and often across different species.

We have already seen that in undifferentiated, proliferating cells random genome alterations can be eliminated from the population if they adversely affect fitness. In highly specialized cell populations this is generally much more difficult, especially when the cells are postmitotic. In such differentiated cells the only genes that are active, apart from housekeeping genes, are those contributing to the particular function they need to carry out to service the organism as a whole. Hence, genome alterations affecting this function will immediately manifest as functional decline at the level of the organism. All cellular tasks are carried out through networks, functionally connecting different genes and proteins into modules. All genes in such networks as well as their regulatory regions are targets for

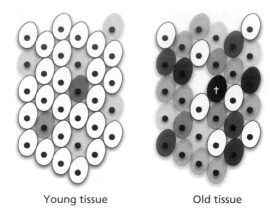

Young tissue   Old tissue

Fig. 7.6 Age-related, cell functional divergence as a consequence of stochastic effects. Functional decline is indicated by increasing darkness of shading. Even in a young tissue, the function of highly differentiated cells in an organ or tissue is never maximized. The old tissue is different in the sense that there are many more cells that have suffered functional decline, with some of them dying (†) or eliminated (open space). Others have grown into hyperplastic or neoplastic lesions (not shown) or have been replaced by fibrosis (not shown).

mutational alteration. The effect will vary depending on the level of redundancy in the pathway and the type of mutational alteration. Because genome alterations are random, the effect on each cell will be different, varying from no functional decline at all to a complete ablation of function. For the organ or tissue as a whole the decline in function would be the average of the effect on each cell in the tissue. This is schematically depicted in Fig. 7.6. Of note, during development in as-yet-undifferentiated, proliferating cell populations, there will be continuous selection against mutations adversely affecting housekeeping function, but probably not always against those negatively influencing the specialized function these cells are destined to perform upon terminal differentiation.

Because mutation accumulation is not restricted to adulthood, in young adults organ function will already be compromised to a varying extent and almost never maximized. Indeed, although we demonstrated that certain types of mutations occur as readily in quiescent cells as in cells that are mitotically active (see above and Fig. 7.5), it is highly likely that overall the mutation rate during development is high due to the large number of DNA transactions that need to take place. This would be in keeping with the hypothesis of Leonid Gavrilov and Natalia Gavrilova (both in Chicago, IL, USA) that early development produces a high load of initial damage, which is comparable with the amount of damage received during the remaining part of the life span[829]. This allowed them to explain the Gompertz model of aging in terms of an exponentially increasing loss of functional redundancy. Of note, such a model is in agreement with our own observations that already as young adults mice suffer from a considerable somatic mutation load (Table 6.2).

The model for age-related functional decline in Fig. 7.6 is compatible with impaired patterns of gene regulation as its proximal cause. Impaired gene regulation is a likely consequence of genome alterations interfering with transcription, for example, methylation changes or chromosomal deletions, resulting in haploinsufficiency, or translocations altering the position of regulatory sequences relative to the transcription unit (Chapter 3). If this is correct one would expect to see that aging is associated with increased transcriptional noise, random molecular fluctuations that create increased variability in the levels of gene expression within a cell population. Transcriptional noise is to some extent inherent in the basic process of transcription, especially for genes expressed at low levels[830,831], but could be greatly enhanced by increased mutational loads in the genome. As we have seen in Chapter 3, gene expression is a complex process that involves chromatin remodeling, transcription-factor binding, and initiation, elongation, processing, and export of RNA from the nucleus to the cytoplasm where mRNA molecules are translated into proteins. Genome alterations could greatly increase stochastic variation in mRNA levels. This has been demonstrated, for example, in cases of haploinsufficiency of transcription-factor genes, which increases the probability of stochastic activation or inactivation of target genes[832]. It is conceivable that random alterations in the somatic genome, for example, as induced by ROS, contribute to increased stochasticity of gene expression.

To directly test for such increased transcriptional noise in aged tissue, Rumana Bahar, a postdoctoral researcher in my laboratory, dissociated single cardiomyocytes from fresh heart samples of young and old mice, followed by the quantitative analysis of mRNA levels of a panel of housekeeping and heart-specific genes. Accurate quantification of mRNA levels in single cells requires the amplification of the entire cell message, with each of the different mRNAs amplified at the same level of efficiency. This should avoid loss of representation; that is, the quantitative relationship between the expression level of the different genes should be the same as in the original sample. A reliable protocol to this end has been developed in the laboratory of our collaborator, Christoph Klein (Munich, Germany)[833]. His method is based on PCR (Chapter 1) with a single poly-C primer, making all amplified sequences equally GC-rich and allowing a high annealing temperature (schematically depicted in Fig. 7.7).

We decided to use this poly-C-based amplification method, but since it had never been applied to cells taken directly from the *in vivo* situation, we designed a procedure to test its reproducibility in maintaining correct representation of individual mRNA levels. We repeatedly amplified the equivalent of a single cell from a pool of lysed cardiomyocytes as compared with independent amplifications from single cells, all from the heart of a 6-month-old mouse. We subsequently quantified for each cell or cell equivalent the expression level of the heart-specific gene encoding the myosin ventricular regulatory light chain (Myl2). This was done by real-time PCR, a variant of the PCR amplification method described in Chapter 1. Real-time PCR quantifies the initial amount of the

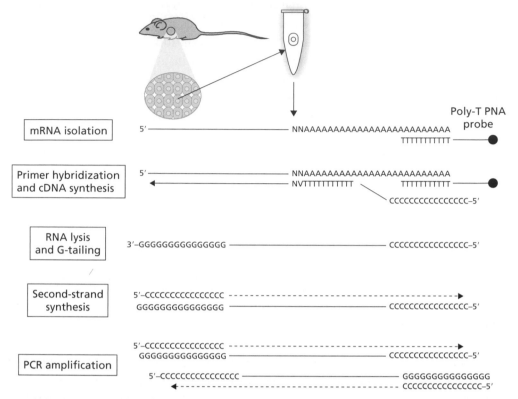

Poly-T PNA probe

| mRNA isolation |

| Primer hybridization and cDNA synthesis |

| RNA lysis and G-tailing |

| Second-strand synthesis |

| PCR amplification |

Fig. 7.7 Schematic depiction of global mRNA amplification using the method developed by Christoph Klein[833] and applied to cells directly taken from animal tissue. Critically important in this method are the use of a peptide nucleic acid (PNA) as a high-affinity probe to bind the poly-A tails of the mRNAs and immobilize them to small magnetic beads, and the single poly-C primer for the PCR-based amplification, which permits unbiased amplification at high annealing temperatures (high specificity). The poly-T primer used for cDNA synthesis (the PNA probe cannot be extended) is anchored immediately after the last A by using random bases (N = any possible base; V = any possible base but T). Because poly-A tails are often very long, such anchoring precludes the incorporation of large stretches of As in the final PCR amplicons. The poly-C tail of this primer in combination with the subsequent G tailing allows the use of the single poly-C primer in PCR amplification.

template by monitoring amplicon production during each PCR cycle, hence the term real time as opposed to the endpoint detection that is common in PCR. Products are detected during the exponential phase of the reaction by using a dye that produces a fluorescent signal each time a double-stranded product is made. The more copies of the template of interest present at the start of the reaction, the fewer amplification cycles needed to cross the detection threshold. Such C_T values (a measure of the copy number of the mRNA amplified) are used to calculate the quantity of the original template (Fig. 7.8). Since it is difficult to obtain absolute values in this way, the results are always expressed as

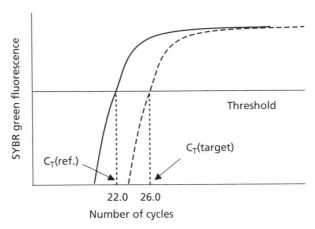

Fig. 7.8 The principle of real-time PCR. The number of PCR cycles when the threshold is reached is a measure of the copy number of the mRNA amplified and is called the C_T value for that mRNA. Relative measures can be obtained by subtracting the C_T value of a reference gene (ref.; also called the normalizer) from the C_T of the target gene, yielding the so-called ΔC_T value. The median ΔC_T value for all cells analyzed in a given sample is calculated and subtracted from the ΔC_T of each cell. These cell-specific values are called $\Delta\Delta C_T$. The results are linearized and plotted as relative quantities on a logarithmic scale.

relative measures; that is, the expression level of the test gene relative to another reference gene in the same sample. Because in our situation any gene had the potential to undergo random changes in its transcript level, we used different reference genes, reasoning that whereas this would not allow us to draw conclusions about absolute differences, our basic hypothesis of a significant age-related increase in cell-to-cell variation in gene expression remained perfectly testable.

In the example in Fig. 7.9, the mitochondrially encoded cytochrome c oxidase subunit 1 gene (*mtCo1*) was used as the reference gene. There was very small variation in the relative quantity of the *Myl2* gene among single-cell equivalents, which was within the range of experimental error when the real-time PCR was performed repeatedly on the same template. By contrast, variation in expression of the *Myl2* gene relative to that of *mtCo1* among single cells was significant. This result indicates that even among seemingly identical cells in the heart of a young mouse the *Myl2* gene is not expressed to the same extent. This was also proved to be true for other genes that are relatively highly expressed in heart cells. For genes expressed at a low level the cell-to-cell variation was in the same range as that in the single-cell-equivalent control. This is most likely due to the variation arising from the stochastic distribution of low-copy-number template molecules, which limits the accuracy of any amplification system sensitive enough to amplify the total mRNA content of a single cell.

To test whether our prediction of increased stochasticity of gene expression at old age was correct, fifteen heart cardiomyocytes were isolated from a similar area of ventricular

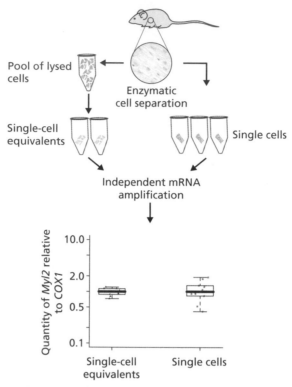

Fig. 7.9 Validation of the accuracy of the procedure for amplifying all mRNAs from a single cell. Repeated global mRNA amplification from a pool of lysed cells, by taking a one-cell equivalent 10 times, yielded virtually identical values when measuring the expression level of the heart-specific gene *Myl2* relative to the mitochondrially encoded *COX1* gene (*mtCo1*; single cell equivalents). By contrast, 15 single cells yielded significantly higher variation ($P = 0.0026$), indicating transcriptional noise of this gene among normal heart cardiomyocytes already in a young mouse. Boxes in the box plots indicate the interquartile range with the median; the whiskers indicate 1.5 times the interquartile range. Re-printed with permission from ref. 834.

heart tissue of three young (6-month-old) and three old (27-month-old) male mice. Each cell was subsequently subjected to the global mRNA amplification procedure, after which the expression level of each target gene was quantified using real-time PCR. For normalization we used different genes in the panel, depending on the expression level. Rumana subsequently quantified mRNA levels of 12 nuclear and three mitochondrial genes in every one of the 15 cardiomyocytes isolated from each animal. The results indicated a highly significant increase in cell-to-cell variation in the expression levels of all nuclear genes in the old animals as compared with the young ones. This increased cell-to-cell variation was observed irrespective of the reference gene used. Examples of the increased cell-to-cell variation in transcript levels in old mouse heart are shown in Fig. 7.10a for β-actin (*Actb*), β-2 microglobulin (*B2m*), lipoprotein lipase (*Lpl*), and cardiac α-actin

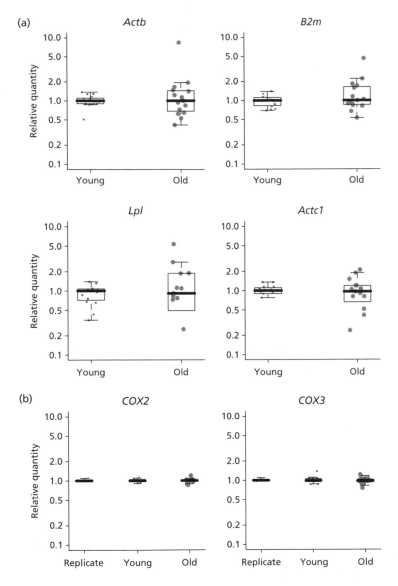

Fig. 7.10 Increased cell-to-cell variation in gene expression among cardiomyocytes from the heart of old as compared to young mice. (a) Examples of four genes showing a statistically significant age-related increase in cell-to-cell variability in expression: β-actin (*Actb*), β-2 microglobulin (*B2m*), lipoprotein lipase (*Lpl*), and cardiac α-actin (*Actc1*), normalized over glyceraldehyde-3-phosphate dehydrogenase (*Gapd*). Box plots are as in Fig. 7.9. (b) Expression of mitochondrially encoded cytochrome c oxidase genes (normalized over *COX1*) in the same cells showed no statistically significant increase in stochastic variation. Ten replicates from the same template were analyzed along with the young and old cells to determine the real-time PCR error. Box plots are as in Fig. 7.9. Re-printed with permission from ref. 834.

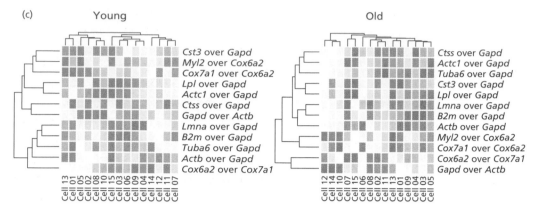

Fig. 7.10 (c) Representative heat maps of one young and one old animal revealed no significant pattern of change in $\Delta\Delta C_T$ (see legend of Fig. 7.8 for an explanation) among the cells (dark = low, white = high), indicating that the observed cell-to-cell variation is random. Re-printed with permission from ref. 834.

(*Actc1*), normalized over glyceraldehyde-3-phosphate dehydrogenase (*Gapd*). Rumana also quantified three mitochondrially encoded cytochrome c oxidase genes (*mtCo1–3*), against each other as the reference gene (Fig. 7.10b). In these cases no significant cell-to-cell variation in gene expression was observed. This is not unexpected since these genes are transcribed as one long RNA molecule, with a poly-A tail to guarantee its capture in our amplification assay (Chapter 3). Hence, relative to each other these mito-chondrially encoded genes should be expressed at the same level in all cells. Indeed, even if they would come from different transcripts, the large number of mitochondrial genomes in each cell would always yield an average value that is likely to be similar from cell to cell. Therefore, the absence of stochastic variation in transcript level for mitochon-drially encoded genes independently confirmed the unbiased nature and reproducibility of the global mRNA amplification method. Of note, the use of a nuclear gene as a refer-ence gene immediately introduced significant cell-to-cell variation in the relative quan-tities of each mitochondrially encoded transcript.

Finally, to test whether the observed cell-to-cell variation in gene expression was com-pletely random and not biased towards, for example, generally low gene-expression levels in cells close to dying, heat plots were constructed for each group of cells; that is, from the three young and three old animals. The results show no specific correlation of the relative gene-expression levels for any of the cells, allowing us to conclude that the observed vari-ation is random and free of systematic bias (Fig. 7.10c).

The results of these experiments established, at least for the heart, that the aging process is associated with a significant increase in stochasticity of gene expression. What is not known, however, is whether this is due to random genome alterations or other factors. This issue is far from resolved and much more experimentation is required. However, we did test whether genotoxic exposure of cells, the ultimate source of most permanent genome alterations, does lead to increased stochasticity of gene expression. For this purpose, Rumana incubated cultured mouse embryonic fibroblasts from the *lacZ*-plasmid reporter mice described in Chapter 6 for 2 hours in medium containing a low dose of hydrogen peroxide, a known generator of oxidative damage. After washing the cells (which grow attached to the culture plates), fresh medium was provided and the incubation continued. At different time points thereafter, Rumana harvested single cells from the population, which she then subjected to the global mRNA-amplification procedure, similar to the cardiomyocytes from old and young heart. In the fibroblasts she assessed the mRNA levels of several housekeeping genes, including the aforementioned *Actb* and *B2m*, with *Gapd* as the normalizer. As compared with single-cell equivalents, untreated cells displayed a trend towards cell-to-cell variation in gene expression, similar to the situation in cardiomyocytes from young animals, albeit of lesser magnitude. At 48 h after the treatment with hydrogen peroxide, cell-to-cell variation in gene expression greatly increased. This is shown in Fig. 7.11a for *Actb* as an example, but a similar increase was observed for *B2m* and tubulin alpha (Tube 6).

To investigate when this hydrogen peroxide-induced transcriptional noise occurred and how long it would last, we studied cell-to-cell variation in the expression of these housekeeping genes at 6 h, 48 h, and 9 days after treatment in several independent experiments. Cell-to-cell variation in gene-expression levels was plotted as the average coefficient of variation for the three genes from at least two independent experiments. As shown in Fig. 7.11b, at 6 h after treatment, cell-to-cell variation in the relative expression level of neither of these genes was increased. This strongly suggests that the increased transcriptional noise induced by hydrogen peroxide is not due to direct chemical damage to DNA, proteins, or lipids, the effects of which are likely to appear almost immediately. At 48 h after treatment, there was significantly increased stochastic variation, as mentioned above. This variation was still present 9 days after treatment (Fig. 7.11b).

A potential complicating factor in these experiments is the induction of replicative senescence by hydrogen peroxide and other genotoxic agents in primary human or mouse cells, as mentioned several times in this and previous chapters. At 48 h after treatment and later, the cell populations started to show an increased number of senescent cells, characterized by the flattened and enlarged morphology typical for senescent fibroblasts irrespective of the species (Fig. 7.4). During this period, when the cells were maintained in the same culture plates, survival after this low dose was high and cell death not apparent. However, at 9 days after treatment almost all cells were senescent, as evident from their flattened and enlarged morphology. At that time, the increased cell-to-cell variation in gene expression among the hydrogen peroxide-treated populations was still highly significant (Fig. 7.11b), suggesting that the senescent response has no influence on the expression level of the three housekeeping genes studied.

Based on these results we concluded that hydrogen peroxide-induced transcriptional noise is a relatively late event, which is virtually absent during the first hours after treatment.

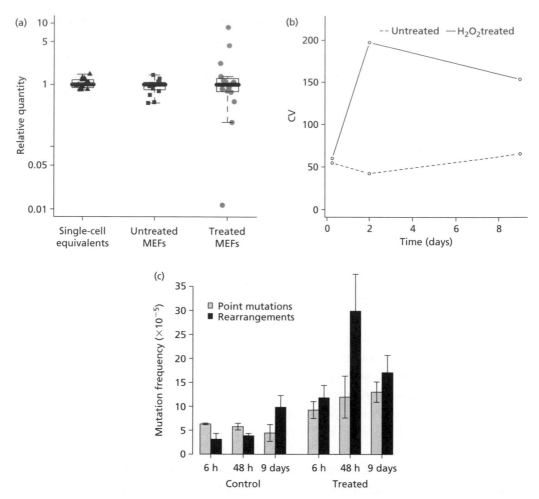

Fig. 7.11 Hydrogen peroxide treatment of mouse embryonic fibroblasts (MEFs) increases cell-to-cell variation in gene expression in parallel to the induction of genome rearrangements. (a) Cell-to-cell variation in relative expression of *Actb* (over *Gapd*) among mouse embryonic fibroblast single-cell equivalents, untreated mouse embryonic fibroblasts, and mouse embryonic fibroblasts at 48 h after treatment with 0.1 mM H_2O_2. The variation among the treated cells is significantly greater than among the untreated cells ($P < 0.0001$). Box plots are as in Fig. 7.9. (b) Coefficient of variation (CV) calculated for all relative quantities of *Actb*, *B2m*, and *Tuba6* at each time point and treatment condition. The 6-h, 48-h, and 9-day time points are based on two, four, and three independent experiments, respectively. Each experiment consisted of 11–15 single-cell determinations for each gene under each treatment condition. (c) *lacZ* mutation frequencies at 6 h, 48 h, and 9 days after hydrogen peroxide treatment compared with control populations. Mutation frequencies are averages from three determinations from each of two parallel experiments. The sub-division into point mutations (light-gray bars) and genome rearrangements (dark-gray bars) was made on the basis of restriction-enzyme analysis of 48 mutants taken from each experiment. Error bars indicate standard deviations. Re-printed with permission from ref. 834.

However, at least in a population of low cell turnover, the induced transcriptional noise is also persistent. To demonstrate that such late-arising, persistent cell-to-cell variation in gene expression induced by hydrogen peroxide treatment at least parallels the induction of genomic mutations, we tested for the presence of mutations in the *lacZ* reporter gene. The results of this analysis indicate that whereas some mutations were already present 6 h after treatment, their frequency increased substantially at 48 h, and declined somewhat at 9 days. Most of these mutations were large genome rearrangements, as indicated by restriction analysis of the mutant *lacZ* plasmids recovered from the cells (Fig. 7.11c). Hence, the observed transcriptional noise appears to follow the kinetics of hydrogen peroxide-induced mutations.

In view of the results described above[834], we can merely speculate that the accumulation of genome rearrangements may contribute to the increased stochasticity of gene expression, for example, through gene-dose and/or position effects. This could be true for both the hydrogen peroxide-treated cells and the aging mouse heart. Indeed, as described in the previous chapter, genome rearrangements at the *lacZ* reporter locus accumulate with age in the mouse heart. However, there are other types of genome alteration that are relatively late events of DNA damage. Increased external noise in transcription could be a consequence of random changes in DNA-methylation or histone-modification patterns, for example, due to incomplete restoration of these patterns after DNA repair or replication. Like genome rearrangements, such epigenomic alterations can be expected to modify transcriptional activity in a stochastic manner. While random genomic alterations would be a logical explanation for the observed transcriptional noise, we can at this stage not exclude other cellular changes unrelated to DNA damage.

Stochastic changes in gene expression have been considered as a possible cause of aging since Leslie Orgel proposed his error catastrophe theory (Chapter 1). After that, there has been a shift in focus towards cell-signaling processes and various programmatic aspects of aging. At present, the role of chance events is increasingly appreciated[835], as also testified by the work on nematode aging described above. Our present results do not allow any further conclusions with respect to increased transcriptional noise as a mechanism of aging. However, stochasticity is ubiquitous in biological systems and increased noise in gene expression has been implicated in reduced organismal fitness[836]. It is conceivable that a variety of persistent forms of damage to biological macromolecules could initiate a gradual increase in transcriptional noise introducing phenotypic variation among cells. Such variation could generally become detrimental to normal cell functioning, which would explain many etiological characteristics of the aging process, most notably its highly variable, progressive decline in organ function. If a gradual loss of the informational integrity of the somatic genome is responsible for increased transcriptional noise and a general deregulation of gene expression, how can we experimentally test whether such events are causally related to aging?

7.3 **Testing the role of genome instability in aging**

To test for the possibility that aging is caused by a gradual loss of genome integrity it is not enough to demonstrate an increasing load of genome alterations in aging cells and tissues and correlate that with aging-related functional loss and disease. We have already seen that even in the much simpler case of mtDNA mutations accumulating to a high level, concomitant with abnormalities in the electron-transport system, it has not been possible to distinguish between cause and effect (Chapter 6). Hence, further exploration of increased genome instability in various tissues of mice or humans will contribute useful new data sets and provide further insight into how increased mutation loads could be causally related to aging, but will not allow any definite conclusions regarding a causal relationship. For this purpose it will be necessary to design experimental interventions in combination with genetic studies. Analogous to the situation in yeast—the organism for which we have come very close in demonstrating that genome instability is a major cause of aging—it should now be possible to design such studies in nematodes, flies, mice, and humans. Two lines of research can be distinguished: (1) genetic modeling of genome instability and (2) population-based genetic studies.

7.3.1 GENETIC MODELING OF GENOME INSTABILITY

If loss of genome integrity is a major cause of aging, then it should be possible to engineer accelerated aging in a model organism by increasing genotoxic stress. We have already seen that a major reason for suspecting DNA mutations as a cause of aging was the discovery in the late 1940s that low, daily doses of ionizing radiation accelerated symptoms of normal aging in rodents[505]. However, the type of damage induced by radiation is unlikely to be exactly the same as the spontaneous DNA damage inflicted during normal aging. An obvious way to enhance such natural damage is through the engineering of genetic defects in genome maintenance. In Chapter 5 I described the genetic modeling of DNA-repair defects in the mouse, based on the discovery that human segmental progeroid syndromes are almost always caused by heritable defects in genome maintenance. Such modeling seemed to have come tantalizingly close to a model of aging in which increased genotoxic stress through the constitutive activation of major players in the DNA-damage response greatly impaired cellular proliferation. This might especially affect stem-cell populations, preventing them from properly rejuvenating aged tissue, which would eventually lead to organ dysfunction and degenerative disease. While a plausible scenario, I closed the chapter by expressing some doubts about this model, based on the absence of any phenotypic evidence that aging is caused by a lack of cells. Hematopoietic stem cells, for example, can outlive their original donor upon repeated transplantation in lethally irradiated recipients for several generations[809] (see above). As mentioned, some do not consider such mice as good models for normal aging in the first place because they never display a full complement of aging phenotypes; hence, the term segmental progeroid syndromes. As outlined in Chapter 5, I disagree with such a generalization, but it cannot be denied that in some of these mouse models the symptoms of premature aging are of doubtful validity.

The strategy of modeling genome-maintenance defects in the mouse to test the hypothesis that aging is due to genomic instability has at least three inherent limitations. First, manipulating single genes almost always leads to unwanted side effects due to the multiple roles most genes play in the physiology of an organism. An example is the *Xpd* mouse mutant, which has defects in both DNA repair and transcription (Chapter 5). In this mouse model it is not known whether the observed symptoms of premature aging are due to the defect in repair or in transcription. Second, due to the several hundred genes directly involved only in DNA repair it is not realistic to assume that one single gene change is sufficient to create an accelerated copy of the normal aging process. As we have seen in the examples of the *WRN* and *ATM* genes, premature aging only becomes apparent in combination with short telomeres. Since it is very difficult to cross multiple traits into a single mouse, the multigenic character of genome maintenance is a serious stumbling block in modeling aging through genetic manipulation at that level. Third, virtually all mouse models of accelerated aging are based on genes that are entirely ablated or severely impaired in their activity. A better approach would be to systematically replace genome-maintenance genes by variants that are only slightly defective, possibly polymorphic gene variants found associated with aging-related phenotypes in human population studies (see below).

One way to address some of the limitations of genetically manipulating genome maintenance in accelerating the normal aging process is to carefully select only those genome-maintenance pathways most likely to affect aging. From Chapter 4, we were able to conclude that DSB repair pathways could be among the ones most critical for aging. Erroneous repair of such lesions could easily lead to large genome rearrangements, mutational events most likely to severely impact patterns of normal gene regulation. In Chapter 5 we have also seen that mice with defects in the key genes acting in this pathway—*Ku80* and *Prkdc* (*DNA PK$_{CS}$*)—all show symptoms of accelerated aging and a shorter lifespan. The *Ku80*-null mice, made by Paul Hasty[458], which are the ones investigated most thoroughly, show these phenotypes at similar points in their survival curves and at the same level of severity as the normal control mice. The involvement of NHEJ in V(D)J recombination may confound some of the results obtained; overall, however, this mouse model appears to reflect the normal aging process more accurately than most other models.

My laboratory, in collaboration with Paul Hasty's laboratory, has now begun to study genome instability in the Ku80-deficient mouse model by crossing it with the *lacZ*-plasmid reporter mice and then analyzing *lacZ* mutations in the hybrids and their littermate controls. The first results show a somewhat complicated picture. Very similar mutation

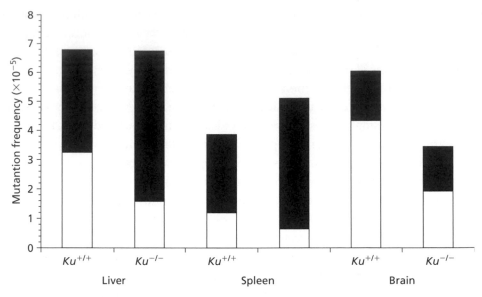

Fig. 7.12 Average mutation frequencies at the *lacZ*-plasmid reporter locus in liver, spleen, and brain of *Ku80*-deficient mice ($Ku^{-/-}$) compared with littermate controls ($Ku^{+/+}$). Point mutations are shown by the white portions of the bars (no change), genome rearrangements by black portions (size change). In both liver ($P = 0.0253$) and spleen ($P = 0.0002$), the frequency of genome rearrangements was significantly higher in the *Ku80*-defective mice than in the controls. In brain ($P = 0.0003$) and liver ($P = 0.0005$), the frequency of point mutations was found to be significantly lower in the *Ku80*-defective mice. R. Busuttil, unpublished work.

frequencies were observed in liver, significantly increased mutation frequencies in spleen ($P = 0.016$) and significantly reduced mutation frequencies in brain ($P = 0.013$; Fig. 7.12). After characterizing the mutations it appeared that while in liver the total mutation frequencies were the same as in the control mice, the *Ku80*-null mice showed a significantly higher number of genome rearrangements and a significantly lower number of point mutations. In spleen, the higher mutation frequency was due to a significantly higher frequency of genome rearrangements in the *Ku80*-null mice. In spleen there was also a trend toward a lower frequency of point mutations. Finally, in brain the lower mutation frequency in the *Ku80*-null mice was entirely due to a significant reduction of the number of point mutations. In brain, of course, almost all spontaneous mutations are point mutations (Chapter 6).

We hypothesize that these results reflect a tendency for increased errors in the repair of DNA DSBs in the *Ku80*-null mice. This confirms earlier observations by others of increased chromosomal aberrations in fibroblasts of these mice[350]. The reduction in point mutations, events typically associated with replication errors (see above and Fig. 7.5), could reflect a reduced number of cell divisions in the *Ku80*-null mice. Indeed, these mice are already very small at birth and it is possible that the increased member of DNA double-strand breaks impairs cellular replication during development. A reduced number of cellular replications would naturally limit the number of errors associated with such replication (replication errors are likely to be independent of NHEJ) as compared with normal mice. There is evidence for increased cellular senescence in primary fibroblasts isolated from Ku80-deficient mouse embryos[458]. Because, as we have seen, unrepaired DNA DSBs are associated with cellular senescence (as well as aging *in vivo*), increased genome rearrangements could explain the premature aging phenotypes observed in this mouse.

The Ku80 story not only suggests that increased genome instability could very well be the main cause of the observed premature aging symptoms in this mouse model, it also underscores the importance of focusing on those genome-maintenance pathways likely to be most critical as determinants of aging. Nevertheless, aging in *Ku80*-mutant mice is not exactly the same as normal aging. Accelerating normal aging would require much more subtle genetic modeling, for which currently the technological basis is not fully available. Is it possible to model the downstream events that may underlie premature aging in the Ku80 mice and, possibly, normal aging? Assuming that large genome rearrangements are a major cause of normal aging it may be possible to gradually increase the frequency of the main lesion thought to underlie these events: DNA DSBs. Such an approach has been taken for the mitochondrial genome. A mouse model was developed in which a mitochondrially targeted restriction enzyme was expressed in skeletal muscle[837]. Such transgenic mice quickly developed a mitochondrial myopathy, at 6 months, with large mtDNA deletions present in muscle. In the absence of a lifespan study it is unknown whether the increased DSBs gave rise to an accelerated-aging phenotype. However, as we have seen in Chapter 6, transgenic mice expressing the gene *twinkle*—a nuclear-encoded mtDNA helicase with a dominant mutation causing adult-onset progressive external ophthalmoplegia in humans—developed myopathy, but its effects were only mild and did not include symptoms of accelerated aging in spite of the demonstrated accumulation of deletions in the mtDNA in tissues such as brain and muscle in these mice[493]. We have already seen that proofreading-deficient mice accumulate point mutations in the mtDNA much faster than the control animals, which was found to be associated with premature aging, albeit most likely a consequence of a DNA-damage response (Chapter 5). The results with *twinkle*, therefore, indicate that mtDNA deletions either do not play a role in normal aging or need to accumulate at a much higher rate to exert an effect. As yet a mouse model with an increased rate of rearrangements in the nuclear genome based on a restriction enzyme targeted to the nucleus has not been generated.

7.3.2 POPULATION-BASED GENETIC STUDIES

A second approach recently made possible by the surge in advanced genetic technologies as a consequence of the Human Genome Project is direct association of the genes acting in genome-maintenance pathways with aging-related phenotypes. In principle such an approach can be applied to outbred populations of all species, but is most eminently suitable for humans. The reasons are that, as mentioned, the aging phenome of humans is the best characterized of all species and large 'natural' populations of outbred individuals that can be easily sampled and clinically evaluated are widely available.

There are two strategies to test the hypothesis that alleles at loci involved in genome maintenance are associated with aging-related phenotypes[838–840]. First, in a so-called global genome approach one would test the entire genome for regions that co-inherit with one or more aging or longevity phenotypes. A second approach is a so-called candidate approach, in which only those genes known to be involved in genome maintenance are tested. The former approach is the better one because it is an objective screen without bias towards genome maintenance.

In the past, the identification of genes, mutational variants of which were causing disease, was cumbersome and only relatively straightforward in the case of monogenic disorders. For monogenic disorders, such as cystic fibrosis, there are often extensive pedigrees available in which co-segregation of a polymorphic DNA marker allele with the disease-causing allele provided an indication as to the chromosomal location of the disease gene. This is called linkage analysis and requires a so-called linkage map of polymorphic markers. The higher the density of the map, the more precisely the location of the disease gene can be mapped, albeit at a higher rate of false positives. Once the disease gene has been mapped to a relatively small chromosomal region other methods, such as positional cloning, have been used effectively to identify the culprit gene on the basis of the presence of a disease-causing mutation in the affected but not in the unaffected individuals. However, since the segment of the genome defined by the marker alleles is still very large—several millions of base pairs—even in this relatively simple approach the final identification of the gene one is looking for is not simple.

The identification of genes involved in complex phenotypes with late onset, such as aging, is even more challenging. In such cases a standard linkage approach is essentially constrained due to the need to collect DNA samples from families of affected individuals. This is seldom a problem with disease phenotypes where onset is during childhood or early adulthood but aging phenotypes arising later in life increase the likelihood that family members will be deceased. Moreover, inherent complexity of phenotypes related to aging suggests involvement of multiple genes with small effect and tight interaction with environmental factors. The segments of the genome containing such genes are known as quantitative trait loci (QTLs), regions containing genes that make a significant contribution to the expression of a complex trait. It has proved notoriously difficult to identify the genes underlying a QTL, hence the low success rate of the linkage and positional-cloning approach in finding genes involved in complex traits.

A conceptually much simpler approach to associating gene variants with phenotypes is to directly compare the frequencies of variant alleles in a group of affected versus unrelated non-affected individuals. Such association studies are essentially case-control studies. Association-based studies are generally considered to be more effective tools for studying complex traits because they have greater statistical power to detect genes of small effect. There are two types of association study: direct association studies and indirect association studies. Direct association studies rely on the aforementioned candidate genes, selected because of *a priori* hypotheses about their role in the phenotype. If a positive association is found, then the gene is immediately known and needs no further identification. At present, our knowledge about the putative functional relevance of candidate genes and its potentially important interactions for a given phenotype is rarely sufficient to warrant direct association studies.

Indirect association studies rely on alleles of marker loci co-inherited with alleles at the loci involved in the phenotype of interest. Indirect association analysis has recently attracted a lot of attention because for the first time this approach can now be taken genome-wide, using markers that are evenly spaced throughout the genome without regard for their function or context in a specific gene. As we have seen, a consensus sequence for the entire human genome as well as for that of related species, such as mouse, rat, and chimpanzee is now available. Although we have still not identified and annotated all the genes in the human genome, we do have extensive maps of individual sequence variation. As we have seen, most of the variation in our genome is due to single-nucleotide polymorphisms (SNPs). Such variants, either in protein-coding sequences or in gene-regulatory regions, could be important determinants of complex traits such as aging.

Genome-wide association analysis may be within reach as a consequence of the completion of the HapMap project[841]. The International HapMap Project, briefly mentioned in Chapter 1, is an international collaborative effort to organize the SNPs in the human genome in a so-called haplotype map. In this map groups of SNPs are organized on the basis of linkage disequilibrium, a consequence of the fact that chromosomes do not pass from each generation to the next as identical copies. When germ cells are being formed, the chromosome pairs undergo recombination, resulting in a hybrid chromosome containing pieces from both members of a chromosome pair. Over the course of many generations, repeated recombination events in an interbreeding population lead to extensive shuffling of chromosomal segments, some of which are still shared by multiple individuals. These segments have not been broken up by recombination and are called haplotypes. Alleles of SNPs in such haplotypes tend to be inherited together.

The HapMap project will define the haplotype block structure of the entire human genome and 'tag' the common haplotypes by so-called haplotype-tagging SNPs. This should organize the 10 million common SNPs in the human genome into a few hundred thousand tag SNPs to represent common haplotypes. Hence, comparing a group of people with a disease to a group without the disease for all the 500 000 or so tagging SNPs should reveal the chromosome regions where the two groups differ in haplotype frequencies. These areas might contain genes affecting the disease. Hence, rather than analyzing each of the approx. 10 million SNP loci, analyzing about 500 000 tagging SNPs is enough to spot all possible genetic differences between the affected and non-affected groups[842].

Technically, we are now able to do whole-genome association studies according to the HapMap principle, for example, using the microarray chips now offered by companies like Affymetrix (www.affymetrix.com) or Illumina (www.illumina.com), containing hundreds of thousands of probes for SNP markers. However, there are some doubts whether the predictions made by the HapMap consortium are entirely correct[843]. It is known that there are multiple chromosome regions where associations among SNPs due to linkage disequilibrium are only weak or fall apart entirely. Ironically, this is true for the human IGF-1 receptor gene, as part of the IGF-1 axis often implicated in aging and aging-associated disease (Y. Suh, personal communication). In such cases it will be necessary to either score all SNPs in the region or potentially miss important information. The HapMap strategy is also based on the assumption that complex diseases are mainly affected by common SNPs. This is called the common disease/common variant (CDCV) hypothesis, which is conveniently based on the fact that all SNPs known were found by comparing only a limited number of unrelated individuals of few ethnic groups. However, this assumption is seriously disputed[843]. The fact that many monogenic disorders are due to a heterogeneous collection of variants that are individually rare has drawn attention to the importance of rare gene variants in common, complex traits. If combinations of rare SNPs turn out to be mainly responsible for the genetic component of such diseases it will be necessary to re-sequence each and every gene with all their regulatory regions for each individual participating in a population study. As mentioned in Chapter 1, such $1000 genome sequencing has been predicted to be a reality within 10 years[844].

Based on the rapid progress in technology development we should know within a few years if the predictions made by the HapMap consortium will be correct. If so, objective, genome-wide association analysis will undoubtedly be carried out to identify genes involved in aging phenotypes. If genome instability is a major cause of aging we should expect to find genome maintenance genes among the genes associated with aging phenotypes. If the HapMap predictions are wrong and most aging and disease-related phenotypes turn out to be characterized by a wide range of common and rare variants, each with minimal individual attributable risk, we may need to proceed with re-sequencing genes acting in candidate pathways (awaiting the $1000 genome). In that case, progress will be slower and biased towards the hypothesis under study, but most definitely providing increased insight into genome maintenance as a possible determinant of aging.

Awaiting a likely future surge of information on the genes specifying aging and longevity, what is the current state of the art? As we have seen, the genetics of aging has been strongly dominated by gene variants conferring increased longevity on mainly invertebrate organisms. However, as discussed in Chapter 2, as we have seen, many if not all of these mutants are artifacts of laboratory conditions and would not survive in the wild. It will be necessary to demonstrate that individual differences in aging and lifespan can be explained by variation at such loci in the wild. Trudy Mackay (Raleigh, NC, USA) and co-workers took a systematic approach to link candidate longevity genes to QTLs that affect variation in longevity in *Drosophila*[845]. It is very difficult to map QTLs to the level of the individual gene, even for *Drosophila*. These workers therefore used quantitative complementation tests to determine whether 16 candidate genes affecting *Drosophila* longevity contributed to naturally segregating variation in lifespan. In quantitative complementation tests inbred strains derived from a natural population are crossed to stocks containing loss-of-function mutations or deletions removing the candidate genes, maintained over a so-called balancer chromosome (carrying inversions that suppress crossing-over with the chromosome that carries the candidate gene mutation). Failure to complement suggests a genetic interaction between the allele of the candidate gene and the naturally occurring QTL and can be attributed to either allelism or epistasis. The results of this study demonstrated, for the first time, that several candidate longevity genes, including the insulin-like receptor locus and at least one gene involved in oxidative stress response, exhibit genetic variation for lifespan in *Drosophila*. Candidate genes for genome maintenance were not tested in this study. These results and those of others suggest that quantitative complementation testing is a promising candidate approach in studying the possible role of genome maintenance in determining lifespan. However, the approach has also been criticized, mainly because it is likely to yield false-positive results[846].

Systematic approaches have also been explored in mice to search for genes explaining natural variation in lifespan. Richard Miller (Ann Arbor, MI, USA) and collaborators used a population of mice obtained by mating the offspring of a cross between two different inbred mouse strains with the offspring of a cross between two other strains. Such a four-way-cross mouse population resembles to some extent a human pedigree. Hence, they can be subjected to linkage analysis using a relatively small number of genetic markers (typically fewer than 500). Using this approach QTLs were identified that modulate longevity[847]. There is no evidence yet for the nature of the genetic influences identified through this analysis. This is not surprising, because the size of the genetic intervals delineated is still extremely large, at 23 cM.

What do we know about gene variants and aging in humans? In humans, the heritability of lifespan is relatively minor[848], but extreme longevity appears to have much stronger heritability. Although the available records do not support biblical or folkloristic claims of the existence of humans with abnormally high longevity, people now, more frequently than in the past, live to be 100. Indeed, the oldest person thus far recorded is Jeanne Calment, who was more than 122 years old when she died in 1997. Thomas Perls (Boston, MA, USA) and co-workers originally found that the survival ratio for siblings of centenarians compared with siblings of 73-year-olds was about 4 for ages 80–94[849]. In a later assessment involving 444 centenarian families, these investigators found that, compared with the US 1900 cohort, male and female siblings of centenarians were at least 17 and eight times as likely to attain age 100, respectively[850]. Similar findings of a much stronger familial basis of unusual longevity have been obtained by Kerber *et al.*, who demonstrated an increased recurrence risk for siblings for surviving to extreme ages[851], whereas Gudmundsson *et al.*[852] found that the first-degree relatives of probands who live to an extremely old age are twice as likely as controls to survive to the same age. Moreover, multiple investigators have observed moderate familial clustering for extreme longevity. For example, the immediate ancestors of the aforementioned Jeanne Calment were shown to have a >10-fold greater likelihood of living 80 years or more, compared with the control ancestors of a reference family[853]. Are the genetic factors contributing to extreme longevity involved in genome maintenance?

Because it is virtually impossible to obtain DNA samples of pedigrees to follow the segregation of extreme longevity, sibling-pair analysis has been adopted as an alternative in the study of such late-onset genetic traits. Results of a genome-wide sibling-pair study of 308 persons belonging to 137 families with exceptional longevity indicated significant evidence for linkage with a locus on chromosome 4[854]. A subsequent haplotype-based fine mapping study of the interval identified a marker within microsomal transfer protein (MTP), which is involved in lipoprotein synthesis, as the possible modifier of human lifespan[855]. These results were not replicated in other populatons of centenarians[856], raising the possibility that the finding was specific for that particular population or was a false-positive result.

A role for lipoprotein genes in longevity would suggest that it is mainly a lower risk for cardiovascular disease that determines the longevity phenotype in human natural populations. This would be in accordance with the demonstration of the importance of large high- and low-density lipoprotein particles for exceptional longevity[857]. Small LDL particle size has been associated with cardiovascular risk and a possible explanation for the long lives of centenarians could very well be their possession of heritable so-called atherosclerotic protective factors. Indeed, results published in the same paper indicated the association of this lipoprotein phenotype of exceptional longevity with an increased frequency of homozygosity of a common functional variant (I405V) in the gene for cholesterol ester transfer protein (CETP), and decreased CETP levels. Moreover, these investigators also found that homozygosity for the APOC3 -641C allele is associated with a favorable lipoprotein profile, cardiovascular health, insulin sensitivity, and longevity[858]. These results are also in agreement with the previously reported decreased frequency of the apoliprotein E-4 (APOE) allelic variant, conferring a high risk for both cardiovascular and Alzheimer's disease, in French centenarians, suggesting a decreased risk for these diseases in individuals with exceptional longevity[859].

While limited mainly to lipoprotein-related genes in candidate-gene studies, these results strongly support the utility of centenarians as a model system in which to study the genetics of longevity and aging in humans. One persistent problem in centenarian studies is the lack of good controls. Nir Barzilai (New York, NY, USA) and collaborators realized that family members of centenarians are likely to inherit genetic factors that modulate aging processes and disease susceptibility. While also harboring many genes that are unrelated to the longevity phenotype, such offspring can be expected to be enriched for longevity genes. Comparing offspring of centenarians in a case-control study with their spouses as the control group, these investigators found significantly lower prevalence of hypertension, diabetes, heart attacks, and strokes among the children of the centenarians[860].

Based on these initial results of linkage and association analysis in the human system, most of which are based on candidate-gene studies involving only very few genes and virtually no regulatory regions, it is clear that no conclusions can be drawn regarding genome maintenance as a possible causal factor in aging. However, the fact that these first studies identified genes involved in lipoprotein metabolism as genetic determinants of extreme longevity suggests that genome instability may not be the main cause of age-related mortality in humans.

7.3.3 FUTURE PERSPECTIVES

In the 1970s, before very much was known about the regulation of cellular senescence, several authors, most notably Alvaro Macieiro-Coelho (Versailles, France)[861], proposed that cellular aging and immortalization both depended on random reorganization of the cell genome, as determined by chance, intrinsic properties of the cell and environmental factors. Macieiro-Coelho argued that such random genomic alterations as genome rearrangements, transpositions, amd methylation changes caused the loss of division potential. However, some cells had a high chance to overcome this genome reorganization and escape from senescence. He took the mouse genome as especially plastic, based on the assessment of a number of variables, such as the rate of sister chromatid exchanges, chromosomal aberrations, and aneuploidy. Whereas this high plasticity quickly leads to immortalization, Macieiro-Coelho observed a stabilization of the genome thereafter. It is now known that replicative senescence is a programmed cellular response against genomic stress, rather than a result of genome alterations per se, but these observations are in striking agreement with what could very well be the major cause of aging. As we have seen, due to natural limitations in somatic maintenance and repair, somatic-cell populations gradually lose their orchestral nature and diverge into cellular mosaics, as is their fate as intermediaries between two germ lines. The accumulation of variation in the somatic genome is inherently stochastic, with genetic and environmental factors playing important roles in modifying patterns of genomic instability in different individuals of a population.

There is abundant evidence that genomic alterations in somatic cells are diverse and can occur at significant rates, but exactly how such random events cause functional decline and disease is not known. A random mutation can exert an effect by being amplified. The best example is of course amplification by clonal expansion and we have seen that this is the most likely mechanism underlying the dramatic increase of cancer at old age. There are other forms of clonal expansion, such as developmental expansion when all mitotic descendants of a mutated cell inherit the mutation, with its altered behavior. Another example is the aforementioned atherosclerotic plaque formation which may involve local hyperproliferation. The small sets of cells involved in such forms of amplification could easily influence distant cell populations, for example, by releasing factors that induce cellular alterations. This could significantly extend the reach of a clonal expansion.

Somatic mutations contributing to disease, based on the amplification of a mutation normally associated with a heritable form of the disease, are now well documented[862]. Such cases are called segmental. For example, a microdeletion in the heritable disease gene, neurofibromin, has been detected in fibroblasts cultured from the *cafe-au-lait* lesions in the affected segment of the body of a patient with segmental neurofibromatosis[863]. The mutant allele was present in a mosaic pattern in the *cafe-au-lait* fibroblasts from the spot lesions, but was absent in fibroblasts from normal skin or peripheral blood lymphocytes. Whereas there are many other examples of such somatic mosaicism, their contribution to normative aging is unclear at present.

Whereas clonal expansion and diffusion of released factors could very well play an important role in translating random mutations into aging-related adverse effects (including non-cancer phenotypes), stochastic alternatives should not be ignored. Indeed, the alternative possibility, proposed in this chapter, of increased transcriptional noise generating cell mosaics of functional decline in highly specialized cell populations, could very well be more generally applicable as a cause of aging.

In this chapter I have provided the alternative, more general possibility that consistent patterns of functional decline in highly specialized cell populations can be a result of increased randomization of gene-regulatory patterns. Irrespective of the exact mechanisms by which random somatic genome alterations act to cause aging and disease, the important question to address is to what extent such events contribute to the causes of aging. Although it is hard to see how genome alterations could underlie each and every symptom of aging, the key issue is whether it is the ultimate limiting condition that we have to consider in designing interventions to attenuate the adverse effects of aging. Indeed, is it possible that random genome alteration essentially constrains attempts to significantly delay or even reverse aging? This is the topic of the final chapter.

8 A genomic limit to life?

In previous chapters I have discussed the possibility that aging is a universal outcome of the natural plasticity of the genome. Random alteration of genomes and the principle of natural selection provided the creative impulse behind the emergence of robust living systems, most of which are simple, unicellular organisms. It is only in the small fraction of complex metazoan species, with the soma held captive by the germ line, that aging first became apparent. As we have seen in Chapter 6, most deleterious mutations are efficiently purged from the germ line through selection during propagation of the germ cells, embryogenesis, and early life, resulting in cell death, spontaneous abortion, still birth, or early death of individuals, preventing them from reproducing. To some extent, stem cells or progenitor cells should also be able to rid themselves from cells harboring mutations with adverse effects. Highly specialized somatic cells, however, can be expected to ultimately succumb to an ever-increasing mutation load.

A model, presented in the previous chapter tentatively explains how random mutations can lead not only to cancer, but also to reduced function of the highly differentiated cells in vertebrates. Aging would then be the result of the ultimate failure of genome maintenance in preventing increased noise in the execution of various cellular programs due to accumulating errors in the information content of the genome. This model remains untested and it could be wrong. Instead, aging could be partially or wholly due to a combination of other processes, most notably increased damage to cellular macromolecules, which may not be restricted to DNA, and the responses to their adverse effects. This possibility has been extensively discussed in previous chapters. Here I will discuss the consequences of a scenario in which aging is predominantly a result of random, irreversible alterations in the somatic genome, and how that would impact our prospects to intervene and possibly retard or reverse aging-related cellular degeneration and organismal death.

8.1 Aiming for immortality

The inevitability of aging and death has been a major theme in philosophy and literature throughout the ages. It first appeared in the oldest written story, the epic of Gilgamesh. It was written in ancient Sumeria on 12 clay tablets in cuneiform script and tells the story of Gilgamesh, the historical King of Uruk, somewhere between 2750 and 2500 BCE. The central theme of the epic is the vanity of all human effort to escape death. Although Gilgamesh ultimately realizes that the best way mortals can achieve immortality is through lasting works of civilization and culture, eternal life would remain on the front burner of human interest. One could argue that all world religions ultimately sprang from the desire for immortality and that science, itself a religious offshoot, was born from early practitioners of alchemy and hermetics and their attempts to find the elixir of youth when not occupied by seeking the philosophers' stone to change base metals into gold. Nowadays, such endeavors are considered as pseudoscience and serious scientists try to stay as far away from such activities as possible.

Nevertheless, anti-aging movements blossom as never before, due to the availability of a multitude of possible treatments to delay the inevitable, including dietary supplements, cosmetics, hormone therapy, and many other interventions that are not part of standard medical care. Although they are often considered as pseudoscience, there is in fact legitimacy in such activities. This is apparent, for example, from the website of the American Academy of Anti-Aging Medicine (A4M). While hyperbole is generally not spared in the information provided by this organization, it does reflect the mainstream scientific literature. Its recommendations may be frequently flawed and its predictions that an age-less society is just around the corner not based on a correct interpretation of the facts, but the organization is nevertheless a genuine product of modern science. For example, I have not found any evidence that A4M supports the use of undefined anti-aging potions, concocted when the full moon is in Gemini. Indeed, while it is impossible to know if any of the proposed treatments extend human lifespan (or have long-term, adverse side effects, as suggested for growth-hormone treatment), for many of A4M's recommendations (such as exercise) there is evidence that they improve health and well-being in old age. At least some of the preventive measures advocated by A4M and others are likely to retard damage accumulation, often considered as the basic cause of aging.

Ironically, some of the various statements by mainstream scientists, claiming that there are no effective anti-aging measures at present and that human lifespan cannot be significantly extended, are hardly more rational than A4M's advocacy of an imminent age-less society. Although it is important to warn against the more outrageous, unsubstantiated claims of some entrepreneurs promoting anti-aging products, there seems no reason to be unduly pessimistic about our future prospects of enjoying an increasingly longer and increasingly healthier lifespan. Predictions made by, for example, Olshansky and colleagues in a 1990 paper[864], that future declines in death rates would not reach the levels required for life expectancy at birth in low-mortality populations to exceed 85 years, were based on past trends and theoretical assumptions that have never been tested. In fact, as described below, there is clear evidence for a continuing mortality decline of elderly populations in developed countries. Arguments that even eliminating all aging-related causes of death currently written on the death certificates of the elderly will not increase human life expectancy by more than 15 years make little sense in the face of the fact that we are still relatively ignorant about the causes of aging-related illnesses and their relationship to basic processes of aging (Chapter 7). To assume that the molecular and cellular basis of such processes is distinct from 'aging-related causes of death' is at the very least premature.

Hence, we are confronted on the one hand by the optimistic statements of an overenthusiastic anti-aging movement, and on the other hand by those who maintain that under no circumstances can processes of aging be modified to significantly extend lifespan. On the basis of the scientific evidence, neither conclusion seems totally secure. The statements made by the anti-aging movement are the easiest to discard, in the sense that there is simply no evidence that any of their advocated practices intervenes in basic processes of aging. Conversely, the possibility that at least some of them actually do retard aging can also not be completely excluded. The pessimists are more difficult to deal with because they often represent mainstream scientists, and their message to reject false claims of entrepreneurs promoting products to slow or reverse aging is a sympathetic one. Nevertheless, their assessment that future increases in lifespan are all but impossible is overly pessimistic in view of the abundant evidence to the contrary, mostly obtained in animal model systems. Recent developments provide some support for potential initiatives to extend human lifespan. There are at least two bodies of knowledge that suggest that lifespan is flexible and can range far above the natural average under certain conditions.

The first line of evidence stems from studies of animal and human populations. As we have seen in Chapter 2, mortality increases exponentially with age, as originally established by Benjamin Gompertz for humans in the early nineteenth century. While valid for a great variety of species, it has never been clear if this exponential rise in mortality would continue at much older ages. We have already concluded that in natural populations older cohorts are rarely expected due to extrinsic mortality. A group of researchers has now studied mortality at advanced ages for a number of species under protected conditions, including flies, worms, and humans. Especially the data on flies, from James Carey's (Davis, CA, USA)[865] and James Curtsinger's (St. Paul, MN, USA)[866] laboratories, some based on studying over 1 million individuals, indicate that mortality decelerates at older ages and that many individual flies live much longer than what was assumed to be their maximum lifespan. Similar findings were made in the nematode worm in Thomas Johnson's laboratory[867]. In humans, work by James Vaupel (Rostock, Germany)[56] and others, based on 70 million elderly people, suggests that mortality decelerates after age 80, reaching perhaps a maximum around age 110. Superficially these results appear to refute the evolutionary basis of aging which predicts that mortality should continue to increase ever more rapidly after the reproductive period. However, populations are never homogenous, even if they are inbred, and it is likely that the results of these studies simply reflect genetic good fortune, chance, and environmental permissiveness. Moreover, in the absence of specific quantitative predictions of the evolutionary theory of aging with respect to extreme old age it is premature to conclude that a deceleration of the increase in mortality with age is not in keeping with its basic tenets.

Human life expectancies are increasing almost everywhere in the world where socioeconomic circumstances are permissive[3,868] and there is no evidence that a limit to life is anywhere near[1,869]. This increase in lifespan effectively rules out a compression of morbidity as proposed by James Fries (Palo Alto, CA, USA)[870]. According to this hypothesis, we can expect to reduce morbidity and delay the onset of chronic disease. Mortality, however, was assumed to be unalterable and fixed according to the species' genetic make-up. While reducing morbidity should of course always be a major goal in aging research and, as we have seen, the dramatic improvements in general nutrition, hygiene and public safety have indeed been responsible for a rectangularization of our survival curve, the evidence is lacking that it is possible to continue along this track. Indeed, it is highly doubtful whether we will be able to eliminate aging-related diseases without substantially increasing maximum lifespan. In view of the absence of any scientific criteria to separate such diseases from an hypothetical intrinsic aging process (see Chapter 7) we have to assume that a successful delay of chronic disease will automatically increase lifespan. This is exactly what we are now witnessing[2]. Older individuals have never been so healthy and further improvements in life style, environmental conditions, and medical care, on a sound scientific basis, are likely to help this trend to continue. It will obviously take decades before we will know the effect of such recent medical advances as the large-scale use of antihypertensive and cholesterol-lowering drugs. Nevertheless, all available evidence points towards a continuing rise in both human health and lifespan.

A second line of evidence supporting the plasticity of animal lifespan and the potential for intervention is provided by the discovery that intervention in growth, energy metabolism, reproduction, or nutrient sensing can greatly increase longevity in a range of model organisms, from worms and flies to mice[871]. As extensively discussed in Chapter 2, dampening such activities activates pathways that promote survival by regulating genes that control cellular maintenance and repair. The realization that aging, like many other biological processes, is subject to regulation by pathways that have been conserved during evolution suggests that it is also possible to intervene in its basic causes in humans. Indeed, it was noted by Cynthia Kenyon that the 6-fold extension of lifespan in *C. elegans*, achieved by removing the reproductive system and decreasing insulin-signaling activity in the same animals, would be the equivalent of healthy, active 500-year-old humans[871]. Pharmacological simulation of these effects may therefore prolong life and youthful vigor. This approach has been discussed by Richard Miller, who concluded that the existence of long-lived mutants in a variety of species together with an expected enormously increased knowledge about aging were a strong basis for predicting a lifespan increase of up to 40% in humans[872]. However, as we have seen, intervention in metabolism carries a price and it remains to be seen whether compounds can be identified that mimic the effect of, for example, calorie restriction, without adverse effects. In addition, since the pro-longevity effects of intervention in these pathways on invertebrates are often dramatic (up to 4-fold), but much less in rodents (up to 50–60%), they may increase lifespan of humans only marginally. Indeed, whereas the first results of calorie restriction on non-human primates do indicate similar beneficial effects on health and function as in rodents, such as reduced body temperature and insulin levels, significantly longer life-spans are not yet apparent. Because these studies are still ongoing definite conclusions cannot be drawn. Nevertheless, it is entirely possible that much smaller increases in life-span in humans or rodents, percentage-wise, than in worms or flies may be a consequence of the increased complexity of pathways such as the insulin/IGF-1 pathway in long-lived mammals. Of note is that these invertebrate organisms are composed entirely of postmitotic cells, whereas mammals have renewable tissues that need to be replenished by cell proliferation. Nevertheless, the widespread existence in the animal world of the option to increase lifespan through metabolic modulation is strong evidence that humans can also be made to live longer, once a suitable strategy has been found.

8.2 SENS, and does it make sense?

We are now confronted with real scientific evidence that animal lifespan is malleable. However, what evidence do we have that this malleability can be exploited to increase human lifespan? Some time ago a group of scientists, spearheaded by Aubrey de Grey (Cambridge, UK), proposed a different approach to slow or reverse aging[873]. Based on the premise that aging is due to the accumulation of somatic damage they argue that intervention can occur at three levels: metabolism, damage, and pathology. Metabolism generates damage, for example, oxidative damage, as an inevitable side effect. A fraction of the damage cannot be removed by endogenous repair processes, which are necessarily imperfect (Chapter 2), and thus accumulates. This accumulating damage ultimately drives age-related degeneration and leads to pathology.

Interventions can be designed at all three stages. I have already discussed the option to mimic the metabolic interventions that proved so successful in animal models, especially invertebrates, to increase somatic maintenance and repair and decrease the accumulation of unrepaired somatic damage. However, intervention at this level is likely to carry a heavy price due to the extreme interconnectedness of metabolic pathways, optimized over eons of natural selection. Similarly, intervention in pathology is necessarily impractical in the long term if the damage that drives it is accumulating unabated. Therefore, so the argument goes, intervention to remove the accumulating damage is the most promising strategy to sever the link between metabolism and pathology and has the potential to postpone aging indefinitely.

Negligible senescence—the absence of a statistically detectable increase in mortality rate with age—has been demonstrated to exist at least in one metazoan, *Hydra*[874]. *Hydra*, a cnidarian, is one of the simplest of metazoans. It can reproduce vegetatively as well as sexually and is basically a bag of stem cells, permeated throughout its body. This virtual absence of a clear distinction between soma and germ line gives the organism almost limitless power to regenerate. This, together with a lack of highly differentiated cells, probably allows it to escape senescence. As discussed in Chapters 2 and 6, the growth centers (meristems) of plants provide an unlimited source of 'stem cells' and a similar open-ended lifespan as in *Hydra*. This does not mean that mutations do not accumulate in such cells. However, in the absence of differentiated cells with highly specialized functions, deleterious mutations can be efficiently purged from the population by intra-organismal selection, whereas others might even be beneficial and a source of adaptive fitness[810].

Other multicellular organisms, especially certain poikilotherm (cold-blooded) vertebrates, may exhibit negligible senescence but this has never been confirmed in a statistically satisfactory way. No warm-blooded animal (homeotherm) has ever been shown to do so. The strategies outlined by Aubrey de Grey and colleagues to emulate *Hydra* and follow in the footsteps of Gilgamesh to stop aging in its tracks are termed strategies for engineered negligible senescence or SENS for short.

SENS includes a number of interventions to reverse aging, first and foremost clearance of damage, including intra- and extracellular aggregates and macromolecular cross-links. The focus here is on protein damage, not on DNA damage. A second type of intervention prominent in SENS is stem-cell therapy to replenish tissues subject to cell loss. As we have already seen, whereas cell loss may contribute to age-related functional decline, its share in the causes of aging is likely to be small, except in disease states. Cancer is recognized as a special problem due to its intrinsic genetic instability, facilitating resistance to any treatment by activating or inactivating specific genes. In this case a rather drastic solution is proposed involving the entire replacement of the body's own stem cells by periodic reseeding with telomere-elongation-incompetent stem cells. The lack of telomerase would effectively prevent unlimited proliferation and therefore block tumor progression, but would allow tissue replenishment for a given amount of time. The two final components involve the removal of senescent cells (which could be important in select tissues) and the rescue of mitochondria from accumulating mutations by the transfer of their genes to the nuclear genome. All these possible interventions have been described in more or less detail in the scientific literature[875].

Aubrey de Grey and colleagues have speculated that the advanced methodology to repair each type of damage could become available in time to increase the expected age at death of a healthy 55-year-old, which is currently about 85, to 115 by the year 2030. Since the longest-lived person ever, Jeanne Calment, reached the age of 122 several years ago, a target of 115 seems rather modest. However, it should be realized that efforts to increase the average lifespan in such a short time to even this age will only succeed by drastically accelerating the current downward trend in mortality. Indeed, even optimistic predictions of average lifespans of 95–100 would require a large deviation from past trends[869].

Predictions of rapid increases in life expectancy to 115 therefore require rather drastic interventions which are unlikely to be accomplished through further optimization of lifestyle, the use of nutritional supplements, and more frequent applications of novel preventive drugs such as the aforementioned antihypertensive and cholesterol-lowering drugs. Instead, it will require preventive and/or therapeutic treatment inhibiting the basic processes that underlie aging and its associated diseases. Hence, SENS relies on technology not yet developed. Is this unrealistic?

Roadmaps towards a desirable goal in science are not new and are frequently applied. They merely serve the purpose of an extended look at the future of a chosen field of enquiry to bridge the space between where we are and where we want to be. A roadmap is not a crystal ball and the reality will deviate from expectations. It is also perfectly legitimate to rely, in part, on new technology expected to be developed. An example of this is the Human Genome Project, which was originally meant to rely heavily on technology that still remained to be developed. In reality, the project was accomplished ahead of time solely based on improvements to the existing Sanger sequencing technology (Chapter 1). Another example is the so-called war on cancer, which has not been particularly successful in terms of clinical significance. However, it focused everybody's attention on a major problem and a desirable outcome and its advances have been enormous in terms of our understanding of the biological basis of cancer; it is still likely that this will ultimately lead to successful therapies. Likewise, our greatly increased knowledge of the possible causes of aging and enormously improved tools for engineering cells and animals make it worthwhile to at least consider the possibility that current trends towards declining mortality can be accelerated considerably. While the time frame for obtaining even modest results is likely to be excessively optimistic (as it was with cancer), SENS certainly provides a welcome step towards formulating a rational goal for aging research. With respect to the SENS agenda, therefore, neither the uncertainty of as-yet-undeveloped technology nor the speculative nature of the predicted time frame should necessarily be considered problematic. What is important, however, is its feasibility, which depends on an accurate definition of the problem. In other words, is SENS in keeping with what we now know about aging?

Unlike cancer, aging is not a clonal phenomenon, but reflects a divergence from a common differentiated lineage towards a mosaic of cells, differing from each other in the frequency and location of the scars of aging. The assumption of SENS is that the types of damage are very similar in all cells of a given tissue and can be specifically targeted. For example, intra- and extracellular aggregates are proposed to be removed through enzymatic digestion and phagocytosis, respectively, macromolecular cross-links will be removed through cross-link breakers, senescent cells by driving them into apoptosis, and mtDNA mutations by incorporating the critical parts of the mitochondrial genome into the presumably much better defended nuclear genome. As mentioned, cancer cells will be prevented from progressing into tumors by depriving all somatic cells of telomerase and tissues suffering from cell loss will be rejuvenated using stem cells. All these damage-control measures together are expected to reverse the adverse effects of aging and restore youthful vigor.

One could of course have some reasonable doubt about the speed with which such therapies would become available, based on the rather disappointing track record of successful medical interventions through the ages. Indeed, in spite of some major break-throughs in health care, such as the development of antibiotics, successful interventions in aging-related diseases have been few and it is not immediately clear why the translation from newly discovered basic biological principles into clinically robust treatments would suddenly dramatically accelerate. However, this is not a decisive objection against SENS, because we have clearly made significant progress in understanding the basic causes of aging and recent progress in biomolecular engineering has been dramatic. Hence, the possibility cannot be excluded that preventive and therapeutic interventions to cleanse the human body of molecular and cellular damage will be successful on a very short time frame. The question is whether the list of proposed causes of aging as provided by SENS is complete. As argued by Huber Warner (St. Paul, MN, USA), it is not clear at pres-ent if these proposed causal factors have an impact on longevity, even in an experimental animal model[876]. Indeed, as mentioned throughout this book, it is uncertain whether factors like cell loss, accumulating senescent cells, or various forms of chemical damage are causes or consequences of the aging process.

If a significant part of the adverse effects associated with aging is the consequence of random alterations in the information content of the genome, corrective intervention would be impossible, or at the very least subject to enormous complications. In contrast to chemical or cellular damage, which share the basic features suitable for therapeutic targeting, random genome alteration creates (epi)genetically different cells, a situation that cannot be easily reversed. In the absence of turnover from an immutable template, altered genomes can also not be liquidated as in protein repair. Hence, every individual cell would now be genetically different and would need individual 'treatment' to correct its mutation load. In principle, such correction would be very similar to current attempts at gene therapy: through the use of vectors transferring the correct piece of DNA sequence into the sick cell to replace the mutated fragment. Editing genes to correct defects has made great strides, but it is unlikely that we will ever be able to correct the collection of random mutations in our genome that result from natural wear and tear. Indeed, it is entirely reasonable to assume that stochastic alterations in the information content of the genome are essentially irreversible. If this proves to be the main underlying cause of aging SENS would have a problem because its strategy ignores the need to correct genomic errors as a potential source of cellular malfunction.

Intriguingly, the SENS program does include a strategy to neutralize the accumulation of mutations in the mtDNA: this potential cause of aging is very difficult to ignore given the numerous observations of high levels of mutated mtDNA copies coinciding with functionally defective cells at old age (see Chapter 6). The proposal to rescue mitochondria from aging through this mechanism by copying the genes residing in these organelles to the nuclear genome depends on the safety of that environment from genomic or epige-nomic alteration. Although it is true that there is as yet no evidence that changes in the information content of the nucleus reach levels high enough to cause cell-degenerative effects, the results mentioned in the previous chapter shed some doubt on the assumption that the nucleus provides the safe haven for genes and their correct expression anticipated by SENS.

If aging is based on random genome alterations that are essentially irreversible because they lack a common signature that could serve as therapeutic target, does this mean that lifespan is ultimately limited? Direct reversal of informational errors is impossible, or at least outside the realm of current or foreseeable technology, but we may consider the option to greatly improve genome-maintenance systems. Because the highly complex, interconnected DNA-repair systems do not lend themselves well to further optimization (see Chapter 4), intervention at the level of DNA-damage signaling to increase apoptotic responses appears a reasonable strategy. As we have seen from the work of Peter Stambrook, increased apoptosis may very well be the main mechanism through which the mouse ES cells are able to maintain the integrity of their genome so dramatically better than normal mouse fibroblasts (Chapter 4). Although excessive apoptosis is likely to increase rather than mitigate age-related functional decline and pathology (due to cell loss, which may not be a major problem during normal aging; see Chapter 5), a slight increase may lead to a more rapid rate of cell replacement and an increased preservation of organ function. Of note, slightly increased apoptosis rates have been found associated with caloric restriction, a condition that extends lifespan[407]. This strategy would heavily depend on regeneration as a mechanism to replace cells eliminated because of their heavy load of genetic damage. While regeneration of the liver after partial hepatectomy has been described in previous chapters, this is not a general characteristic of human tissues. Nevertheless, the capacity to regenerate exists in a wide variety of species and may still be in place in humans. Hence, as speculative as this is, the possibility cannot be excluded that we will be able at some point in time to harness the power of regeneration to more rapidly cleanse our organs and tissues from damaged cells.

Although further optimization of genome maintenance may be a logical and possibly the only feasible strategy for SENS, a gradual loss of genome integrity is ultimately unavoidable. Indeed, maximization of genome maintenance could cause severe meta-bolic imbalances and may not be tolerated by the organism. A logical way out of this dilemma would be to use stem cells to replace cells worn out by high mutation loads. This strategy has its own potential limitations, which make it even more speculative than increasing cell turnover and regeneration. Stem cells may only be able to undertake very limited repairs of organs like liver and heart. Indeed, fibrosis (scarring) may be a fall-back mechanism put in place when the natural stem-cell capacities are exceeded. Moreover, in spite of the more robust genome-maintenance systems in stem cells than in differentiated cells, there is no evidence that stem-cell populations do not accumulate genomic or epigenomic alterations. On the contrary, preserving the genetic integrity of human stem cells is a major concern, as exemplified by the reported increased frequency of karyotypic abnormalities in human ES cell lines after extended passage in culture[877]. Whereas this was found to be dependent on the way the cell lines were propagated—manually or by bulk-passage methods—and therefore to some extent subject to control[878], accumulation of genomic mutations will hinder stem-cell therapy and a strong focus on designing novel strategies to further stabilize genomes is critically important.

In spite of the above and the rather fitful advances in tissue engineering and stem-cell nuclear transfer of late, it is conceivable that major improvements over the next decades will allow transplantation of immunologically matched organs grown from stem cells. Some tissues, especially the brain and skeletal muscle, are clearly not amenable to whole-sale replacement by transplantation, but may be amenable to slow and steady cell replacement. Hence, cell replacement is an essential component of SENS-like roadmaps.

In spite of widespread skepticism from the scientific community, I find it not unlikely that the essence of the predictions made in SENS will prove to be correct. Indeed, it seems reasonable that novel interventions will appear, including some of those outlined in SENS, and that their application will significantly extend human lifespan, possibly even beyond the target of 115 years by the year 2030, as speculated by Aubrey de Grey. There certainly are a host of proximal causes of aging that lend themselves well to intervention. Most of these may have emerged, as dictated by the evolutionary logic of aging, during evolution through genetic drift or as adverse by-products of genetic traits with a beneficial effect early in life (Chapter 2). This will give us a grace period, as it were, for intervention, which we have already begun to explore. It is difficult to know the length of this grace period, but there can be no doubt that it will end when mutation loads become too heavy and cell replacement is no longer enough to compensate for the functional losses. Irrespective of our progress in developing new measures for prevention or therapy, genome instability will be the final curtain and the ultimate limit to life. Genomes cannot be cleansed of all genetic damage, because it is their nature to change and undergo mutation. Indeed, to keep genomes free of change would be to tamper with the logic of life itself.

In 1998, the controversial French novelist, Michel Houellebecq, published his nihilistic classic, *The Elementary Particles* (*Les Particules élémentaires*; Flammarion, Paris, 1998), an indictment of our individualistic ('atomized') society. In this novel, the brilliant scientist Michel Djerzinski develops the theoretical basis for re-writing our genetic code into a new form, no longer susceptible to mutation, with the capacity to replicate indefinitely without error. In the novel's epilogue this rather undigested chunk of science is the basis for the creation of a new species of genetically identical individuals of one sex without the capacity to reproduce and no longer susceptible to aging or disease. While philosophically this reflection on the rise of such a serene successor to humankind (with gratitude to the French philosopher Michel Foucault) may meet approval, I think most of us would prefer to live a bit shorter, with our imperfect selves and an aging genome.

■ EPILOGUE

The history of science is characterized by a quest for fundamental theories. This is true, for example, in physics with quantum theory, theory of relativity, and superstring theory, but no less so in biology, which is dominated by the theory of evolution. There is probably no subspecialty in science in which the formulation of theories has been as pervasive as in the science of aging[51,879]. To see aging as a universal process based on principles that are valid in a wide range of species is only natural in view of the common structural basis of life. By now most of us have come to accept that aging is a random process that falls outside the realm of natural selection. As such it is merely a by-product of natural selection. A random process of wear and tear kept in check for different periods of time, depending on the life history of the species, is a logical explanation for aging, based both on what we see and on what we know. We can all witness the process of wear and tear, its many faces, and its enormous individual and species-specific variation. It is the accomplishments of a number of great minds in science, some of whom have died and some of whom are still my colleagues, that we now know the rational basis behind this wear and tear and understand its relationship with the evolutionary logic of life.

Mutation accumulation as a universal mechanism of aging and death is in keeping with the basic tenets of the theory of why we age and perhaps the only fully convincing explanation of how we age, as I have tried to point out in this book. However, it is by no means the only possible explanation and there is a lot that we do not know, both within and outside the somatic mutation theory. What I have tried to accomplish with this book is to lay down the possible arguments for somatic mutation as a universal theory of aging, critically discuss the available evidence on both sides of the debate, and predict its consequences for possible strategies to halt or reverse aging in humans.

Meanwhile, we have barely scratched the surface in our attempts to test the somatic mutation theory, which can only happen through a painstaking process of unraveling the impact of genomic instability on cellular function. This is part of the much broader task of linking genotype to phenotype. As we now know, this is not the same as studying single genes and their products. Although molecular biology has in fact never been as guilty of genetic determinism as sometimes implied, we have now definitely reached a stage where we can begin exploring the function of each gene and its multiple protein products in the context of all the others. There are many problems on this road, the end of which is not in sight. In the face of an overwhelming amount of data we still have insufficient information in our current databases to predict the behavior of complex systems, assuming that this is not inherently impossible. We also lack the computational tools to work with gigantic data-sets, analyze them in an integrated manner, make predictions, and formulate hypotheses.

The study of the molecular basis of aging requires all of the above and a great deal more. Thus far, the most dramatic progress in the science of aging recently has been the discovery of conserved genes and functional pathways that control lifespan in multiple organisms. While this field is still very much alive and systematic genetic screens, now also including screens *in silico*, are still yielding increasingly detailed genetic information as to how lifespan is controlled, the next major development may be around the corner. Based on strong evidence that the survival pathways, identified first in nematodes and yeast, then in flies and mice, work by limiting the accumulation of unrepaired somatic damage, we will now need to return from questions about longevity to questions about aging. We will need to know the mechanisms that lead from random, unrepaired somatic damage to functional decline and disease. We need to understand how such a stochastic process of damage accumulation can lead to reproducible patterns of phenotypic change and how to explain the numerous observations of phenotypic variation in isogenic or clonal populations or among different, seemingly identical cells. It is clear that to understand aging we will need to understand how its stochastic component is linked to its reproducible outcome.

Organismic function is based on networks of coupled biochemical reactions and feedback signals. Within those regulatory networks gene activity is controlled by molecular signals that determine when, where, and how frequently a given gene is transcribed. Regulatory proteins may act in combination with environmental signals to control multiple genes, which results in complex, branching networks of interaction. It is highly unlikely that the outcome of such regulatory networks is completely deterministic. On the contrary, there is now conclusive evidence for a major stochastic component in gene expression already at the level of transcription. As outlined in Chapter 7 of this book, such transcriptional noise could gradually increase with age, for example, as a consequence of accumulating genetic damage.

Increased noise in gene expression can be offset by such factors as redundancy between genes and pathways, integration of expression over time or relatively stable protein levels. Eventually, however, it will increase probability as a factor in determining the outcome of a gene-regulatory event. This will result in increased fluctuations in protein levels, with proteins no longer arriving at the right moment in the right amount. In turn, this will adversely affect cell signaling with signals no longer arriving in time. This could explain many examples of delayed cellular responses in old individuals. It could be the reason for a decline in temperature control, immune response, and glucose tolerance. Genes encoding components of multi-protein complexes are likely to be especially sensitive to random fluctuations in gene expression. Adverse effects on the timing for the subunits to come together would compromise the integrity of the complex and may lead to increased protein aggregation.

So, increased stochasticity, beginning at the level of the genome, may work its way through the system and eventually translate into aging as a pattern of cells and tissues acting increasingly out of tune. It is now possible to explore a wealth of ever-more sophisticated tools in both cell and molecular biology and the computational sciences to systematically study these cascades of stochastic variation in the various model systems for aging that are now available. This is likely to finally unlock the door to a full understanding of how we grow old.

GLOSSARY

acentric fragment A fragment of a chromosome generated by breakage that lacks a centromere and is lost during cell division.

aging The time-dependent loss of viability and increase in vulnerability to disease. At the population level, aging is defined as an exponentially increasing probability of death after maturity. *See also* **senescence**.

allele An alternative form or variant of a DNA sequence at a genetic locus. Of the two alleles per locus in an individual, one is inherited from the father, the other from the mother.

alternative splicing Different ways of combining exons of a gene resulting in protein variants.

Alu family A set of about 1 million dispersed repeats in the human genome, each about 300 bp long. Named after the restriction endonuclease *Alu*I that cleaves it.

amplification An increase in the number of copies of a specific DNA fragment.

amyloid An extracellular protein deposit found to increase in diverse tissues with age, most notably in the brain, where it has been associated with Alzheimer's disease.

antagonistic pleiotropy When there is a beneficial effect of a gene early in life, but an adverse effect later.

anticipation The increased severity of a genetic disorder through the generations. Can be caused by, for example, expansion of microsatellite repeats.

antigen A foreign substance provoking an immune response after entering the body of an organism.

aneuploidy When a cell has lost or gained one or more entire chromosomes.

apoptosis Programmed cell death to get rid of cells that are not needed or heavily damaged.

arteriosclerosis The hardening of artery walls and the consequent narrowing of the vessel lumen.

atherosclerosis The most common form of arteriosclerosis, caused by plaque formation in the wall of arteries or arterioles.

autoimmune disease A pathological condition in which the immune response is directed to some of the organism's own molecules. According to Frank Macfarlane Burnet somatic mutations accumulating during aging giving rise to 'forbidden' lymphocyte clones are the most likely explanation for such diseases.

autoradiography A technique that uses X-ray film to visualize radioactively labeled molecules.

autosome A chromosome not involved in sex determination.

bacterial artificial chromosome (BAC) A vector used to clone DNA fragments of 100–200 kb in *E. coli*.

bacteriophage A virus for which the natural host is a bacterial cell. Bacteriophage λ is often used as a vector for cloning DNA fragments.

base The part of a DNA or RNA molecule that determines sequence specificity.

base pair (bp) Two bases (i.e. A and T or G and C) held together by hydrogen bonds.

B cell A cell of the immune system involved in adaptive immunity by producing specific antibodies.

bioinformatics The management and analysis of biological data using advanced computing techniques.

caloric restriction (CR) Intake of a low-calorie diet, demonstrated to extend lifespan in multiple species.

cancer A type of disease in which cells become abnormal and multiply unchecked.

candidate gene A gene located in a chromosome region suspected of being involved in a particular phenotype.

carcinogen An agent or compound which causes cancer, often by damaging the DNA.

cell The basic unit of all living organisms.

centenarian A person who has lived for at least 100 years.

centimorgan (cM) A measure of recombination frequency. One centimorgan is the equivalent, on average, of 1 million bp.

central dogma The flow of genetic information, from DNA via RNA to protein. There are several important exceptions to the central dogma, such as reverse transcription, where DNA is synthesized from an RNA template.

centromere The part of the chromosome to which spindle fibers attach during cell division.

chromatid pair A chromosome and its copy produced by replication just before they separate during mitosis.

chromatin The ordered complex of DNA, RNA, and proteins that forms the uncondensed eukaryotic chromosomes.

chromosome A self-replicating complex of DNA and protein that carries the genetic information of a cell. Human cells have two sets of 22 chromosomes plus two sex chromosomes.

clone An exact copy of a biological entity, such as a DNA segment, a cell, or an organism.

cloning Producing multiple, exact copies of a DNA fragment (in *E. coli*), a cell, or an organism.

co-dominance When both alleles from the same locus are expressed.

codon A sequence of three nucleotides in the DNA, specifying an amino acid. The sequence of codons in the mRNA defines the primary structure of the polypeptide chain.

collagen A fibrous glycoprotein and the major component of connective tissue, providing strength and support to the skin, ligaments, tendons, bones, and other parts of the body.

complementary DNA (cDNA) DNA that is synthesized in the laboratory from a mRNA template.

complex trait A trait that does not follow strict Mendelian inheritance, for example, because it involves multiple different genes as well as environmental factors.

conjugation The exchange of genetic material between two cells during contact.

conserved DNA sequence A DNA fragment that has remained essentially unchanged throughout evolution.

CpG islands Stretches of DNA, about 1–2 kb long, with an average GC content of about 60%, as compared with 40% for the genome overall. CpG islands often surround the promoters of constitutively expressed genes.

crisis When further replication of primary cells in culture is prevented by a wave of end-to-end fusions of chromosomes, apoptosis, and replicative senescence due to extremely short telomeres. Sometimes a few cells can bypass the telomere block and become immortal cell lines.

crossing over Exchange of corresponding sections of the DNA in a chromosome through breaking and rejoining during, for example, meiosis.

cytogenetics The study of the physical appearance of chromosomes.

deletion The loss of a segment of DNA from a chromosome.

diploid A complete set of chromosome pairs, one from each parent.

disposable soma theory First proposed by Tom Kirkwood, this theory states that lifespan is the outcome of a balance between investments in somatic maintenance or growth and reproduction. It follows that aging is caused by the accumulation of unrepaired somatic damage.

DNA (deoxyribonucleic acid) The molecule that encodes genetic information.

DNA damage Changes in the DNA chemical structure induced by a variety of different chemical, physical, and biological agents.

DNA repair The complex of enzymatic systems to remove DNA damage and restore the original situation.

DNA replication The process of copying the entire complement of genetic information.

DNA transcription The process of copying the genetic information encoded in a gene into mRNA.

domain I. Part of a protein with its own function. II. The region of a chromosome in which the DNA has a similar higher-order structure, independent of other such domains on that chromosome.

dominant allele An allele with an effect that is seen in the phenotype if it is homozygous or heterozygous.

elastin A fibrous protein providing resilience and elasticity to skin, ligaments, and artery walls.

electrophoresis A method of separating DNA or protein molecules on the basis of their size or specific sequence using their electrical charge.

embryonic stem cell (ES cell) An embryonic cell that can replicate indefinitely and is pluripotent; that is, it can give rise to many types of differentiated cells.

endonuclease An enzyme that cleaves within a nucleic acid chain.

enzyme A protein that acts as catalyst by increasing the rate of a biochemical reaction.

epigenetic Describing a change that influences the phenotype without altering the genotype.

epistasis When the action of one gene interferes with that of another.

error catastrophe An hypothesis, first proposed by Leslie Orgel, in the wake of the central dogma, stating that aging is a result of random errors in macromolecular synthesis. A positive-feedback loop driven by errors in the macromolecules that are themselves involved in producing DNA, RNA, and proteins would lead to a general breakdown of cellular information transfer. *See also* **central dogma**.

euchromatin The part of the genome with an open chromatin configuration which usually contains the active or potentially active genes. *See also* **heterochromatin**.

eukaryote An organism with cells that have a structurally discrete nucleus.

exon Part of the protein-coding DNA sequence of a gene.

exonuclease An enzyme that cleaves nucleotides from the end of a nucleic acid molecule, from either the 5' or 3' end.

fecundity The physiological ability to reproduce. Different from fertility which is the actual production of live offspring.

fertility Reproductive potential; for example, in humans it is the number of births per woman.

fibrosis Deposition of fibrous, scar-like connective tissue, for example, after injury, inflammation, or infection.

free radicals Molecules with unpaired electrons that react readily with other molecules. Free radicals, which in aerobic organisms are mainly adverse by-products of oxidative phosphorylation, are thought to be a major cause of the aging process.

functional genomics The integrated study of all genes, their resulting proteins, and their interactions in relation to their function in the organism.

gamete Male or female reproductive cell (sperm or ovum) with a haploid set of chromosomes (23 for humans).

gene The basic unit of the genetic material. Physically, a gene is the segment of DNA encoding a polypeptide chain; it includes regions immediately preceding and following the coding sequence as well as intervening sequences (introns) between individual coding sequences.

gene conversion The non-reciprocal transfer of genetic information between two homologous DNA fragments. The process is not an exchange of genetic information between two double-strand DNA molecules, but the conversion of the genotype of one strand of a heteroduplex to the genotype of the other strand.

gene expression The conversion of the information encoded in a gene into proteins that provide structure and function to the cell and the organism.

genetic code Sequence of nucleotides that in the form of a triplet code (codons) determines the sequence of amino acids in a protein.

genetic mosaic An organism or tissue in which different cells contain different genetic sequence, for example, due to mutation accumulation.

genetic polymorphism Difference in DNA sequence at a specific locus among individuals, such as a single-nucleotide polymorphism (SNP).

genetics The study of the inheritance of specific traits.

genome The complete set of genetic information of an organism. The human genome consists of approximately 6 billion bp of DNA distributed among 46 DNA–protein complexes, termed chromosomes.

genotype The genetic constitution of an organism, as distinct from its phenotype.

geriatrics The part of medicine concerned with the care and treatment of older patients.

gerontology The multidisciplinary study of aging. It ranges from basic biology to sociology and can even involve architecture to suit dwellings for aged people.

germ cell A sperm or egg cell and its precursors.

germ line The genetic information that flows from one generation to the next.

glycation The non-enzymatic addition of glucose residues to proteins or DNA. This so-called Maillard reaction eventually leads to advanced glycation endproducts (AGEs). Glycation should be distinguished from glycosylation, which is an enzyme-directed site-specific addition of saccharides to proteins and lipids.

Gompertz function A relationship in which mortality rates increase exponentially during adult life.

haploid The single set of chromosomes as it is present in the egg and sperm cells of animals.

haploinsufficiency A reduction in phenotype due to the loss of one of the turn gene copies at a particular locus.

haplotype The genotype of a number of closely linked loci on the same chromosome. A gene haplotype is essentially the same as an allele.

Hayflick limit The maximum number of times a normal somatic cell can divide in culture. That normal human cells have a limited division potential was first discovered by Leonard Hayflick. *See also* **senescence**.

helicase An enzyme that uses ATP to separate the strands of a nucleic acid duplex.

heterochromatin Regions of the genome that are highly condensed and generally not containing transcribed sequences.

heterozygosity When different alleles are present at a locus on the two homologous chromosomes.

histone A protein found associated with DNA in chromatin.

Holliday structure An intermediate structure in homologous recombination, with the two duplexes still connected by the exchange points on two of the four strands. Nicks or single-strand breaks resolve these structures and restore two separate duplexes. The structure was first proposed by Robin Holliday in 1964.

homologous chromosome The other chromosome of the pair; that is, the one derived from the other parent.

homologous recombination The exchange of DNA fragments between paired chromosomes.

homology Similarity in DNA or protein sequence between different species.

homozygosity When the same alleles are present at a locus on the two homologous chromosomes.

hybridization The process of joining two complementary strands of DNA or RNA.

imprinting The phenomenon of an allele inherited from one of the parents but not expressed in offspring. This 'memory' of a chromosome as to which parent it was inherited from is not determined by the DNA sequence but most likely from its pattern of methylation at CpG sites.

in vitro Outside a living organism, as in cultured cells or in protein extracts.

in vivo In living organisms, as in mice, but sometimes also understood as in cells.

intron A transcribed but non-coding part of a gene, separating one exon (the coding part) from another. Introns are removed from the transcript by splicing together the exons on either side of it.

L1 retrotransposon The L1 (LINE-1, long interspersed nuclear element) repeat family of about 6 kb long, non-long terminal repeat (LTR) retrotransposons makes up 27% of the human genome. L1 elements transpose by first reverse transcribing their RNA into DNA, which is then inserted at a new genomic locus. About 100 L1 retrotransposons are still thought to be active in the human genome.

lagging strand The strand of DNA that grows from 3' to 5', but is replicated discontinuously as short fragments synthesized from 5' to 3' and later covalently connected.

leading strand The strand of DNA synthesized continuously from 5' to 3'.

life expectancy The period an individual organism can be expected to live, as determined by statistics. Often expressed as the age at which 50% of a cohort has died. The same as average or mean lifespan.

life history The features of the life cycle of an organism, as determined by its specific strategies for aging, survival, and reproduction.

ligament Mass of parallel collagen fibers that strengthens a joint or supports an organ.

linkage When two or more genes or polymorphic marker loci are close together on a chromosome.

linkage disequilibrium When two alleles occur together more often than predicted by chance.

lipofuscin An insoluble, brown pigment found in granules representing lipid-containing residues of lysosomal digestion. Its occurrence in organs and tissue increases with age and it is often called age pigment.

locus The place on a chromosome where a gene or DNA sequence marker is located.

longevity The length or duration of life.

lysosome A membrane-bound organelle of the cytoplasm containing enzymes for intracellular digestion.

karyotype The number, size, and shape of an individual's chromosomes.

knockin Targeted insertion of a transgene at a selected locus of a cell or animal germ line. This often involves replacing a gene or part of a gene by a variant sequence, for example, in the mouse germ line.

knockout When a gene is ablated, for example, in the mouse germ line.

macrophage A phagocytic cell of the innate immune system that acts as a scavenger to engulf dead cells, foreign substances, and other debris.

maximum lifespan The age at death of the oldest individual of a population or species.

meiosis The process of two consecutive cell divisions generating the four gametes.

messenger RNA (mRNA) Single-strand RNA copy of a gene that serves as a template for protein synthesis.

microarray Miniaturized set of DNA fragments, antibodies, cells, or pieces of tissue on a solid support to screen biological samples at high throughput.

mitochondrial DNA (mtDNA) The genetic material found in mitochondria.

mitosis The process of nuclear division that produces two identical daughter cells from one parent cell.

monogenic disorder A heritable disease caused by mutation of a single gene.

mutagen An agent that causes a permanent DNA sequence alteration in a cell.

mutation A permanent change in the sequence of DNA in a genome that can involve a single base-pair position (point mutation) or a rearrangement (deletion, insertion, translocation, transposition, or amplification). The mutation frequency *in vivo* is defined as the frequency at which a mutated variant of the gene is found among non-mutated (wild-type) copies.

necrosis Cell death as a consequence of traumatic injury. Necrosis is different from apoptosis, in that it invariably induces an inflammatory reaction.

nuclear transfer The removal of a cell's nucleus and its transfer into another cell, such as an oocyte that has had its own nucleus removed.

nucleolar organizing region (NOR) NORs are regions on different chromosomes containing rRNA genes. These regions get together to form the nucleolus in the cell nucleus to transcribe ribosomal RNA genes.

nucleotide A subunit of DNA or RNA consisting of a base, a phosphate, and a sugar molecule.

nucleus The cellular organelle in eukaryotes that contains most of the genetic material.

oligonucleotide A small DNA or RNA molecule, usually single-stranded and composed of 25 or fewer nucleotides.

oncogene A gene that when activated by a mutation causes the cell to become malignant.

operon A set of genes, usually in prokaryotes, transcribed under the control of an operator sequence.

PCR amplification An *in vitro* process used as a laboratory tool for copying DNA fragments in an amount sufficient for their analysis.

phenotype Outward characteristics of an organism or cell.

plasmid An extra-chromosomal circular DNA molecule that replicates autonomously in *E. coli* and other bacteria.

pleiotropy Multiple effects of a single gene.

pluripotency The potential of a cell to develop into more than one differentiated cell type.

poly(A) tail About 200 adenine nucleotides added to the 3' end of an mRNA following its synthesis.

polygenic trait A heritable trait resulting from the combined action of more than one gene.

polymorphism A difference in DNA sequence among individuals that may underlie differences in phenotype. Irrespective of a possible function, polymorphic variants are used as markers to connect genes to phenotypes.

positional cloning The identification of genes based on their location on a chromosome.

primer Single-strand oligonucleotide that can be hybridized to a known part of a DNA molecule which can then be extended by a polymerase in the presence of new deoxyribonucleotides.

probe A specific DNA or RNA molecule labeled with a radioactive isotope, a fluorescent molecule, or an enzyme that is used to detect a similar base sequence by hybridization to the separated strands in a sample containing nucleic acids.

progeroid A phenotype with features resembling accelerated aging.

prokaryote Cell that lacks a structurally discrete nucleus, such as bacteria.

promoter The portion of a gene to which RNA polymerase must bind before transcription of the gene can begin.

protein translation The process of producing a protein based on the mRNA code.

proteome Total protein complement expressed by a cell or organ at a particular time and under specific conditions.

proteomics The study of proteomes.

pseudogene A disabled copy of a functional gene that was once active in the ancient genome. The human genome contains thousands of pseudogenes.

rate of living theory The theory according to which lifespan is inversely related to metabolic rate. Hence, animals with the most rapid metabolism should have the shortest lifespans. Although this theory has been important for our thinking about aging, it contains many flaws and is certainly insufficient to explain the aging process.

recessive allele An allele with an effect that is only seen in the phenotype if it is homozygous.

regeneration The ability to restore lost or damaged tissues by reactivating already differentiated cells to start dividing again. This is in contrast to fibrosis, which replaces damaged tissue by non-functional scar-like tissue. It is also different from tissue renewal by stem cells, which derives from separate compartments in some organs and tissues.

regulatory DNA sequence A DNA sequence that controls gene expression.

repetitive DNA A collective term for all the DNA sequences that occur more than once in the genome of an organism. Examples include minisatellites, ribosomal RNA genes, and Alu repeat sequences.

replicative senescence The stage at which a cell has permanently stopped dividing.

reverse transcriptase An enzyme catalyzing the synthesis of DNA from an RNA template.

ribosome The cellular structure composed of specialized ribosomal RNA and protein that functions as the site of protein synthesis.

RNA (ribonucleic acid) Similar molecule to DNA, with the exception that it is usually single-stranded and has a uracil instead of a thymine base.

Sanger sequencing The most widely used method of determining the order of bases in DNA.

senescence Those irreversible, deteriorative changes causing functional decline, disease, and death in aged organisms. In this respect senescence is somewhat similar to aging although aging also includes positive changes over time. Senescence is often used to indicate cells that have ceased to divide and reached their Hayflick limit. To make things even more confusing, a senescent cell can also mean a cell that has undergone functional decline in a tissue of an aged organism.

sex chromosome The chromosomes determining the sex of an individual. In humans females have two X chromosomes; males have one X and one Y chromosome.

somatic cell Any cell in the body except gametes and their precursors.

stem cell An undifferentiated cell that can be totipotent (for example, a human zygote), pluripotent (for example, human stem cells from the blastocyst), or multipotent (for example, adult stem cells).

T cell A cell of the immune system involved in adaptive immunity in a variety of cellular interactions.

telomere The end of a chromosome.

transcription factor A protein that binds to gene-regulatory regions in the control of transcription.

transcriptome The complete set of mRNAs of a given cell or cell type.

transgenic mouse A mouse in which a foreign gene or DNA fragment has been artificially introduced in some or all of its somatic cells, often including the germ line.

transposable element A DNA sequence that has the capacity to move from one chromosomal site to another.

trisomy Three copies of a particular chromosome instead of the normal two copies.

X chromosome The female sex chromosome. Males have only one copy of this chromosome; females have two.

Y chromosome The male sex chromosome. Females do not have a copy of this chromosome. Males have one X and one Y chromosome.

wild type The 'normal' version of an organism; the non-mutated variant.

zygote The cell formed by the fusion of two gametes; the start of a new individual.

REFERENCES

1 Oeppen, J. & Vaupel, J.W. Demography. Broken limits to life expectancy. Science 296, 1029–1031. 2002.

2 Manton, K.G. & Gu, X. Changes in the prevalence of chronic disability in the United States black and non-black population above age 65 from 1982 to 1999. Proc. Natl. Acad. Sci. USA 98, 6354–6359. 2001.

3 Wilmoth, J.R., Deegan, L.J., Lundstrom, H., & Horiuchi, S. Increase of maximum life-span in Sweden, 1861–1999. Science 289, 2366–2368. 2000.

4 Farber, P.L. Finding Order in Nature: the Naturalist Tradition from Linnaeus to E.O. Wilson. The Johns Hopkins University Press, Baltimore, 2000.

5 Darwin, C. On the Origin of Species a Facsimile of the First. Harvard University Press, Cambridge, MA, 1975.

6 Harman, O.S. Cyril Dean Darlington: the man who 'invented' the chromosome. Nat. Rev. Genet. 6, 79–85. 2005.

7 Kirkwood, T.B. & Cremer, T. Cytogerontology since 1881: a reappraisal of August Weismann and a review of modern progress. Hum. Genet. 60, 101–121. 1982.

8 Dawkins, R. The Selfish Gene. Oxford University Press, Oxford, 1989.

9 Kirkwood, T.B. Evolution of ageing. Nature 270, 301–304. 1977.

10 Charlesworth, B. Fisher, Medawar, Hamilton and the evolution of aging. Genetics 156, 927–931. 2000.

11 Avery, O.T., MacLeod, C.M., & McCarty, M. Studies on the chemical nature of the substance inducing transformation of pneumococcal types. J. Exp. Med. 98, 451–460. 1944.

12 Watson, J. & Crick, F. Molecular structure of nucleic acids. Nature 171, 737–738. 1953.

13 Meselson, M. & Stahl, F. The replication of DNA in E. coli. Proc. Natl. Acad. Sci. USA 44, 671–682. 1958.

14 Jacob, F. & Monod, J. Genetic regulatory mechanisms in the synthesis of proteins. J. Mol. Biol. 3, 318–356. 1961.

15 Crick, F.H.C., Barnett, L., Brenner, S., & Watts-Tobin, R.J. General nature of the genetic code for proteins. Nature 192, 1227–1232. 1961.

16 Nirenberg, M.W. & Matthaei, J.H. The dependence of cell-free protein synthesis in E. coli upon naturally occurring or synthetic polyribonucleotides. Proc. Natl. Acad. Sci. USA 47, 1588–1602. 1961.

17 Khorana, H.G., Buchi, H., Ghosh, H., Gupta, N., Jacob, T.M., Kossel, H. et al. Polynucleotide synthesis and the genetic code. Cold Spring Harb. Symp. Quant. Biol. 31, 39–49. 1966.

18 Hoagland, M.B., Zamecnik, P.C., & Stephenson, M.L. Intermediate reactions in protein biosynthesis. Biochim. Biophys. Acta 24, 215–216. 1957.

19 Holley, R.W., Apgar, J., Everett, G.A., Madison, J.T., Marquisee, M. et al. Structure of a ribonucleic acid. Science 147, 1462–1465. 1965.

20 Muller, H.J. Artificial transmutation of the gene. Science 66, 84–87. 1927.

21 Boorstin, D. The Discoverers: a History of Man's Search to Know His World and Himself. Vintage Books, New York, 1985.

22 Mitchell, P. Coupling of phosphorylation to electron and hydrogen transfer by a chemiosmotic type of mechanism. Nature 191, 144–148. 1961.

23 Harman, D. Aging: a theory based on free radical and radiation chemistry. J. Gerontol. 2, 298–300. 1956.

24 Russo, E. The birth of biotechnology. Nature 421, 456–457. 2003.

25 Gordon, J.W., Scangos, G.A., Plotkin, D.J., Barbosa, J.A., & Ruddle, F.H. Genetic transformation of mouse embryos by microinjection of purified DNA. Proc. Natl. Acad. Sci. USA 77, 7380–7384. 1980.

26 Palmiter, R.D., Brinster, R.L., Hammer, R.E., Trumbauer, M.E., Rosenfeld, M.G. *et al*. Dramatic growth of mice that develop from eggs microinjected with metallothionein-growth hormone fusion genes. Nature 300, 611–615. 1982.

27 Delehanty, J., White, R.L., & Mendelsohn, M.L. International Commission for Protection Against Environmental Mutagens and Carcinogens. ICPEMC Meeting Report No. 2. Approaches to determining mutation rates in human DNA. Mutat. Res. 167, 215–232. 1986.

28 Wadman, M. Company aims to beat NIH human genome efforts. Nature 393, 101. 1998.

29 Lander, E.S., Linton, L.M., Birren, B., Nusbaum, C., Zody, M.C., Baldwin, J. *et al*. Initial sequencing and analysis of the human genome. Nature 409, 860–921. 2001.

30 Venter, J.C., Adams, M.D., Myers, E.W., Li, P.W., Mural, R.J., Sutton, G.G. *et al*. The sequence of the human genome. Science 291, 1304–1351. 2001.

31 International Human Genome Sequencing Consortium. Finishing the euchromatic sequence of the human genome. Nature 431, 931–945. 2004.

32 Sanger, F., Nicklen, S., & Coulson, A.R. DNA sequencing with chain-terminating inhibitors. Proc. Natl. Acad. Sci. USA 74, 5463–5467. 1977.

33 Smith, L.M., Sanders, J.Z., Kaiser, R.J., Hughes, P., Dodd, C., Connell, C.R. *et al*. Fluorescence detection in automated DNA sequence analysis. Nature 321, 674–679. 1986.

34 Braslavsky, I., Hebert, B., Kartalov, E., & Quake, S.R. Sequence information can be obtained from single DNA molecules. Proc. Natl. Acad. Sci. USA 100, 3960–3964. 2003.

35 Hood, L. Systems biology: integrating technology, biology, and computation. Mech. Ageing Dev. 124, 9–16. 2003.

36 Blattner, F.R., Plunkett, G., 3rd, Bloch, C.A., Perna, N.T., Burland, V., Riley, M. *et al*. The complete genome sequence of Escherichia coli K-12. Science 277, 1453–1474. 1997.

37 Waterston, R.H., Lindblad-Toh, K., Birney, E., Rogers, J., Abril, J.F., Agarwal, P. *et al*. Initial sequencing and comparative analysis of the mouse genome. Nature 420, 520–562. 2002.

38 Wolf, Y.I., Rogozin, I.B., Grishin, N.V., & Koonin, E.V. Genome trees and the tree of life. Trends Genet. 18, 472–479. 2002.

39 Ureta-Vidal, A., Ettwiller, L., & Birney, E. Comparative genomics: genome-wide analysis in metazoan eukaryotes. Nat. Rev. Genet. 4, 251–262. 2003.

40 Collins, F.S., Green, E.D., Guttmacher, A.E., & Guyer, M.S. A vision for the future of genomics research. Nature 422, 835–847. 2003.

41 Hartwell, L.H., Hopfield, J.J., Leibler, S., & Murray, A.W. From molecular to modular cell biology. Nature 402, C47–C52. 1999.

42 Fire, A., Xu, S., Montgomery, M.K., Kostas, S.A., Driver, S.E., & Mello, C.C. Potent and specific genetic interference by double-stranded RNA in Caenorhabditis elegans. Nature 391, 806–811. 1998.

43 Dillon, C.P., Sandy, P., Nencioni, A., Kissler, S., Rubinson, D.A., & Van Parijs, L. Rnai as an experimental and therapeutic tool to study and regulate physiological and disease processes. Annu. Rev. Physiol. 67, 147–173. 2005.

44 Ptacek, J., Devgan, G., Michaud, G., Zhu, H., Zhu, X., Fasolo, J. *et al*. Global analysis of protein phosphorylation in yeast. Nature 438, 679–684. 2005.

45 Buchman, T.G. The community of the self. Nature 420, 246–251. 2002.

46 Hunter, P.J. & Borg, T.K. Integration from proteins to organs: the Physiome Project. Nat. Rev. Mol. Cell. Biol. 4, 237–243. 2003.

47 Crampin, E.J., Halstead, M., Hunter, P., Nielsen, P., Noble, D., Smith, N. *et al.* Computational physiology and the Physiome Project. Exp. Physiol. 89, 1–26. 2004.

48 Vijg, J. & Suh, Y. Functional genomics of ageing. Mech. Ageing Dev. 124, 3–8. 2003.

49 Strehler, B.L. Deletional mutations are the basic cause of aging: historical perspectives. Mutat. Res. 338, 3–17. 1995.

50 Friedman, D.B. & Johnson, T.E. A mutation in the age-1 gene in Caenorhabditis elegans lengthens life and reduces hermaphrodite fertility. Genetics 118, 75–86. 1988.

51 Medvedev, Z.A. An attempt at a rational classification of theories of ageing. Biol. Rev. Camb. Phil. Soc. 65, 375–398. 1990.

52 Tatar, M., Bartke, A., & Antebi, A. The endocrine regulation of aging by insulin-like signals. Science 299, 1346–1351. 2003.

53 Martin, G.M. Keynote: mechanisms of senescence–complificationists versus simplificationists. Mech. Ageing Dev. 123, 65–73. 2002.

54 Brunet-Rossinni, A.K. & Austad, S.N. Senescence in natural populations of mammals and birds. In Handbook of the Biology of Aging, 6th edn, pp. 243–265 (eds. Masoro, E.J. & Austad, S.N.). Elsevier Academic Press, San Diego, CA, 2006.

55 Kirkwood, T.B. Understanding the odd science of aging. Cell 120, 437–447. 2005.

56 Vaupel, J.W., Carey, J.R., Christensen, K., Johnson, T.E., Yashin, A.I., Holm, N.V. *et al.* Biodemographic trajectories of longevity. Science 280, 855–860. 1998.

57 Finch, C.E. Longevity, Senescence, and the Genome. University of Chicago Press, Chicago, 1990.

58 Austad, S.N. Why We Age: What Science is Discovering about the Body's Journey Through Life. Wiley, New York, 1997.

59 Kirkwood, T.B. Time of Our Lives: the Science of Human Ageing. Oxford University Press, Oxford, 1999.

60 Medawar, P.B. An Unsolved Problem in Biology. H.K. Lewis, London, 1952.

61 Clare, M.J. & Luckinbill, L.S. The effects of gene-environment interaction on the expression of longevity. Heredity 55, 19–26. 1985.

62 Stearns, S.C., Ackermann, M., Doebeli, M., & Kaiser, M. Experimental evolution of aging, growth, and reproduction in fruitflies. Proc. Natl. Acad. Sci. USA 97, 3309–3313. 2000.

63 Austad, S.N. Retarded aging rate in an insular population of opossums. J. Zool. 229, 695–708. 1993.

64 Reznick, D.N., Bryant, M.J., Roff, D., Ghalambor, C.K., & Ghalambor, D.E. Effect of extrinsic mortality on the evolution of senescence in guppies. Nature 431, 1095–1099. 2004.

65 Abrams, P.A. Evolutionary biology: mortality and lifespan. Nature 431, 1048. 2004.

66 Arking, R. Gene expression and regulation in the extended longevity phenotypes of Drosophila. Ann. NY Acad. Sci. 928, 157–167. 2001.

67 Ackermann, M., Stearns, S.C., & Jenal, U. Senescence in a bacterium with asymmetric division. Science 300, 1920. 2003.

68 Stewart, E.J., Madden, R., Paul, G., & Taddei, F. Aging and death in an organism that reproduces by morphologically symmetric division. PLoS Biol. 3, e45. 2005.

69 Bell, G. Sex and Death in Protozoa. Cambridge University Press, Cambridge, 1988.

70 Sonneborn, T.M. The relation of autogamy to senescence and rejuvenescence in P. aurelia. J. Protozool. 1, 36–53. 1954.

71 Cui, Y., Chen, R.S., & Wong, W.H. The coevolution of cell senescence and diploid sexual reproduction in unicellular organisms. Proc. Natl. Acad. Sci. USA 97, 3330–3335. 2000.

72 Kirkwood, T.B. Sex and ageing. Exp. Gerontol. 36, 413–418. 2001.

73 Muller, H.J. The relation of recombination to mutational advance. Mut. Res. 1, 2–9. 1964.

74 Martin, G.M., Austad, S.N., & Johnson, T.E. Genetic analysis of ageing: role of oxidative damage and environmental stresses. Nat. Genet. 13, 25–34. 1996.

75 Martin, G.M. Genetic syndromes in man with potential relevance to the pathobiology of aging. Birth Defects Orig. Artic. Ser. 14, 5–39. 1978.

76 Williams, G.C. Pleiotropy, natural selection, and the evolution of senescence. Evolution 11, 398–411. 1957.

77 Franceschi, C., Bonafe, M., Valensin, S., Olivieri, F., De Luca, M., Ottaviani, E. *et al.* Inflamm-aging. An evolutionary perspective on immunosenescence. Ann. NY Acad. Sci. 908, 244–254. 2000.

78 Campisi, J. Cellular senescence and apoptosis: how cellular responses might influence aging phenotypes. Exp. Gerontol. 38, 5–11. 2003.

79 Campisi, J. Cancer and ageing: rival demons? Nat. Rev. Cancer 3, 339–349. 2003.

80 Sgro, C.M. & Partridge, L. A delayed wave of death from reproduction in Drosophila. Science 286, 2521–2524. 1999.

81 Partridge, L., Gems, D., & Withers, D.J. Sex and death: what is the connection? Cell 120, 461–472. 2005.

82 Maynard Smith, J. The effects of temperature and of egg-laying on the longevity of Drosophila subob-scura. J. Exp. Biol. 35, 832–842. 1958.

83 Guarente, L. & Kenyon, C. Genetic pathways that regulate ageing in model organisms. Nature 408, 255–262. 2000.

84 Morris, J.Z., Tissenbaum, H.A., & Ruvkun, G. A phosphatidylinositol-3-OH kinase family member regulating longevity and diapause in Caenorhabditis elegans. Nature 382, 536–539. 1996.

85 Kenyon, C., Chang, J., Gensch, E., Rudner, A., & Tabtiang, R. A C. elegans mutant that lives twice as long as wild type. Nature 366, 461–464. 1993.

86 Kimura, K.D., Tissenbaum, H.A., Liu, Y., & Ruvkun, G. daf-2, an insulin receptor-like gene that regulates longevity and diapause in Caenorhabditis elegans. Science 277, 942–946. 1997.

87 Lin, K., Dorman, J.B., Rodan, A., & Kenyon, C. daf-16: An HNF-3/forkhead family member that can function to double the life-span of Caenorhabditis elegans. Science 278, 1319–1322. 1997.

88 Ogg, S., Paradis, S., Gottlieb, S., Patterson, G.I., Lee, L., Tissenbaum, H.A. *et al.* The Fork head transcription factor DAF-16 transduces insulin-like metabolic and longevity signals in C. elegans. Nature 389, 994–999. 1997.

89 Libina, N., Berman, J.R., & Kenyon, C. Tissue-specific activities of C. elegans DAF-16 in the regulation of lifespan. Cell 115, 489–502. 2003.

90 Hsin, H. & Kenyon, C. Signals from the reproductive system regulate the lifespan of C. elegans. Nature 399, 362–366. 1999.

91 Hekimi, S. & Guarente, L. Genetics and the specificity of the aging process. Science 299, 1351–1354. 2003.

92 Burgess, J., Hihi, A.K., Benard, C.Y., Branicky, R., & Hekimi, S. Molecular mechanism of maternal rescue in the clk-1 mutants of Caenorhabditis elegans. J. Biol. Chem. 278, 49555–49562. 2003.

93 Larsen, P.L. & Clarke, C.F. Extension of life-span in Caenorhabditis elegans by a diet lacking coenzyme Q. Science 295, 120–123. 2002.

94 Hansen, M., Hsu, A.L., Dillin, A., & Kenyon, C. New genes tied to endocrine, metabolic, and dietary regulation of lifespan from a Caenorhabditis elegans genomic RNAi screen. PLoS Genet. 1, 119–128. 2005.

95 Hamilton, B., Dong, Y., Shindo, M., Liu, W., Odell, I., Ruvkun, G. *et al.* A systematic RNAi screen for longevity genes in C. elegans. Genes Dev. 19, 1544–1555. 2005.

96 Kapahi, P., Zid, B.M., Harper, T., Koslover, D., Sapin, V., & Benzer, S. Regulation of lifespan in Drosophila by modulation of genes in the TOR signaling pathway. Curr. Biol. 14, 885–890. 2004.

97 Lin, Y.J., Seroude, L., & Benzer, S. Extended life-span and stress resistance in the Drosophila mutant methuselah. Science 282, 943–946. 1998.

98 Cvejic, S., Zhu, Z., Felice, S.J., Berman, Y., & Huang, X.Y. The endogenous ligand Stunted of the GPCR Methuselah extends lifespan in Drosophila. Nat. Cell Biol. 6, 540–546. 2004.

99 Knauf, F., Rogina, B., Jiang, Z., Aronson, P.S., & Helfand, S.L. Functional characterization and immunolocalization of the transporter encoded by the life-extending gene Indy. Proc. Natl. Acad. Sci. USA 99, 14315–14319. 2002.

100 McCay, C.M., Crowell, M.F., & Maynard, L.A. The effect of retarded growth upon the length of life span and upon the ultimate body size. J. Nutr. 10, 63–79. 1935.

101 Masoro, E. Caloric Restriction: a Key to Understanding and Modulating Aging. Elsevier, Amsterdam, 2002.

102 Weindruch, R. & Walford, R.L. Dietary restriction in mice beginning at 1 year of age: effect on life-span and spontaneous cancer incidence. Science 215, 1415–1418. 1982.

103 Yu, B.P., Masoro, E.J., & McMahan, C.A. Nutritional influences on aging of Fischer 344 rats: I. Physical, metabolic, and longevity characteristics. J. Gerontol. 40, 657–670. 1985.

104 Mattison, J.A., Lane, M.A., Roth, G.S., & Ingram, D.K. Calorie restriction in rhesus monkeys. Exp. Gerontol. 38, 35–46. 2003.

105 Van Voorhies, W.A. Live fast–live long? A commentary on a recent paper by Speakman *et al.* Aging Cell 3, 327–330. 2004.

106 Pearl, R. The Rate of Living. Alfred A. Knopf, New York, 1928.

107 Yu, B.P. Aging and oxidative stress: modulation by dietary restriction. Free Radic. Biol. Med. 21, 651–668. 1996.

108 Masoro, E.J. & McCarter, R.J. Aging as a consequence of fuel utilization. Aging (Milano) 3, 117–128. 1991.

109 Van Voorhies, W.A., Khazaeli, A.A., & Curtsinger, J.W. Testing the "rate of living" model: further evidence that longevity and metabolic rate are not inversely correlated in Drosophila melanogaster. J. Appl. Physiol. 97, 1915–1922. 2004.

110 Houthoofd, K., Braeckman, B.P., Lenaerts, I., Brys, K., De Vreese, A., Van Eygen, S. *et al.* No reduction of metabolic rate in food restricted Caenorhabditis elegans. Exp. Gerontol. 37, 1359–1369. 2002.

111 Hulbert, A.J., Clancy, D.J., Mair, W., Braeckman, B.P., Gems, D., & Partridge, L. Metabolic rate is not reduced by dietary-restriction or by lowered insulin/IGF-1 signalling and is not correlated with individual lifespan in Drosophila melanogaster. Exp. Gerontol. 39, 1137–1143. 2004.

112 Brown-Borg, H.M., Borg, K.E., Meliska, C.J., & Bartke, A. Dwarf mice and the ageing process. Nature 384, 33. 1996.

113 Ikeno, Y., Bronson, R.T., Hubbard, G.B., Lee, S., & Bartke, A. Delayed occurrence of fatal neoplastic diseases in ames dwarf mice: correlation to extended longevity. J. Gerontol. A Biol. Sci. Med. Sci. 58, 291–296. 2003.

114 Flurkey, K., Papaconstantinou, J., Miller, R.A., & Harrison, D.E. Lifespan extension and delayed immune and collagen aging in mutant mice with defects in growth hormone production. Proc. Natl. Acad. Sci. USA 98, 6736–6741. 2001.

115 Hsieh, C.C., DeFord, J.H., Flurkey, K., Harrison, D.E., & Papaconstantinou, J. Effects of the Pit1 mutation on the insulin signaling pathway: implications on the longevity of the long-lived Snell dwarf mouse. Mech. Ageing Dev. 123, 1245–1255. 2002.

116 Coschigano, K.T., Holland, A.N., Riders, M.E., List, E.O., Flyvbjerg, A., & Kopchick, J.J. Deletion, but not antagonism, of the mouse growth hormone receptor results in severely decreased body weights, insulin, and insulin-like growth factor I levels and increased life span. Endocrinology 144, 3799–3810. 2003.

117 Holzenberger, M., Dupont, J., Ducos, B., Leneuve, P., Geloen, A., Even, P.C. et al. IGF-1 receptor regulates lifespan and resistance to oxidative stress in mice. Nature 421, 182–187. 2003.

118 Bluher, M., Kahn, B.B., & Kahn, C.R. Extended longevity in mice lacking the insulin receptor in adipose tissue. Science 299, 572–574. 2003.

119 Masoro, E.J. A forum for commentaries on recent publications. FIRKO mouse report: important new model—but questionable interpretation. J. Gerontol. A Biol. Sci. Med. Sci. 58, B871–B872. 2003.

120 Harrison, D.E., Archer, J.R., & Astle, C.M. Effects of food restriction on aging: separation of food intake and adiposity. Proc. Natl. Acad. Sci. USA 81, 1835–1838. 1984.

121 Liu, X., Jiang, N., Hughes, B., Bigras, E., Shoubridge, E., & Hekimi, S. Evolutionary conservation of the clk-1-dependent mechanism of longevity: loss of mclk1 increases cellular fitness and lifespan in mice. Genes Dev. 19, 2424–2434. 2005.

122 Jenkins, N.L., McColl, G., & Lithgow, G.J. Fitness cost of extended lifespan in Caenorhabditis elegans. Proc. Biol. Sci. 271, 2523–2526. 2004.

123 Walker, D.W., McColl, G., Jenkins, N.L., Harris, J., & Lithgow, G.J. Evolution of lifespan in C. elegans. Nature 405, 296–297. 2000.

124 Gershon, H. & Gershon, D. Caenorhabditis elegans—a paradigm for aging research: advantages and limitations. Mech. Ageing Dev. 123, 261–274. 2002.

125 Spencer, C.C., Howell, C.E., Wright, A.R., & Promislow, D.E. Testing an 'aging gene' in long-lived drosophila strains: increased longevity depends on sex and genetic background. Aging Cell 2, 123–130. 2003.

126 Austad, S.N. Does caloric restriction in the laboratory simply prevent overfeeding and return house mice to their natural level of food intake? Sci. Aging Knowledge Environ. 2001, pe3. 2001.

127 Barabasi, A.L. & Oltvai, Z.N. Network biology: understanding the cell's functional organization. Nat. Rev. Genet. 5, 101–113. 2004.

128 Promislow, D.E. Protein networks, pleiotropy and the evolution of senescence. Proc. Biol. Sci. 271, 1225–1234. 2004.

129 Ferrarini, L., Bertelli, L., Feala, J., McCulloch, A.D., & Paternostro, G. A more efficient search strategy for aging genes based on connectivity. Bioinformatics 21, 338–348. 2005.

130 Jeong, H., Mason, S.P., Barabasi, A.L., & Oltvai, Z.N. Lethality and centrality in protein networks. Nature 411, 41–42. 2001.

131 Gandhi, T.K., Zhong, J., Mathivanan, S., Karthick, L., Chandrika, K.N., Mohan, S.S. et al. Analysis of the human protein interactome and comparison with yeast, worm and fly interaction datasets. Nat. Genet. 38, 285–293. 2006.

132 Johnson, T.E., Henderson, S., Murakami, S., de Castro, E., de Castro, S.H., Cypser, J. *et al.* Longevity genes in the nematode Caenorhabditis elegans also mediate increased resistance to stress and prevent disease. J. Inherit. Metab. Dis. 25, 197–206. 2002.

133 Murakami, S. & Johnson, T.E. The OLD-1 positive regulator of longevity and stress resistance is under DAF-16 regulation in Caenorhabditis elegans. Curr. Biol. 11, 1517–1523. 2001.

134 Walker, G.A. & Lithgow, G.J. Lifespan extension in C. elegans by a molecular chaperone dependent upon insulin-like signals. Aging Cell 2, 131–139. 2003.

135 Morley, J.F. & Morimoto, R.I. Regulation of longevity in Caenorhabditis elegans by heat shock factor and molecular chaperones. Mol. Biol. Cell 15, 657–664. 2004.

136 Hsu, A.L., Murphy, C.T., & Kenyon, C. Regulation of aging and age-related disease by DAF-16 and heat-shock factor. Science 300, 1142–1145. 2003.

137 Chavous, D.A., Jackson, F.R., & O'Connor, C.M. Extension of the Drosophila lifespan by overexpression of a protein repair methyltransferase. Proc. Natl. Acad. Sci. USA 98, 14814–8. 2001.

138 Ruan, H., Tang, X.D., Chen, M.L., Joiner, M.L., Sun, G., Brot, N. *et al.* High-quality life extension by the enzyme peptide methionine sulfoxide reductase. Proc. Natl. Acad. Sci. USA 99, 2748–2753. 2002.

139 Sun, J. & Tower, J. FLP recombinase-mediated induction of Cu/Zn-superoxide dismutase transgene expression can extend the life span of adult Drosophila melanogaster flies. Mol. Cell. Biol. 19, 216–228. 1999.

140 Sun, J., Folk, D., Bradley, T.J., & Tower, J. Induced overexpression of mitochondrial Mn-superoxide dismutase extends the life span of adult Drosophila melanogaster. Genetics 161, 661–672. 2002.

141 Bartke, A. & Brown-Borg, H. Life extension in the dwarf mouse. Curr. Top. Dev. Biol. 63, 189–225. 2004.

142 Sinclair, D.A. Toward a unified theory of caloric restriction and longevity regulation. Mech. Ageing Dev. 126, 987–1002. 2005.

143 Murakami, S., Salmon, A., & Miller, R.A. Multiplex stress resistance in cells from long-lived dwarf mice. FASEB J. 17, 1565–1566. 2003.

144 Harper, J.M., Salmon, A.B., Chang, Y., Bonkowski, M., Bartke, A., & Miller, R.A. Stress resistance and aging: influence of genes and nutrition. Mech. Ageing Dev. 127, 687–694. 2006.

145 Harrison, D.E. & Archer, J.R. Natural selection for extended longevity from food restriction. Growth Dev. Aging 53, 3. 1989.

146 Holliday, R. Food, reproduction and longevity: is the extended lifespan of calorie-restricted animals an evolutionary adaptation? Bioessays 10, 125–127. 1989.

147 Blander, G. & Guarente, L. The Sir2 family of protein deacetylases. Annu. Rev. Biochem. 73, 417–435. 2004.

148 Shore, D. The Sir2 protein family: a novel deacetylase for gene silencing and more. Proc. Natl. Acad. Sci. USA 97, 14030–14032. 2000.

149 Sinclair, D.A. & Guarente, L. Extrachromosomal rDNA circles–a cause of aging in yeast. Cell 91, 1033–1042. 1997.

150 Guarente, L. SIR2 and aging–the exception that proves the rule. Trends Genet. 17, 391–392. 2001.

151 McMurray, M.A. & Gottschling, D.E. An age-induced switch to a hyper-recombinational state. Science 301, 1908–1911. 2003.

152 Gaubatz, J.W. Extrachromosomal circular DNAs and genomic sequence plasticity in eukaryotic cells. Mutat. Res. 237, 271–292. 1990.

153 Tissenbaum, H.A. & Guarente, L. Increased dosage of a sir-2 gene extends lifespan in Caenorhabditis elegans. Nature 410, 227–230. 2001.

154 Greer, E.L. & Brunet, A. FOXO transcription factors at the interface between longevity and tumor suppression. Oncogene 24, 7410–7425. 2005.

155 Daitoku, H., Hatta, M., Matsuzaki, H., Aratani, S., Ohshima, T., Miyagishi, M. *et al.* Silent information regulator 2 potentiates Foxo1-mediated transcription through its deacetylase activity. Proc. Natl. Acad. Sci. USA 101, 10042–10047. 2004.

156 Motta, M.C., Divecha, N., Lemieux, M., Kamel, C., Chen, D., Gu, W. *et al.* Mammalian SIRT1 represses forkhead transcription factors. Cell 116, 551–563. 2004.

157 Brunet, A., Sweeney, L.B., Sturgill, J.F., Chua, K.F., Greer, P.L., Lin, Y. *et al.* Stress-dependent regulation of FOXO transcription factors by the SIRT1 deacetylase. Science 303, 2011–2015. 2004.

158 Luo, J., Nikolaev, A.Y., Imai, S., Chen, D., Su, F., Shiloh, A. *et al.* Negative control of p53 by Sir2alpha promotes cell survival under stress. Cell 107, 137–148. 2001.

159 Tran, H., Brunet, A., Grenier, J.M., Datta, S.R., Fornace, A.J., Jr, DiStefano, P.S. *et al.* DNA repair pathway stimulated by the forkhead transcription factor FOXO3a through the Gadd45 protein. Science 296, 530–534. 2002.

160 Lin, S.J., Defossez, P.A., & Guarente, L. Requirement of NAD and SIR2 for life-span extension by calorie restriction in Saccharomyces cerevisiae. Science 289, 2126–2128. 2000.

161 Anderson, R.M., Bitterman, K.J., Wood, J.G., Medvedik, O., & Sinclair, D.A. Nicotinamide and PNC1 govern lifespan extension by calorie restriction in Saccharomyces cerevisiae. Nature 423, 181–185. 2003.

162 Kaeberlein, M., Kirkland, K.T., Fields, S., & Kennedy, B.K. Sir2-independent life span extension by calorie restriction in yeast. PLoS Biol. 2, E296. 2004.

163 Cohen, H.Y., Miller, C., Bitterman, K.J., Wall, N.R., Hekking, B., Kessler, B. *et al.* Calorie restriction promotes mammalian cell survival by inducing the SIRT1 deacetylase. Science 305, 390–392. 2004.

164 Cox, M.M. A path for coevolution of recombinational DNA repair, transposition, and the common nucleotides. Mutat. Res. 384, 15–22. 1997.

165 Bernstein, H., Hopf, F.A., & Michod, R.E. The molecular basis of the evolution of sex. Adv. Genet. 24, 323–370. 1987.

166 Kibota, T.T. & Lynch, M. Estimate of the genomic mutation rate deleterious to overall fitness in E. coli. Nature 381, 694–696. 1996.

167 Elena, S.F. & Lenski, R.E. Test of synergistic interactions among deleterious mutations in bacteria. Nature 390, 395–398. 1997.

168 Bolotin, A., Quinquis, B., Renault, P., Sorokin, A., Ehrlich, S.D., Kulakauskas, S. *et al.* Complete sequence and comparative genome analysis of the dairy bacterium Streptococcus thermophilus. Nat. Biotechnol. 22, 1554–1558. 2004.

169 Bjedov, I., Tenaillon, O., Gerard, B., Souza, V., Denamur, E., Radman, M. *et al.* Stress-induced mutagenesis in bacteria. Science 300, 1404–1409. 2003.

170 Jackson, A.L. & Loeb, L.A. The contribution of endogenous sources of DNA damage to the multiple mutations in cancer. Mutat. Res. 477, 7–21. 2001.

171 Walbot, V. & Evans, M.M. Unique features of the plant life cycle and their consequences. Nat. Rev. Genet. 4, 369–379. 2003.

172 Evans, H.J. Mutation as a cause of genetic disease. Phil. Trans. R. Soc. Lond. B Biol. Sci. 319, 325–340. 1988.

173 Crow, J.F. The origins, patterns and implications of human spontaneous mutation. Nat. Rev. Genet. 1, 40–47. 2000.

174 Clark, A.M. & Rubin, M.A. The modification by X-irradiation of the life span of haploids and diploids of the wasp, Habrobracon SP. Radiat. Res. 15, 244–253. 1961.

175 Morley, A.A. Is ageing the result of dominant and co-dominant mutations? J. Theor. Biol. 98, 469–474. 1982.

176 Lewin, B. Genes VIII. Pearson Prentice Hall, Upper Saddle River, NJ, 2004.

177 Latchman, D.S. Eukaryotic Trancription Factors. Elsevier, London, 2004.

178 Wang, A.H., Quigley, G.J., Kolpak, F.J., Crawford, J.L., van Boom, J.H., van der Marel, G. *et al.* Molecular structure of a left-handed double helical DNA fragment at atomic resolution. Nature 282, 680–686. 1979.

179 Lynch, M. & Conery, J.S. The origins of genome complexity. Science 302, 1401–1404. 2003.

180 Shabalina, S.A. & Spiridonov, N.A. The mammalian transcriptome and the function of non-coding DNA sequences. Genome Biol. 5, 105. 2004.

181 Dover, G. Molecular drive: a cohesive mode of species evolution. Nature 299, 111–117. 1982.

182 Dermitzakis, E.T., Reymond, A., & Antonarakis, S.E. Conserved non-genic sequences—an unexpected feature of mammalian genomes. Nat. Rev. Genet. 6, 151–157. 2005.

183 Wilson, A.C., Bush, G.L., Case, S.M., & King, M.C. Social structuring of mammalian populations and rate of chromosomal evolution. Proc. Natl. Acad. Sci. USA 72, 5061–5065. 1975.

184 Cutler, R.G. Evolution of human longevity and the genetic complexity governing aging rate. Proc. Natl. Acad. Sci. USA 72, 4664–4668. 1975.

185 Gilbert, W. Why genes in pieces? Nature 271, 501. 1978.

186 Roy, S.W. & Gilbert, W. The evolution of spliceosomal introns: patterns, puzzles and progress. Nat. Rev. Genet. 7, 211–221. 2006.

187 McClintock, B. The origin and behavior of mutable loci in maize. Proc. Natl. Acad. Sci. USA 36, 344–355. 1950.

188 Kazazian, H.H., Jr. Mobile elements and disease. Curr. Opin. Genet. Dev. 8, 343–350. 1998.

189 Tomilin, N.V. Control of genes by mammalian retroposons. Int. Rev. Cytol. 186, 1–48. 1999.

190 Han, J.S. & Boeke, J.D. LINE-1 retrotransposons: modulators of quantity and quality of mammalian gene expression? Bioessays 27, 775–784. 2005.

191 Lyon, M.F. The Lyon and the LINE hypothesis. Semin. Cell Dev. Biol. 14, 313–318. 2003.

192 Muotri, A.R., Chu, V.T., Marchetto, M.C., Deng, W., Moran, J.V., & Gage, F.H. Somatic mosaicism in neuronal precursor cells mediated by L1 retrotransposition. Nature 435, 903–910. 2005.

193 Debrauwere, H., Gendrel, C.G., Lechat, S., & Dutreix, M. Differences and similarities between various tandem repeat sequences: minisatellites and microsatellites. Biochimie 79, 577–586. 1997.

194 Pearson, C.E., Nichol Edamura, K., & Cleary, J.D. Repeat instability: mechanisms of dynamic mutations. Nat. Rev. Genet. 6, 729–742. 2005.

195 Rudd, M.K. & Willard, H.F. Analysis of the centromeric regions of the human genome assembly. Trends Genet. 20, 529–533. 2004.

196 Henikoff, S. & Dalal, Y. Centromeric chromatin: what makes it unique? Curr. Opin. Genet. Dev. 15, 177–184. 2005.

197 de Lange, T. T-loops and the origin of telomeres. Nat. Rev. Mol. Cell. Biol. 5, 323–329. 2004.

198 Griffith, J.D., Comeau, L., Rosenfield, S., Stansel, R.M., Bianchi, A., Moss, H. *et al.* Mammalian telomeres end in a large duplex loop. Cell 97, 503–514. 1999.

199 Morey, C. & Avner, P. Employment opportunities for non-coding RNAs. FEBS Lett. 567, 27–34. 2004.

200 Chow, J.C., Yen, Z., Ziesche, S.M., & Brown, C.J. Silencing of the mammalian X chromosome. Annu. Rev. Genomics Hum. Genet. 6, 69–92. 2005.

201 Lang, B.F., Gray, M.W., & Burger, G. Mitochondrial genome evolution and the origin of eukaryotes. Annu. Rev. Genet. 33, 351–397. 1999.

202 Belmont, A.S. & Bruce, K. Visualization of G1 chromosomes: a folded, twisted, supercoiled chromonema model of interphase chromatid structure. J. Cell Biol. 127, 287–302. 1994.

203 Adkins, N.L., Watts, M., & Georgel, P.T. To the 30-nm chromatin fiber and beyond. Biochim. Biophys. Acta 1677, 12–23. 2004.

204 Bernstein, B.E., Humphrey, E.L., Liu, C.L., & Schreiber, S.L. The use of chromatin immunoprecipitation assays in genome-wide analyses of histone modifications. Methods Enzymol. 376, 349–360. 2004.

205 Bernstein, B.E., Liu, C.L., Humphrey, E.L., Perlstein, E.O., & Schreiber, S.L. Global nucleosome occupancy in yeast. Genome Biol. 5, R62. 2004.

206 Gilbert, N., Boyle, S., Fiegler, H., Woodfine, K., Carter, N.P., & Bickmore, W.A. Chromatin architecture of the human genome: gene-rich domains are enriched in open chromatin fibers. Cell 118, 555–566. 2004.

207 Jaenisch, R. & Bird, A. Epigenetic regulation of gene expression: how the genome integrates intrinsic and environmental signals. Nat. Genet. 33 (suppl), 245–254. 2003.

208 Dion, M.F., Altschuler, S.J., Wu, L.F., & Rando, O.J. Genomic characterization reveals a simple histone H4 acetylation code. Proc. Natl. Acad. Sci. USA 102, 5501–5506. 2005.

209 Olovnikov, A.M. Telomeres, telomerase, and aging: origin of the theory. Exp. Gerontol. 31, 443–448. 1996.

210 Polo, S.E. & Almouzni, G. Histone metabolic pathways and chromatin assembly factors as proliferation markers. Cancer Lett 220, 1–9. 2005.

211 Bestor, T., Laudano, A., Mattaliano, R., & Ingram, V. Cloning and sequencing of a cDNA encoding DNA methyltransferase of mouse cells. The carboxyl-terminal domain of the mammalian enzymes is related to bacterial restriction methyltransferases. J Mol Biol. 203, 971–983. 1988.

212 Morgan, H.D., Santos, F., Green, K., Dean, W., & Reik, W. Epigenetic reprogramming in mammals. Hum. Mol. Genet. 14, R47–R58. 2005.

213 Allegrucci, C., Denning, C., Priddle, H., & Young, L. Stem-cell consequences of embryo epigenetic defects. Lancet 364, 206–208. 2004.

214 Gruenbaum, Y., Margalit, A., Goldman, R.D., Shumaker, D.K., & Wilson, K.L. The nuclear lamina comes of age. Nat. Rev. Mol. Cell. Biol. 6, 21–31. 2005.

215 Goldman, R.D., Gruenbaum, Y., Moir, R.D., Shumaker, D.K., & Spann, T.P. Nuclear lamins: building blocks of nuclear architecture. Genes Dev. 16, 533–547. 2002.

216 Vlcek, S., Dechat, T., & Foisner, R. Nuclear envelope and nuclear matrix: interactions and dynamics. Cell. Mol. Life Sci. 58, 1758–1765. 2001.

217 Muchir, A. & Worman, H.J. The nuclear envelope and human disease. Physiology (Bethesda) 19, 309–314. 2004.

218 Eng, C. & Mulligan, L.M. Mutations of the RET proto-oncogene in the multiple endocrine neoplasia type 2 syndromes, related sporadic tumours, and hirschsprung disease. Hum. Mutat. 9, 97–109. 1997.

219 Scaffidi, P. & Misteli, T. Lamin A-dependent nuclear defects in human aging. Science 312, 1059–1063. 2006.

220 Haithcock, E., Dayani, Y., Neufeld, E., Zahand, A.J., Feinstein, N., Mattout, A. et al. Age-related changes of nuclear architecture in Caenorhabditis elegans. Proc. Natl. Acad. Sci. USA 102, 16690–16695. 2005.

221 Lemon, K.P. & Grossman, A.D. Localization of bacterial DNA polymerase: evidence for a factory model of replication. Science 282, 1516–1519. 1998.

222 Pederson, T. Half a century of "the nuclear matrix". Mol. Biol. Cell 11, 799–805. 2000.

223 Nickerson, J. Experimental observations of a nuclear matrix. J. Cell Sci. 114, 463–474. 2001.

224 Cremer, T. & Cremer, C. Chromosome territories, nuclear architecture and gene regulation in mammalian cells. Nat. Rev. Genet. 2, 292–301. 2001.

225 Parada, L.A., McQueen, P.G., & Misteli, T. Tissue-specific spatial organization of genomes. Genome Biol. 5, R44. 2004.

226 Misteli, T. Spatial positioning; a new dimension in genome function. Cell 119, 153–156. 2004.

227 Roix, J.J., McQueen, P.G., Munson, P.J., Parada, L.A., & Misteli, T. Spatial proximity of translocation-prone gene loci in human lymphomas. Nat. Genet. 34, 287–291. 2003.

228 Cleveland, D.W., Mao, Y., & Sullivan, K.F. Centromeres and kinetochores: from epigenetics to mitotic checkpoint signaling. Cell 112, 407–421. 2003.

229 Blumenthal, T., Evans, D., Link, C.D., Guffanti, A., Lawson, D., Thierry-Mieg, J. *et al.* A global analysis of Caenorhabditis elegans operons. Nature 417, 851–854. 2002.

230 Jacob, F., Ullman, A., & Monod, J. [The Promotor, a Genetic Element Necessary to the Expression of an Operon.]. CR Hebd. Seances Acad. Sci. 258, 3125–3128. 1964.

231 Latchman, D.S. Eukaryotic Transcription Factors. Elsevier, San Diego, CA, 2004.

232 Ma, J. Crossing the line between activation and repression. Trends Genet. 21, 54–59. 2005.

233 Krek, A., Grun, D., Poy, M.N., Wolf, R., Rosenberg, L., Epstein, E.J. *et al.* Combinatorial microRNA target predictions. Nat. Genet. 37, 495–500. 2005.

234 Chuang, C.H. & Belmont, A.S. Close encounters between active genes in the nucleus. Genome Biol. 6, 237. 2005.

235 Huber, M.C., Kruger, G., & Bonifer, C. Genomic position effects lead to an inefficient reorganization of nucleosomes in the 5'-regulatory region of the chicken lysozyme locus in transgenic mice. Nucleic Acids Res. 24, 1443–1452. 1996.

236 Setlow, R.B. & Carrier, W.L. The disappearance of thymine dimers from DNA: an error-correcting mechanism. Proc. Natl. Acad. Sci. USA 51, 226–231. 1964.

237 Boyce, R.P. & Howard-Flanders, P. Release of ultraviolet light-induced thymine dimers from DNA in E. coli K-12. Proc. Natl. Acad. Sci. USA 51, 293–300. 1964.

238 Hart, R.W. & Setlow, R.B. Correlation between deoxyribonucleic acid excision-repair and life-span in a number of mammalian species. Proc. Natl. Acad. Sci. USA 71, 2169–2173. 1974.

239 Sacher, G.A. Evolutionary theory in gerontology. Perspect. Biol. Med. 25, 339–353. 1982.

240 Friedberg, E.C., Walker, G.R., Siede, W., Wood, R.D., Schultz, R.A., & Ellenberger, T. DNA Repair and Mutagenesis. 2nd Edition. ASM Press, Washington, 2005.

241 Aravind, L., Walker, D.R., & Koonin, E.V. Conserved domains in DNA repair proteins and evolution of repair systems. Nucleic Acids Res. 27, 1223–1242. 1999.

242 de Duve, C. The onset of selection. Nature 433, 581–582. 2005.

243 Nobrega, M.A., Zhu, Y., Plajzer-Frick, I., Afzal, V., & Rubin, E.M. Megabase deletions of gene deserts result in viable mice. Nature 431, 988–993. 2004.

244 White, O., Eisen, J.A., Heidelberg, J.F., Hickey, E.K., Peterson, J.D., Dodson, R.J. *et al.* Genome sequence of the radioresistant bacterium Deinococcus radiodurans R1. Science 286, 1571–1577. 1999.

245 Ames, B.N., Shigenaga, M.K., & Hagen, T.M. Oxidants, antioxidants, and the degenerative diseases of aging. Proc. Natl. Acad. Sci. USA 90, 7915–7922. 1993.

246 Jaruga, P. & Dizdaroglu, M. Repair of products of oxidative DNA base damage in human cells. Nucleic Acids Res. 24, 1389–1394. 1996.

247 Simic, M.G. Urinary biomarkers and the rate of DNA damage in carcinogenesis and anticarcinogenesis. Mutat. Res. 267, 277–290. 1992.

248 Hanawalt, P.C. & Haynes, R.H. Repair replication of DNA in bacteria: irrelevance of chemical nature of base defect. Biochem. Biophys. Res. Commun. 19, 462–467. 1965.

249 Yang, W. Poor base stacking at DNA lesions may initiate recognition by many repair proteins. DNA Repair (Amst.) 5, 654–666. 2006.

250 Hartwell, L.H. & Weinert, T.A. Checkpoints: controls that ensure the order of cell cycle events. Science 246, 629–634. 1989.

251 Kruman, II, Wersto, R.P., Cardozo-Pelaez, F., Smilenov, L., Chan, S.L., Chrest, F.J. et al. Cell cycle activation linked to neuronal cell death initiated by DNA damage. Neuron 41, 549–561. 2004.

252 Bakkenist, C.J. & Kastan, M.B. DNA damage activates ATM through intermolecular autophosphorylation and dimer dissociation. Nature 421, 499–506. 2003.

253 Lee, J.H. & Paull, T.T. ATM activation by DNA double-strand breaks through the Mre11-Rad50-Nbs1 complex. Science 308, 551–554. 2005.

254 Zou, L., Liu, D., & Elledge, S.J. Replication protein A-mediated recruitment and activation of Rad17 complexes. Proc. Natl. Acad. Sci. USA 100, 13827–13832. 2003.

255 Helt, C.E., Cliby, W.A., Keng, P.C., Bambara, R.A., & O'Reilly, M.A. Ataxia telangiectasia mutated (ATM) and ATM and Rad3-related protein exhibit selective target specificities in response to different forms of DNA damage. J. Biol. Chem. 280, 1186–1192. 2005.

256 Lobrich, M. & Jeggo, P.A. Harmonising the response to DSBs: a new string in the ATM bow. DNA Repair (Amst.) 4, 749–759. 2005.

257 Qi, L., Strong, M.A., Karim, B.O., Armanios, M., Huso, D.L., & Greider, C.W. Short telomeres and ataxia-telangiectasia mutated deficiency cooperatively increase telomere dysfunction and suppress tumorigenesis. Cancer Res. 63, 8188–8196. 2003.

258 Dip, R. & Naegeli, H. More than just strand breaks: the recognition of structural DNA discontinuities by DNA-dependent protein kinase catalytic subunit. FASEB J. 19, 704–715. 2005.

259 Lees-Miller, S.P., Sakaguchi, K., Ullrich, S.J., Appella, E., & Anderson, C.W. Human DNA-activated protein kinase phosphorylates serines 15 and 37 in the amino-terminal transactivation domain of human p53. Mol. Cell. Biol. 12, 5041–5049. 1992.

260 Budzowska, M., Jaspers, I., Essers, J., de Waard, H., van Drunen, E., Hanada, K. et al. Mutation of the mouse Rad17 gene leads to embryonic lethality and reveals a role in DNA damage-dependent recombination. EMBO J. 23, 3548–3558. 2004.

261 Helt, C.E., Wang, W., Keng, P.C., & Bambara, R.A. Evidence that DNA damage detection machinery participates in DNA repair. Cell Cycle 4, 529–532. 2005.

262 Chambon, P., Weill, J.D., & Mandel, P. Nicotinamide mononucleotide activation of new DNA-dependent polyadenylic acid synthesizing nuclear enzyme. Biochem. Biophys. Res. Commun. 11, 39–43. 1963.

263 Pero, R.W., Holmgren, K., & Persson, L. Gamma-radiation induced ADP-ribosyl transferase activity and mammalian longevity. Mutat. Res. 142, 69–73. 1985.

264 Grube, K. & Burkle, A. Poly(ADP-ribose) polymerase activity in mononuclear leukocytes of 13 mammalian species correlates with species-specific life span. Proc. Natl. Acad. Sci. USA 89, 1759–11763. 1992.

265 Ljungman, M. & Lane, D.P. Transcription—guarding the genome by sensing DNA damage. Nat. Rev. Cancer 4, 727–737. 2004.

266 Ljungman, M. & Zhang, F. Blockage of RNA polymerase as a possible trigger for u.v. light-induced apoptosis. Oncogene 13, 823–831. 1996.

267 Stojic, L., Brun, R., & Jiricny, J. Mismatch repair and DNA damage signalling. DNA Repair (Amst.) 3, 1091–1101. 2004.

268 Karran, P. & Bignami, M. DNA damage tolerance, mismatch repair and genome instability. Bioessays 16, 833–839. 1994.

269 Fishel, R. The selection for mismatch repair defects in hereditary nonpolyposis colorectal cancer: revising the mutator hypothesis. Cancer Res. 61, 7369–7374. 2001.

270 Foster, E.R. & Downs, J.A. Histone H2A phosphorylation in DNA double-strand break repair. FEBS J. 272, 3231–3240. 2005.

271 Celeste, A., Difilippantonio, S., Difilippantonio, M.J., Fernandez-Capetillo, O., Pilch, D.R., Sedelnikova, O.A. *et al.* H2AX haploinsufficiency modifies genomic stability and tumor susceptibility. Cell 114, 371–383. 2003.

272 Sedelnikova, O.A., Horikawa, I., Zimonjic, D.B., Popescu, N.C., Bonner, W.M., & Barrett, J.C. Senescing human cells and ageing mice accumulate DNA lesions with unrepairable double-strand breaks. Nat. Cell Biol. 6, 168–170. 2004.

273 Cline, S.D. & Hanawalt, P.C. Who's on first in the cellular response to DNA damage? Nat. Rev. Mol. Cell. Biol. 4, 361–372. 2003.

274 Pfeifer, G.P. Mutagenesis at methylated CpG sequences. Curr. Top. Microbiol. Immunol. 301, 259–281. 2006.

275 Dollé, M.E., Snyder, W.K., Dunson, D.B., & Vijg, J. Mutational fingerprints of aging. Nucleic Acids Res. 30, 545–549. 2002.

276 Nouspikel, T. & Hanawalt, P.C. DNA repair in terminally differentiated cells. DNA Repair (Amst.) 1, 59–75. 2002.

277 Vijg, J., Mullaart, E., Berends, F., Lohman, P.H., & Knook, D.L. UV-induced DNA excision repair in rat fibroblasts during immortalization and terminal differentiation in vitro. Exp. Cell Res. 167, 517–530. 1986.

278 Hsu, G.W., Ober, M., Carell, T., & Beese, L.S. Error-prone replication of oxidatively damaged DNA by a high-fidelity DNA polymerase. Nature 431, 217–221. 2004.

279 Goodman, M.F. Error-prone repair DNA polymerases in prokaryotes and eukaryotes. Annu. Rev. Biochem. 71, 17–50. 2002.

280 Friedberg, E.C., Wagner, R., & Radman, M. Specialized DNA polymerases, cellular survival, and the genesis of mutations. Science 296, 1627–1630. 2002.

281 Heller, R.C. & Marians, K.J. Replication fork reactivation downstream of a blocked nascent leading strand. Nature 439, 557–562. 2006.

282 Rupp, W.D. & Howard-Flanders, P. Discontinuities in the DNA synthesized in an excision-defective strain of Escherichia coli following ultraviolet irradiation. J. Mol. Biol. 31, 291–304. 1968.

283 Hoeijmakers, J.H. Genome maintenance mechanisms for preventing cancer. Nature 411, 366–374. 2001.

284 Taddei, F., Matic, I., & Radman, M. cAMP-dependent SOS induction and mutagenesis in resting bacterial populations. Proc. Natl. Acad. Sci. USA 92, 11736–40. 1995.

285 Cairns, J. & Foster, P.L. Adaptive reversion of a frameshift mutation in Escherichia coli. Genetics 128, 695–701. 1991.

286 Diaz, M. & Lawrence, C. An update on the role of translesion synthesis DNA polymerases in Ig hypermutation. Trends Immunol. 26, 215–220. 2005.

287 Blanc, V. & Davidson, N.O. C-to-U RNA editing: mechanisms leading to genetic diversity. J. Biol. Chem. 278, 1395–1398. 2003.

288 Saxowsky, T.T. & Doetsch, P.W. RNA polymerase encounters with DNA damage: transcription-coupled repair or transcriptional mutagenesis? Chem. Rev. 106, 474–488. 2006.

289 van Leeuwen, F.W., Fischer, D.F., Kamel, D., Sluijs, J.A., Sonnemans, M.A., Benne, R. et al. Molecular misreading: a new type of transcript mutation expressed during aging. Neurobiol. Aging 21, 879–891. 2000.

290 Martin, G.M. & Bressler, S.L. Transcriptional infidelity in aging cells and its relevance for the Orgel hypothesis. Neurobiol. Aging 21, 897–900. 2000.

291 Setlow, J.K. & Setlow, R.B. Nature of the photoreactivable ultraviolet lesion in DNA. Nature 197, 560–562. 1963.

292 Thoma, F. Repair of UV lesions in nucleosomes—intrinsic properties and remodeling. DNA Repair (Amst.) 4, 855–869. 2005.

293 Schul, W., Jans, J., Rijksen, Y.M., Klemann, K.H., Eker, A.P., de Wit, J. et al. Enhanced repair of cyclobutane pyrimidine dimers and improved UV resistance in photolyase transgenic mice. EMBO J. 21, 4719–4729. 2002.

294 Aas, P.A., Otterlei, M., Falnes, P.O., Vagbo, C.B., Skorpen, F., Akbari, M. et al. Human and bacterial oxidative demethylases repair alkylation damage in both RNA and DNA. Nature 421, 859–863. 2003.

295 Kawate, H., Sakumi, K., Tsuzuki, T., Nakatsuru, Y., Ishikawa, T., Takahashi, S. et al. Separation of killing and tumorigenic effects of an alkylating agent in mice defective in two of the DNA repair genes. Proc. Natl. Acad. Sci. USA 95, 5116–5120. 1998.

296 Zhou, Z.Q., Manguino, D., Kewitt, K., Intano, G.W., McMahan, C.A., Herbert, D.C. et al. Spontaneous hepatocellular carcinoma is reduced in transgenic mice overexpressing human O6-methylguanine-DNA methyltransferase. Proc. Natl. Acad. Sci. USA 98, 12566–12571. 2001.

297 Samson, L. & Cairns, J. A new pathway for DNA repair in Escherichia coli. Nature 267, 281–283. 1977.

298 Barnes, D.E. & Lindahl, T. Repair and genetic consequences of endogenous DNA base damage in mammalian cells. Annu. Rev. Genet. 38, 445–476. 2004.

299 Sancar, A., Lindsey-Boltz, L.A., Unsal-Kacmaz, K., & Linn, S. Molecular mechanisms of mammalian DNA repair and the DNA damage checkpoints. Annu. Rev. Biochem. 73, 39–85. 2004.

300 Caldecott, K.W. XRCC1 and DNA strand break repair. DNA Repair (Amst.) 2, 955–969. 2003.

301 Bohr, V.A., Stevnsner, T., & de Souza-Pinto, N.C. Mitochondrial DNA repair of oxidative damage in mammalian cells. Gene 286, 127–134. 2002.

302 Osterod, M., Hollenbach, S., Hengstler, J.G., Barnes, D.E., Lindahl, T., & Epe, B. Age-related and tissue-specific accumulation of oxidative DNA base damage in 7,8-dihydro-8-oxoguanine-DNA glycosylase (Ogg1) deficient mice. Carcinogenesis 22, 1459–1463. 2001.

303 Cleaver, J.E. Mending human genes: a job for a lifetime. DNA Repair (Amst.) 4, 635–638. 2005.

304 Pettijohn, D. & Hanawalt, P. Evidence for repair-replication of ultraviolet damaged DNA in bacteria. J. Mol. Biol. 93, 395–410. 1964.

305 Rasmussen, R.E. & Painter, R.B. Evidence for repair of ultra-violet damaged deoxyribonucleic acid in cultured mammalian cells. Nature 203, 1360–1362. 1964.

306 Cleaver, J.E. Defective repair replication of DNA in xeroderma pigmentosum. Nature 218, 652–656. 1968.

307 De Weerd-Kastelein, E.A., Keijzer, W., & Bootsma, D. Genetic heterogeneity of xeroderma pigmentosum demonstrated by somatic cell hybridization. Nat. New Biol. 238, 80–83. 1972.

308 Bohr, V.A., Smith, C.A., Okumoto, D.S., & Hanawalt, P.C. DNA repair in an active gene: removal of pyrimidine dimers from the DHFR gene of CHO cells is much more efficient than in the genome overall. Cell 40, 359–369. 1985.

309 de Waard, H., de Wit, J., Andressoo, J.O., van Oostrom, C.T., Riis, B., Weimann, A. *et al.* Different effects of CSA and CSB deficiency on sensitivity to oxidative DNA damage. Mol. Cell. Biol. 24, 7941–7948. 2004.

310 Cooper, P.K., Nouspikel, T., & Clarkson, S.G. Retraction. Science 308, 1740. 2005.

311 Brooks, P.J., Wise, D.S., Berry, D.A., Kosmoski, J.V., Smerdon, M.J., Somers, R.L. *et al.* The oxidative DNA lesion 8,5′-(S)-cyclo-2′-deoxyadenosine is repaired by the nucleotide excision repair pathway and blocks gene expression in mammalian cells. J. Biol. Chem. 275, 22355–22362. 2000.

312 Branum, M.E., Reardon, J.T., & Sancar, A. DNA repair excision nuclease attacks undamaged DNA. A potential source of spontaneous mutations. J. Biol. Chem. 276, 25421–25426. 2001.

313 Fitch, M.E., Cross, I.V., Turner, S.J., Adimoolam, S., Lin, C.X., Williams, K.G. *et al.* The DDB2 nucleotide excision repair gene product p48 enhances global genomic repair in p53 deficient human fibroblasts. DNA Repair (Amst.) 2, 819–826. 2003.

314 Friedberg, E.C. How nucleotide excision repair protects against cancer. Nat. Rev. Cancer 1, 22–33. 2001.

315 Andressoo, J.O. & Hoeijmakers, J.H. Transcription-coupled repair and premature ageing. Mutat. Res. 577, 179–194. 2005.

316 Dianov, G.L., Houle, J.F., Iyer, N., Bohr, V.A., & Friedberg, E.C. Reduced RNA polymerase II transcription in extracts of cockayne syndrome and xeroderma pigmentosum/Cockayne syndrome cells. Nucleic Acids Res. 25, 3636–3642. 1997.

317 Mellon, I. Transcription-coupled repair: A complex affair. Mutat. Res. 577, 155–161. 2005.

318 de Vries, A., van Oostrom, C.T., Hofhuis, F.M., Dortant, P.M., Berg, R.J., de Gruijl, F.R. *et al.* Increased susceptibility to ultraviolet-B and carcinogens of mice lacking the DNA excision repair gene XPA. Nature 377, 169–173. 1995.

319 Nakane, H., Takeuchi, S., Yuba, S., Saijo, M., Nakatsu, Y., Murai, H. *et al.* High incidence of ultraviolet-B- or chemical-carcinogen-induced skin tumours in mice lacking the xeroderma pigmentosum group A gene. Nature 377, 165–168. 1995.

320 Hanawalt, P.C. Revisiting the rodent repairadox. Environ. Mol. Mutagen. 38, 89–96. 2001.

321 Vijg, J., Mullaart, E., van der Schans, G.P., Lohman, P.H., & Knook, D.L. Kinetics of ultraviolet induced DNA excision repair in rat and human fibroblasts. Mutat. Res. 132, 129–138. 1984.

322 Ganesan, A.K., Spivak, G. & Hanawalt, P.C. Expression of DNA repair genes in mammalian cells. In Manipulation and Expression of Genes in Eukaryotes, pp. 45–54 (eds. Nagley, P., Linnane, A.W., Peacock, W.J., & Pateman, J.A.). Academic Press, Sydney, Australia. 1983.

323 Wijnhoven, S.W., Kool, H.J., Mullenders, L.H., van Zeeland, A.A., Friedberg, E.C., van der Horst, G.T. *et al.* Age-dependent spontaneous mutagenesis in Xpc mice defective in nucleotide excision repair. Oncogene 19, 5034–5037. 2000.

324 Hollander, M.C., Philburn, R.T., Patterson, A.D., Velasco-Miguel, S., Friedberg, E.C., Linnoila, R.I. *et al.* Deletion of XPC leads to lung tumors in mice and is associated with early events in human lung carcinogenesis. Proc. Natl. Acad. Sci. USA 102, 13200–5. 2005.

325 Dollé, M.E., Busuttil, R.A., Garcia, A.M., Wijnhoven, S., van Drunen, E., Niedernhofer, L.J. *et al.* Increased genomic instability is not a prerequisite for shortened lifespan in DNA repair deficient mice. Mutat. Res. 596, 22–35. 2006.

326 Kato, H., Harada, M., Tsuchiya, K., & Moriwaki, K. Absence of correlation between DNA repair in ultraviolet irradiated mammalian cells and life-span of the donor species. Jpn J. Genet. 55, 99–108. 1980.

327 Dzantiev, L., Constantin, N., Genschel, J., Iyer, R.R., Burgers, P.M., & Modrich, P. A defined human system that supports bidirectional mismatch-provoked excision. Mol Cell 15, 31–41. 2004.

328 Owen, B.A., Yang, Z., Lai, M., Gajek, M., Badger, J.D., 2nd, Hayes, J.J. *et al.* (CAG)(n)-hairpin DNA binds to Msh2-Msh3 and changes properties of mismatch recognition. Nat. Struct. Mol. Biol. 12, 663–670. 2005.

329 Wheeler, V.C., Lebel, L.A., Vrbanac, V., Teed, A., te Riele, H., & MacDonald, M.E. Mismatch repair gene Msh2 modifies the timing of early disease in Hdh(Q111) striatum. Hum. Mol. Genet. 12, 273–281. 2003.

330 MacPhee, D.G. Mismatch repair, somatic mutations, and the origins of cancer. Cancer Res. 55, 5489–5492. 1995.

331 De la Chapelle, A. Genetic predisposition to colorectal cancer. Nat. Rev. Cancer 4, 769–780. 2004.

332 Edelmann, L. & Edelmann, W. Loss of DNA mismatch repair function and cancer predisposition in the mouse: animal models for human hereditary nonpolyposis colorectal cancer. Am. J. Med. Genet. C Semin. Med. Genet. 129, 91–99. 2004.

333 Narayanan, L., Fritzell, J.A., Baker, S.M., Liskay, R.M., & Glazer, P.M. Elevated levels of mutation in multiple tissues of mice deficient in the DNA mismatch repair gene Pms2. Proc. Natl. Acad. Sci. USA 94, 3122–3127. 1997.

334 Scherer, S.J., Avdievich, E., & Edelmann, W. Functional consequences of DNA mismatch repair missense mutations in murine models and their impact on cancer predisposition. Biochem. Soc. Trans. 33, 689–693. 2005.

335 Wijnhoven, S.W., Beems, R.B., Roodbergen, M., van den Berg, J., Lohman, P.H., Diderich, K. *et al.* Accelerated aging pathology in ad libitum fed Xpd (TTD) mice is accompanied by features suggestive of caloric restriction. DNA Repair (Amst.) 4, 1314–1324. 2005.

336 Tanaka, H., Arakawa, H., Yamaguchi, T., Shiraishi, K., Fukuda, S., Matsui, K. *et al.* A ribonucleotide reductase gene involved in a p53-dependent cell-cycle checkpoint for DNA damage. Nature 404, 42–49. 2000.

337 Jackson, S.P. Sensing and repairing DNA double-strand breaks. Carcinogenesis 23, 687–696. 2002.

338 van Gent, D.C., Hoeijmakers, J.H., & Kanaar, R. Chromosomal stability and the DNA double-stranded break connection. Nat. Rev. Genet. 2, 196–206. 2001.

339 Dronkert, M.L. & Kanaar, R. Repair of DNA interstrand cross-links. Mutat. Res. 486, 217–247. 2001.

340 Rosselli, F., Briot, D., & Pichierri, P. The Fanconi anemia pathway and the DNA interstrand cross-links repair. Biochimie 85, 1175–1184. 2003.

341 Szostak, J.W., Orr-Weaver, T.L., Rothstein, R.J., & Stahl, F.W. The double-strand-break repair model for recombination. Cell 33, 25–35. 1983.

342 Shao, C., Deng, L., Henegariu, O., Liang, L., Raikwar, N., Sahota, A. *et al.* Mitotic recombination produces the majority of recessive fibroblast variants in heterozygous mice. Proc. Natl. Acad. Sci. USA 96, 9230–9235. 1999.

343 Shao, C., Stambrook, P.J., & Tischfield, J.A. Mitotic recombination is suppressed by chromosomal divergence in hybrids of distantly related mouse strains. Nat. Genet. 28, 169–172. 2001.

344 Levitus, M., Joenje, H., & de Winter, J.P. The Fanconi anemia pathway of genomic maintenance. Cell Oncol. 28, 3–29. 2006.

345 Bassing, C.H. & Alt, F.W. The cellular response to general and programmed DNA double strand breaks. DNA Repair (Amst.) 3, 781–796. 2004.

346 Ma, Y., Schwarz, K., & Lieber, M.R. The Artemis:DNA-PKcs endonuclease cleaves DNA loops, flaps, and gaps. DNA Repair (Amst.) 4, 845–851. 2005.

347 Lieber, M.R. & Karanjawala, Z.E. Ageing, repetitive genomes and DNA damage. Nat. Rev. Mol. Cell. Biol. 5, 69–75. 2004.

348 Richardson, C. & Jasin, M. Frequent chromosomal translocations induced by DNA double-strand breaks. Nature 405, 697–700. 2000.

349 Aten, J.A., Stap, J., Krawczyk, P.M., van Oven, C.H., Hoebe, R.A., Essers, J. *et al.* Dynamics of DNA double-strand breaks revealed by clustering of damaged chromosome domains. Science 303, 92–95. 2004.

350 Difilippantonio, M.J., Zhu, J., Chen, H.T., Meffre, E., Nussenzweig, M.C., Max, E.E. *et al.* DNA repair protein Ku80 suppresses chromosomal aberrations and malignant transformation. Nature 404, 510–514. 2000.

351 Smogorzewska, A., Karlseder, J., Holtgreve-Grez, H., Jauch, A., & de Lange, T. DNA ligase IV-dependent NHEJ of deprotected mammalian telomeres in G1 and G2. Curr. Biol. 12, 1635–1644. 2002.

352 Greenberg, R.A., Allsopp, R.C., Chin, L., Morin, G.B., & DePinho, R.A. Expression of mouse telomerase reverse transcriptase during development, differentiation and proliferation. Oncogene 16, 1723–1730. 1998.

353 Parrinello, S., Samper, E., Krtolica, A., Goldstein, J., Melov, S., & Campisi, J. Oxygen sensitivity severely limits the replicative lifespan of murine fibroblasts. Nat. Cell Biol. 5, 741–747. 2003.

354 Gonzalez-Suarez, E., Samper, E., Flores, J.M., & Blasco, M.A. Telomerase-deficient mice with short telomeres are resistant to skin tumorigenesis. Nat. Genet. 26, 114–117. 2000.

355 Greenberg, R.A., Chin, L., Femino, A., Lee, K.H., Gottlieb, G.J., Singer, R.H. *et al.* Short dysfunctional telomeres impair tumorigenesis in the INK4a(delta2/3) cancer-prone mouse. Cell 97, 515–525. 1999.

356 Lundblad, V. & Blackburn, E.H. An alternative pathway for yeast telomere maintenance rescues est1-senescence. Cell 73, 347–360. 1993.

357 Le, S., Moore, J.K., Haber, J.E., & Greider, C.W. RAD50 and RAD51 define two pathways that collaborate to maintain telomeres in the absence of telomerase. Genetics 152, 143–152. 1999.

358 Bryan, T.M., Englezou, A., Dalla-Pozza, L., Dunham, M.A., & Reddel, R.R. Evidence for an alternative mechanism for maintaining telomere length in human tumors and tumor-derived cell lines. Nat. Med. 3, 1271–1274. 1997.

359 Tarsounas, M., Munoz, P., Claas, A., Smiraldo, P.G., Pittman, D.L., Blasco, M.A. *et al.* Telomere maintenance requires the RAD51D recombination/repair protein. Cell 117, 337–347. 2004.

360 Maser, R.S. & DePinho, R.A. Telomeres and the DNA damage response: why the fox is guarding the henhouse. DNA Repair (Amst.) 3, 979–988. 2004.

361 Beausejour, C.M., Krtolica, A., Galimi, F., Narita, M., Lowe, S.W., Yaswen, P. *et al.* Reversal of human cellular senescence: roles of the p53 and p16 pathways. EMBO J. 22, 4212–4222. 2003.

362 Kim, S.H., Han, S., You, Y.H., Chen, D.J., & Campisi, J. The human telomere-associated protein TIN2 stimulates interactions between telomeric DNA tracts in vitro. EMBO Rep. 4, 685–691. 2003.

363 Karlseder, J., Kachatrian, L., Takai, H., Mercer, K., Hingorani, S., Jacks, T. *et al.* Targeted deletion reveals an essential function for the telomere length regulator Trf1. Mol. Cell. Biol. 23, 6533–6541. 2003.

364 Munoz, P., Blanco, R., Flores, J.M., & Blasco, M.A. XPF nuclease-dependent telomere loss and increased DNA damage in mice overexpressing TRF2 result in premature aging and cancer. Nat. Genet. 37, 1063–1071. 2005.

365 Bradshaw, P.S., Stavropoulos, D.J., & Meyn, M.S. Human telomeric protein TRF2 associates with genomic double-strand breaks as an early response to DNA damage. Nat. Genet. 37, 193–197. 2005.

366 Flores, I., Cayuela, M.L., & Blasco, M.A. Effects of telomerase and telomere length on epidermal stem cell behavior. Science 309, 1253–1256. 2005.

367 Sarin, K.Y., Cheung, P., Gilison, D., Lee, E., Tennen, R.I., Wang, E. *et al.* Conditional telomerase induction causes proliferation of hair follicle stem cells. Nature 436, 1048–1052. 2005.

368 Allsopp, R.C., Morin, G.B., Horner, J.W., DePinho, R., Harley, C.B., & Weissman, I.L. Effect of TERT over-expression on the long-term transplantation capacity of hematopoietic stem cells. Nat. Med. 9, 369–371. 2003.

369 Mohaghegh, P. & Hickson, I.D. Premature aging in RecQ helicase-deficient human syndromes. Int. J. Biochem. Cell. Biol. 34, 1496–1501. 2002.

370 Wu, L., Lung Chan, K., Ralf, C., Bernstein, D.A., Garcia, P.L., Bohr, V.A. *et al.* The HRDC domain of BLM is required for the dissolution of double Holliday junctions. EMBO J. 24, 2679–2687. 2005.

371 Chester, N., Kuo, F., Kozak, C., O'Hara, C.D., & Leder, P. Stage-specific apoptosis, developmental delay, and embryonic lethality in mice homozygous for a targeted disruption in the murine Bloom's syndrome gene. Genes Dev. 12, 3382–3393. 1998.

372 Goss, K.H., Risinger, M.A., Kordich, J.J., Sanz, M.M., Straughen, J.E., Slovek, L.E. *et al.* Enhanced tumor formation in mice heterozygous for Blm mutation. Science 297, 2051–2053. 2002.

373 Crabbe, L., Verdun, R.E., Haggblom, C.I., & Karlseder, J. Defective telomere lagging strand synthesis in cells lacking WRN helicase activity. Science 306, 1951–1953. 2004.

374 Li, B. & Comai, L. Functional interaction between Ku and the werner syndrome protein in DNA end processing. J. Biol. Chem. 275, 28349–28352. 2000.

375 Chen, L., Huang, S., Lee, L., Davalos, A., Schiestl, R.H., Campisi, J. *et al.* WRN, the protein deficient in Werner syndrome, plays a critical structural role in optimizing DNA repair. Aging Cell 2, 191–199. 2003.

376 Wallis, J.W., Chrebet, G., Brodsky, G., Rolfe, M., & Rothstein, R. A hyper-recombination mutation in S. cerevisiae identifies a novel eukaryotic topoisomerase. Cell 58, 409–419. 1989.

377 Gangloff, S., McDonald, J.P., Bendixen, C., Arthur, L., & Rothstein, R. The yeast type I topoisomerase Top3 interacts with Sgs1, a DNA helicase homolog: a potential eukaryotic reverse gyrase. Mol. Cell. Biol. 14, 8391–8398. 1994.

378 Laine, J.P., Opresko, P.L., Indig, F.E., Harrigan, J.A., von Kobbe, C., & Bohr, V.A. Werner protein stimulates topoisomerase I DNA relaxation activity. Cancer Res. 63, 7136–7146. 2003.

379 Ridgway, P. & Almouzni, G. CAF-1 and the inheritance of chromatin states: at the crossroads of DNA replication and repair. J. Cell Sci. 113, 2647–2658. 2000.

380 Linger, J. & Tyler, J.K. The yeast histone chaperone chromatin assembly factor 1 protects against double-strand DNA-damaging agents. Genetics 171, 1513–1522. 2005.

381 Ramey, C.J., Howar, S., Adkins, M., Linger, J., Spicer, J., & Tyler, J.K. Activation of the DNA damage checkpoint in yeast lacking the histone chaperone anti-silencing function 1. Mol. Cell. Biol. 24, 10313–27. 2004.

382 Reik, W., Dean, W., & Walter, J. Epigenetic reprogramming in mammalian development. Science 293, 1089–1093. 2001.

383 Kastan, M.B., Gowans, B.J., & Lieberman, M.W. Methylation of deoxycytidine incorporated by excision-repair synthesis of DNA. Cell 30, 509–516. 1982.

384 Mortusewicz, O., Schermelleh, L., Walter, J., Cardoso, M.C., & Leonhardt, H. Recruitment of DNA methyltransferase I to DNA repair sites. Proc. Natl. Acad. Sci. USA 102, 8905–8909. 2005.

385 Toyota, M. & Issa, J.P. CpG island methylator phenotypes in aging and cancer. Semin. Cancer Biol. 9, 349–357. 1999.

386 Evans, M.D., Dizdaroglu, M., & Cooke, M.S. Oxidative DNA damage and disease: induction, repair and significance. Mutat. Res. 567, 1–61. 2004.

387 Chen, X., Mele, J., Giese, H., Van Remmen, H., Dolle, M.E., Steinhelper, M. *et al*. A strategy for the ubiquitous overexpression of human catalase and CuZn superoxide dismutase genes in transgenic mice. Mech. Ageing Dev. 124, 219–227. 2003.

388 Frei, B. Efficacy of dietary antioxidants to prevent oxidative damage and inhibit chronic disease. J. Nutr. 134, 3196S–3198S. 2004.

389 Meydani, M., Lipman, R.D., Han, S.N., Wu, D., Beharka, A., Martin, K.R. *et al*. The effect of long-term dietary supplementation with antioxidants. Ann. NY Acad. Sci. 854, 352–360. 1998.

390 Queitsch, C., Sangster, T.A., & Lindquist, S. Hsp90 as a capacitor of phenotypic variation. Nature 417, 618–624. 2002.

391 Rutherford, S.L. Between genotype and phenotype: protein chaperones and evolvability. Nat. Rev. Genet. 4, 263–274. 2003.

392 Szilard, L. On the nature of the aging process. Proc. Natl. Acad. Sci. USA 45, 30–45. 1959.

393 Mullaart, E., Boerrigter, M.E., Lohman, P.H., & Vijg, J. Age-related induction and disappearance of carcinogen-DNA-adducts in livers of rats exposed to low levels of 2-acetylaminofluorene. Chem. Biol. Interact. 69, 373–384. 1989.

394 Boerrigter, M.E., Wei, J.Y., & Vijg, J. Induction and repair of benzo[a]pyrene-DNA adducts in C57BL/6 and BALB/c mice: association with aging and longevity. Mech. Ageing Dev. 82, 31–50. 1995.

395 Vijg, J., Mullaart, E., Lohman, P.H., & Knook, D.L. UV-induced unscheduled DNA synthesis in fibroblasts of aging inbred rats. Mutat. Res. 146, 197–204. 1985.

396 Cabelof, D.C., Raffoul, J.J., Yanamadala, S., Ganir, C., Guo, Z., & Heydari, A.R. Attenuation of DNA polymerase beta-dependent base excision repair and increased DMS-induced mutagenicity in aged mice. Mutat. Res. 500, 135–145. 2002.

397 Intano, G.W., Cho, E.J., McMahan, C.A., & Walter, C.A. Age-related base excision repair activity in mouse brain and liver nuclear extracts. J. Gerontol. A Biol. Sci. Med. Sci. 58, 205–211. 2003.

398 Intano, G.W., McMahan, C.A., McCarrey, J.R., Walter, R.B., McKenna, A.E., Matsumoto, Y. *et al*. Base excision repair is limited by different proteins in male germ cell nuclear extracts prepared from young and old mice. Mol. Cell. Biol. 22, 2410–2418. 2002.

399 Imam, S.Z., Karahalil, B., Hogue, B.A., Souza-Pinto, N.C., & Bohr, V.A. Mitochondrial and nuclear DNA-repair capacity of various brain regions in mouse is altered in an age-dependent manner. Neurobiol. Aging 27, 1129–1136. 2005.

400 Krishna, T.H., Mahipal, S., Sudhakar, A., Sugimoto, H., Kalluri, R., & Rao, K.S. Reduced DNA gap repair in aging rat neuronal extracts and its restoration by DNA polymerase beta and DNA-ligase. J. Neurochem. 92, 818–823. 2005.

401 Vyjayanti, V.N. & Rao, K.S. DNA double strand break repair in brain: reduced NHEJ activity in aging rat neurons. Neurosci. Lett. 393, 18–22. 2006.

402 Hanawalt, P.C. On the role of DNA damage and repair processes in aging: evidence for and against. In Modern Biological Theories of Aging, pp. 183–198 (eds. Warner, H.R., Butler, R.N., Sprott, R.L., & Schneider, E.L.). Raven Press, New York, 1987.

403 Tice, R.R. & Setlow, R.B. DNA repair and replication in aging organisms and cells. In Handbook of the Biology of Aging, pp.173–224 (eds. Finch, C.E. & Schneider, E.L.). Van Nostrand Reinhold Company, New York, 1985.

404 Mullaart, E., Lohman, P.H., Berends, F., & Vijg, J. DNA damage metabolism and aging. Mutat. Res. 237, 189–210. 1990.

405 Suh, Y., Kang, U.G., Kim, Y.S., Kim, W.H., Park, S.C., & Park, J.B. Differential activation of c-Jun NH2-terminal kinase and p38 mitogen-activated protein kinases by methyl methanesulfonate in the liver and brain of rats: implication for organ-specific carcinogenesis. Cancer Res. 60, 5067–5073. 2000.

406 Suh, Y., Lee, K.A., Kim, W.H., Han, B.G., Vijg, J., & Park, S.C. Aging alters the apoptotic response to genotoxic stress. Nat. Med. 8, 3–4. 2001.

407 Muskhelishvili, L., Hart, R.W., Turturro, A., & James, S.J. Age-related changes in the intrinsic rate of apoptosis in livers of diet-restricted and ad libitum-fed B6C3F1 mice. Am. J. Pathol. 147, 20–24. 1995.

408 Camplejohn, R.S., Gilchrist, R., Easton, D., McKenzie-Edwards, E., Barnes, D.M., Eccles, D.M. *et al.* Apoptosis, ageing and cancer susceptibility. Br. J. Cancer 88, 487–490. 2003.

409 Ramalho-Santos, M., Yoon, S., Matsuzaki, Y., Mulligan, R.C., & Melton, D.A. "Stemness": transcriptional profiling of embryonic and adult stem cells. Science 298, 597–600. 2002.

410 Cervantes, R.B., Stringer, J.R., Shao, C., Tischfield, J.A., & Stambrook, P.J. Embryonic stem cells and somatic cells differ in mutation frequency and type. Proc. Natl. Acad. Sci. USA 99, 3586–3590. 2002.

411 Frosina, G. Overexpression of enzymes that repair endogenous damage to DNA. Eur. J. Biochem. 267, 2135–2149. 2000.

412 Cairns, J. Mutation selection and the natural history of cancer. Nature 255, 197–200. 1975.

413 Potten, C.S., Owen, G., & Booth, D. Intestinal stem cells protect their genome by selective segregation of template DNA strands. J. Cell Sci. 115, 2381–2388. 2002.

414 Potten, C.S. Extreme sensitivity of some intestinal crypt cells to X and gamma irradiation. Nature 269, 518–521. 1977.

415 Chimpanzee Sequencing and Analysis Consortium. Initial sequence of the chimpanzee genome and comparison with the human genome. Nature 437, 69–87. 2005.

416 Varki, A. & Altheide, T.K. Comparing the human and chimpanzee genomes: searching for needles in a haystack. Genome Res. 15, 1746–1758. 2005.

417 Todaro, G.J. & Green, H. Quantitative studies of the growth of mouse embryo cells in culture and their development into established lines. J. Cell Biol. 17, 299–313. 1963.

418 Hayflick, L. & Moorhead, P.S. The serial cultivation of human diploid cell strains. Exp. Cell Res. 25, 585–621. 1961.

419 Harvey, D.M. & Levine, A.J. p53 alteration is a common event in the spontaneous immortalization of primary BALB/c murine embryo fibroblasts. Genes Dev. 5, 2375–2385. 1991.

420 Harnden, D.G., Benn, P.A., Oxford, J.M., Taylor, A.M., & Webb, T.P. Cytogenetically marked clones in human fibroblasts cultured from normal subjects. Somatic Cell Genet. 2, 55–62. 1976.

421 Campisi, J. Senescent cells, tumor suppression, and organismal aging: good citizens, bad neighbors. Cell 120, 513–522. 2005.

422 Shay, J.W. & Wright, W.E. Aging. When do telomeres matter? Science 291, 839–840. 2001.

423 Kapahi, P., Boulton, M.E., & Kirkwood, T.B. Positive correlation between mammalian life span and cellular resistance to stress. Free Radic. Biol. Med. 26, 495–500. 1999.

424 Martin, G.M. Genetic modulation of senescent phenotypes in Homo sapiens. Cell 120, 523–532. 2005.

425 Epstein, C.J., Martin, G.M., & Motulsky, A.G. Werner's syndrome; caricature of aging. A genetic model for the study of degenerative diseases. Trans. Assoc. Am. Physicians 78, 73–81. 1965.

426 Pollex, R.L. & Hegele, R.A. Hutchinson-Gilford progeria syndrome. Clin. Genet. 66, 375–381. 2004.

427 Yu, C.E., Oshima, J., Fu, Y.H., Wijsman, E.M., Hisama, F., Alisch, R. et al. Positional cloning of the Werner's syndrome gene. Science 272, 258–262. 1996.

428 Eriksson, M., Brown, W.T., Gordon, L.B., Glynn, M.W., Singer, J., Scott, L. et al. Recurrent de novo point mutations in lamin A cause Hutchinson-Gilford progeria syndrome. Nature 423, 293–298. 2003.

429 Shiloh, Y. ATM and related protein kinases: safeguarding genome integrity. Nat. Rev. Cancer 3, 155–168. 2003.

430 Lehmann, A.R. DNA repair-deficient diseases, xeroderma pigmentosum, Cockayne syndrome and trichothiodystrophy. Biochimie 85, 1101–1111. 2003.

431 Lindor, N.M., Furuichi, Y., Kitao, S., Shimamoto, A., Arndt, C., & Jalal, S. Rothmund-Thomson syndrome due to RECQ4 helicase mutations: report and clinical and molecular comparisons with Bloom syndrome and Werner syndrome. Am. J. Med. Genet. 90, 223–228. 2000.

432 Hasty, P., Campisi, J., Hoeijmakers, J., van Steeg, H., & Vijg, J. Aging and genome maintenance: lessons from the mouse? Science 299, 1355–1359. 2003.

433 Kitado, H., Higuchi, K., & Takeda, T. Molecular genetic characterization of the senescence-accelerated mouse (SAM) strains. J. Gerontol. 49, B247–B254. 1994.

434 Harrison, D.E. Potential misinterpretations using models of accelerated aging. J. Gerontol. 49, B245–B246. 1994.

435 Odagiri, Y., Uchida, H., Hosokawa, M., Takemoto, K., Morley, A.A., & Takeda, T. Accelerated accumulation of somatic mutations in the senescence-accelerated mouse. Nat. Genet. 19, 116–117. 1998.

436 Gutmann, D.H., Hunter-Schaedle, K., & Shannon, K.M. Harnessing preclinical mouse models to inform human clinical cancer trials. J. Clin. Invest. 116, 847–852. 2006.

437 DePinho, R.A. The age of cancer. Nature 408, 248–254. 2000.

438 Hasty, P. & Vijg, J. Accelerating aging by mouse reverse genetics: a rational approach to understanding longevity. Aging Cell 3, 55–65. 2004.

439 Bronson, R.T. & Lipman, R.D. Reduction in rate of occurrence of age-related lesions in dietary restricted laboratory mice. Growth Devel. Aging 55, 169–184. 1991.

440 Bronson, R.T. Rate of occurrence of lesions in 20 inbred and hybrid genotypes of rats and mice sacrificed at 6 month intervals during the first years of life. In Genetic Effects of Aging II, pp. 279–358 (ed. Harrison, D.E.). The Telford Press, Caldwell, NJ, 1990.

441 Mahler, J.F., Stokes, W., Mann, P.C., Takaoka, M., & Maronpot, R.R. Spontaneous lesions in aging FVB/N mice. Toxicol. Pathol. 24, 710–716. 1996.

442 Mohr, U., Dungworth D.L., Capen C.C., Carlton, W.W., Sundberg, J.P., Ward, J.M. (eds) Pathobiology of the Aging Mouse. ILSI Press, Washington DC, 1996.

443 Vijg, J. & Suh, Y. Genetics of longevity and aging. Annu. Rev. Med. 56, 193–212. 2005.

444 Castrillon, D.H., Miao, L., Kollipara, R., Horner, J.W., & DePinho, R.A. Suppression of ovarian follicle activation in mice by the transcription factor Foxo3a. Science 301, 215–218. 2003.

445 Hosaka, T., Biggs, W.H., 3rd, Tieu, D., Boyer, A.D., Varki, N.M., Cavenee, W.K. *et al.* Disruption of forkhead transcription factor (FOXO) family members in mice reveals their functional diversification. Proc. Natl. Acad. Sci. USA 101, 2975–2980. 2004.

446 Cheng, H.L., Mostoslavsky, R., Saito, S., Manis, J.P., Gu, Y., Patel, P. *et al.* Developmental defects and p53 hyperacetylation in Sir2 homolog (SIRT1)-deficient mice. Proc. Natl. Acad. Sci. USA 100, 10794–10799. 2003.

447 Mostoslavsky, R., Chua, K.F., Lombard, D.B., Pang, W.W., Fischer, M.R., Gellon, L. *et al.* Genomic instability and aging-like phenotype in the absence of mammalian SIRT6. Cell 124, 315–329. 2006.

448 Bartke, A. & Brown-Borg, H. Life extension in the dwarf mouse. Curr. Top. Dev. Biol. 63, 189–225. 2004.

449 Migliaccio, E., Giorgio, M., Mele, S., Pelicci, G., Reboldi, P., Pandolfi, P.P. *et al.* The p66shc adaptor protein controls oxidative stress response and life span in mammals. Nature 402, 309–313. 1999.

450 Bartke, A., Chandrashekar, V., Bailey, B., Zaczek, D., & Turyn, D. Consequences of growth hormone (GH) overexpression and GH resistance. Neuropeptides 36, 201–208. 2002.

451 Kurosu, H., Yamamoto, M., Clark, J.D., Pastor, J.V., Nandi, A., Gurnani, P. *et al.* Suppression of aging in mice by the hormone Klotho. Science 309, 1829–1833. 2005.

452 Kile, B.T., Hentges, K.E., Clark, A.T., Nakamura, H., Salinger, A.P., Liu, B. *et al.* Functional genetic analysis of mouse chromosome 11. Nature 425, 81–86. 2003.

453 Lombard, D.B., Chua, K.F., Mostoslavsky, R., Franco, S., Gostissa, M., & Alt, F.W. DNA repair, genome stability, and aging. Cell 120, 497–512. 2005.

454 de Boer, J., Andressoo, J.O., de Wit, J., Huijmans, J., Beems, R.B., van Steeg, H. *et al.* Premature aging in mice deficient in DNA repair and transcription. Science 296, 1276–1279. 2002.

455 Weeda, G., Donker, I., de Wit, J., Morreau, H., Janssens, R., Vissers, C.J. *et al.* Disruption of mouse ERCC1 results in a novel repair syndrome with growth failure, nuclear abnormalities and senescence. Curr. Biol. 7, 427–439. 1997.

456 de Boer, J., de Wit, J., van Steeg, H., Berg, R.J., Morreau, H., Visser, P. *et al.* A mouse model for the basal transcription/DNA repair syndrome trichothiodystrophy. Mol. Cell 1, 981–990. 1998.

457 Vijg, J. & Suh, Y. Ageing: chromatin unbound. Nature 440, 874–875. 2006.

458 Vogel, H., Lim, D.S., Karsenty, G., Finegold, M., & Hasty, P. Deletion of Ku86 causes early onset of senescence in mice. Proc. Natl. Acad. Sci. USA 96, 10770–10775. 1999.

459 Espejel, S., Martin, M., Klatt, P., Martin-Caballero, J., Flores, J.M., & Blasco, M.A. Shorter telomeres, accelerated ageing and increased lymphoma in DNA-PKcs-deficient mice. EMBO Rep. 5, 503–509. 2004.

460 Cao, L., Li, W., Kim, S., Brodie, S.G., & Deng, C.X. Senescence, aging, and malignant transformation mediated by p53 in mice lacking the Brca1 full-length isoform. Genes Dev. 17, 201–213. 2003.

461 Gu, Y., Seidl, K.J., Rathbun, G.A., Zhu, C., Manis, J.P., van der Stoep, N. *et al.* Growth retardation and leaky SCID phenotype of Ku70-deficient mice. Immunity 7, 653–665. 1997.

462 Lebel, M. & Leder, P. A deletion within the murine Werner syndrome helicase induces sensitivity to inhibitors of topoisomerase and loss of cellular proliferative capacity. Proc. Natl. Acad. Sci. USA 95, 13097–13102. 1998.

463 Lombard, D.B., Beard, C., Johnson, B., Marciniak, R.A., Dausman, J., Bronson, R. *et al.* Mutations in the WRN gene in mice accelerate mortality in a p53-null background. Mol. Cell. Biol. 20, 3286–3291. 2000.

464 Goto, M. Hierarchical deterioration of body systems in Werner's syndrome: implications for normal ageing. Mech. Ageing Dev. 98, 239–254. 1997.

465 Fukuchi, K., Martin, G.M., & Monnat, R.J., Jr. Mutator phenotype of Werner syndrome is characterized by extensive deletions. Proc. Natl. Acad. Sci. USA 86, 5893–5897. 1989.

466 Robanus-Maandag, E., Dekker, M., van der Valk, M., Carrozza, M.L., Jeanny, J.C., Dannenberg, J.H. *et al.* p107 is a suppressor of retinoblastoma development in pRb-deficient mice. Genes Dev. 12, 1599–1609. 1998.

467 Blackburn, E.H. Telomeres and telomerase: their mechanisms of action and the effects of altering their functions. FEBS Lett. 579, 859–862. 2005.

468 Blasco, M.A., Lee, H.W., Hande, M.P., Samper, E., Lansdorp, P.M., DePinho, R.A. *et al.* Telomere shortening and tumor formation by mouse cells lacking telomerase RNA. Cell 91, 25–34. 1997.

469 Artandi, S.E., Chang, S., Lee, S.L., Alson, S., Gottlieb, G.J., Chin, L. *et al.* Telomere dysfunction promotes non-reciprocal translocations and epithelial cancers in mice. Nature 406, 641–645. 2000.

470 Rudolph, K.L., Chang, S., Lee, H.W., Blasco, M., Gottlieb, G.J., Greider, C. *et al.* Longevity, stress response, and cancer in aging telomerase-deficient mice. Cell 96, 701–712. 1999.

471 Chun, H.H. & Gatti, R.A. Ataxia-telangiectasia, an evolving phenotype. DNA Repair (Amst.) 3, 1187–1196. 2004.

472 Xu, Y., Ashley, T., Brainerd, E.E., Bronson, R.T., Meyn, M.S., & Baltimore, D. Targeted disruption of ATM leads to growth retardation, chromosomal fragmentation during meiosis, immune defects, and thymic lymphoma. Genes Dev. 10, 2411–2422. 1996.

473 Wong, K.K., Maser, R.S., Bachoo, R.M., Menon, J., Carrasco, D.R., Gu, Y. *et al.* Telomere dysfunction and Atm deficiency compromises organ homeostasis and accelerates ageing. Nature 421, 643–648. 2003.

474 Chang, S., Multani, A.S., Cabrera, N.G., Naylor, M.L., Laud, P., Lombard, D. *et al.* Essential role of limiting telomeres in the pathogenesis of Werner syndrome. Nat. Genet. 36, 877–882. 2004.

475 Tarsounas, M. & West, S.C. Recombination at mammalian telomeres: an alternative mechanism for telomere protection and elongation. Cell Cycle 4, 672–674. 2005.

476 Laursen, L.V., Bjergbaek, L., Murray, J.M., & Andersen, A.H. RecQ helicases and topoisomerase III in cancer and aging. Biogerontology 4, 275–287. 2003.

477 Lebel, M., Cardiff, R.D., & Leder, P. Tumorigenic effect of nonfunctional p53 or p21 in mice mutant in the Werner syndrome helicase. Cancer Res. 61, 1816–1819. 2001.

478 Poot, M., Gollahon, K.A., & Rabinovitch, P.S. Werner syndrome lymphoblastoid cells are sensitive to camptothecin-induced apoptosis in S-phase. Hum. Genet. 104, 10–14. 1999.

479 Kwan, K.Y. & Wang, J.C. Mice lacking DNA topoisomerase IIIbeta develop to maturity but show a reduced mean lifespan. Proc. Natl. Acad. Sci. USA 98, 5717–5721. 2001.

480 Bergo, M.O., Gavino, B., Ross, J., Schmidt, W.K., Hong, C., Kendall, L.V. *et al.* Zmpste24 deficiency in mice causes spontaneous bone fractures, muscle weakness, and a prelamin A processing defect. Proc. Natl. Acad. Sci. USA 99, 13049–13054. 2002.

481 Varela, I., Cadinanos, J., Pendas, A.M., Gutierrez-Fernandez, A., Folgueras, A.R., Sanchez, L.M. *et al.* Accelerated ageing in mice deficient in Zmpste24 protease is linked to p53 signalling activation. Nature 437, 564–568. 2005.

482 Mounkes, L.C., Kozlov, S., Hernandez, L., Sullivan, T., & Stewart, C.L. A progeroid syndrome in mice is caused by defects in A-type lamins. Nature 423, 298–301. 2003.

483 Liu, B., Wang, J., Chan, K.M., Tjia, W.M., Deng, W., Guan, X. *et al.* Genomic instability in laminopathy-based premature aging. Nat. Med. 11, 780–785. 2005.

484 Hanks, S. & Rahman, N. Aneuploidy-cancer predisposition syndromes: a new link between the mitotic spindle checkpoint and cancer. Cell Cycle 4, 225–227. 2005.

485 Baker, D.J., Jeganathan, K.B., Cameron, J.D., Thompson, M., Juneja, S., Kopecka, A. *et al.* BubR1 insufficiency causes early onset of aging-associated phenotypes and infertility in mice. Nat. Genet. 36, 744–749. 2004.

486 Baker, D.J., Jeganathan, K.B., Malureanu, L., Perez-Terzic, C., Terzic, A., & van Deursen, J.M. Early aging-associated phenotypes in Bub3/Rae1 haploinsufficient mice. J. Cell Biol. 172, 529–540. 2006.

487 Arnheim, N. & Cortopassi, G. Deleterious mitochondrial DNA mutations accumulate in aging human tissues. Mutat. Res. 275, 157–167. 1992.

488 Vijg, J. & Dollé, M.E. Large genome rearrangements as a primary cause of aging. Mech. Ageing Dev. 123, 907–915. 2002.

489 Trifunovic, A., Wredenberg, A., Falkenberg, M., Spelbrink, J.N., Rovio, A.T., Bruder, C.E. *et al.* Premature ageing in mice expressing defective mitochondrial DNA polymerase. Nature 429, 417–423. 2004.

490 Kujoth, G.C., Hiona, A., Pugh, T.D., Someya, S., Panzer, K., Wohlgemuth, S.E. *et al.* Mitochondrial DNA mutations, oxidative stress, and apoptosis in mammalian aging. Science 309, 481–484. 2005.

491 Schriner, S.E., Linford, N.J., Martin, G.M., Treuting, P., Ogburn, C.E., Emond, M. *et al.* Extension of murine life span by overexpression of catalase targeted to mitochondria. Science 308, 1909–1911. 2005.

492 Khrapko, K., Kraytsberg, Y., de Grey, A.D., Vijg, J., & Schon, E.A. Does premature aging of the mtDNA mutator mouse prove that mtDNA mutations are involved in natural aging? Aging Cell 5, 279–282. 2006.

493 Tyynismaa, H., Mjosund, K.P., Wanrooij, S., Lappalainen, I., Ylikallio, E., Jalanko, A. *et al.* Mutant mitochondrial helicase Twinkle causes multiple mtDNA deletions and a late-onset mitochondrial disease in mice. Proc. Natl. Acad. Sci. USA 102, 17687–17692. 2005.

494 Vogelstein, B., Lane, D., & Levine, A.J. Surfing the p53 network. Nature 408, 307–310. 2000.

495 Maier, B., Gluba, W., Bernier, B., Turner, T., Mohammad, K., Guise, T. *et al.* Modulation of mammalian life span by the short isoform of p53. Genes Dev. 18, 306–319. 2004.

496 Tyner, S.D., Venkatachalam, S., Choi, J., Jones, S., Ghebranious, N., Igelmann, H. *et al.* p53 mutant mice that display early ageing-associated phenotypes. Nature 415, 45–53. 2002.

497 Levine, A.J., Feng, Z., Mak, T.W., You, H., & Jin, S. Coordination and communication between the p53 and IGF-1-AKT-TOR signal transduction pathways. Genes Dev. 20, 267–275. 2006.

498 Wise, P.M., Kashon, M.L., Krajnak, K.M., Rosewell, K.L., Cai, A., Scarbrough, K. *et al.* Aging of the female reproductive system: a window into brain aging. Recent Prog. Horm. Res. 52, 279–303. 1997.

499 Garcia-Cao, I., Garcia-Cao, M., Martin-Caballero, J., Criado, L.M., Klatt, P., Flores, J.M. *et al.* "Super p53" mice exhibit enhanced DNA damage response, are tumor resistant and age normally. EMBO J. 21, 6225–6235. 2002.

500 Dumble, M., Gatza, C., Tyner, S., Venkatachalam, S., & Donehower, L.A. Insights into aging obtained from p53 mutant mouse models. Ann. NY Acad. Sci. 1019, 171–177. 2004.

501 Dimri, G.P., Lee, X., Basile, G., Acosta, M., Scott, G., Roskelley, C. *et al.* A biomarker that identifies senescent human cells in culture and in aging skin in vivo. Proc. Natl. Acad. Sci. USA 92, 9363–9367. 1995.

502 Lim, D.S., Vogel, H., Willerford, D.M., Sands, A.T., Platt, K.A., & Hasty, P. Analysis of ku80-mutant mice and cells with deficient levels of p53. Mol. Cell. Biol. 20, 3772–3780. 2000.

503 Hamilton, M.L., Van Remmen, H., Drake, J.A., Yang, H., Guo, Z.M., Kewitt, K. *et al.* Does oxidative damage to DNA increase with age? Proc. Natl. Acad. Sci. USA 98, 10469–10474. 2001.

504 Vijg, J. Impact of genome instability on transcription regulation of aging and senescence. Mech. Ageing Dev. 125, 747–753. 2004.

505 Henshaw, P.S., Riley, E.F., & Stapleton, G.E. The biologic effects of pile radiations. Radiology 49, 349–364. 1947.

506 Failla, G. The aging process and carcinogenesis. Ann. NY Acad. Sci. 71, 1124–1135. 1958.

507 Klar, A.J. Propagating epigenetic states through meiosis: where Mendel's gene is more than a DNA moiety. Trends Genet. 14, 299–301. 1998.

508 Fry, J.D. Rapid mutational declines of viability in Drosophila. Genet. Res. 77, 53–60. 2001.

509 Vassilieva, L.L., Hook, A.M., & Lynch, M. The fitness effects of spontaneous mutations in Caenorhabditis elegans. Evol. Int. J. Org. Evol. 54, 1234–1246. 2000.

510 Maynard Smith, J. Review lectures in senescence. I. The causes of ageing. Proc. Roy. Soc. B. 157, 115–127 (1962).

511 Feig, D.I., Reid, T.M., & Loeb, L.A. Reactive oxygen species in tumorigenesis. Cancer Res. 54, 890s–1894s. 1994.

512 Lindahl, T. Instability and decay of the primary structure of DNA. Nature 362, 709–715. 1993.

513 Lindahl, T. & Nyberg, B. Rate of depurination of native deoxyribonucleic acid. Biochemistry 11, 3610–3618. 1972.

514 Lindahl, T. & Andersson, A. Rate of chain breakage at apurinic sites in double-stranded deoxyribonucleic acid. Biochemistry 11, 3618–3623. 1972.

515 Crine, P. & Verly, W.G. A study of DNA spontaneous degradation. Biochim. Biophys. Acta 442, 50–57. 1976.

516 Saul, R.L. & Ames, B.N. Background levels of DNA damage in the population. In Mechanisms of DNA Damage and Repair, pp. 529–536 (ed. Simic, M., Grossman, L., & Upton, A.). Plenum Press, New York, 1985.

517 Lindahl, T. & Nyberg, B. Heat-induced deamination of cytosine residues in deoxyribonucleic acid. Biochemistry 13, 3405–3410. 1974.

518 Karran, P. & Lindahl, T. Hypoxanthine in deoxyribonucleic acid: generation by heat-induced hydrolysis of adenine residues and release in free form by a deoxyribonucleic acid glycosylase from calf thymus. Biochemistry 19, 6005–6011. 1980.

519 Lindahl, T. DNA repair enzymes. Annu. Rev. Biochem. 51, 61–87. 1982.

520 Nakamura, J., Walker, V.E., Upton, P.B., Chiang, S.Y., Kow, Y.W., & Swenberg, J.A. Highly sensitive apurinic/apyrimidinic site assay can detect spontaneous and chemically induced depurination under physiological conditions. Cancer Res. 58, 222–225. 1998.

521 Beckman, K.B. & Ames, B.N. The free radical theory of aging matures. Physiol. Rev. 78, 547–581. 1998.

522 Lim, P., Wuenschell, G.E., Holland, V., Lee, D.H., Pfeifer, G.P., Rodriguez, H. et al. Peroxyl radical mediated oxidative DNA base damage: implications for lipid peroxidation induced mutagenesis. Biochemistry 43, 15339–15348. 2004.

523 Freeman, B.A. & Crapo, J.D. Biology of disease: free radicals and tissue injury. Lab. Invest. 47, 412–426. 1982.

524 Danial, N.N. & Korsmeyer, S.J. Cell death: critical control points. Cell 116, 205–219. 2004.

525 Napoli, C., Martin-Padura, I., de Nigris, F., Giorgio, M., Mansueto, G., Somma, P. et al. Deletion of the p66Shc longevity gene reduces systemic and tissue oxidative stress, vascular cell apoptosis, and early atherogenesis in mice fed a high-fat diet. Proc. Natl. Acad. Sci. USA 100, 2112–2116. 2003.

526 Giorgio, M., Migliaccio, E., Orsini, F., Paolucci, D., Moroni, M., Contursi, C. *et al.* Electron transfer between cytochrome c and p66Shc generates reactive oxygen species that trigger mitochondrial apoptosis. Cell 122, 221–233. 2005.

527 Roots, R. & Okada, S. Estimation of life times and diffusion distances of radicals involved in x-ray-induced DNA strand breaks of killing of mammalian cells. Radiat. Res. 64, 306–320. 1975.

528 Repine, J.E., Pfenninger, O.W., Talmage, D.W., Berger, E.M., & Pettijohn, D.E. Dimethyl sulfoxide prevents DNA nicking mediated by ionizing radiation or iron/hydrogen peroxide-generated hydroxyl radical. Proc. Natl. Acad. Sci. USA 78, 1001–1003. 1981.

529 Cadet, J. & Berger, M. Radiation-induced decomposition of the purine bases within DNA and related model compounds. Int. J. Radiat. Biol. Relat. Stud. Phys. Chem. Med. 47, 127–143. 1985.

530 Teoule, R. Radiation-induced DNA damage and its repair. Int. J. Radiat. Biol. Relat. Stud. Phys. Chem. Med. 51, 573–589. 1987.

531 Cathcart, R., Schwiers, E., Saul, R.L., & Ames, B.N. Thymine glycol and thymidine glycol in human and rat urine: a possible assay for oxidative DNA damage. Proc. Natl. Acad. Sci. USA 81, 5633–5637. 1984.

532 Saul, R.L., Gee, P., & Ames, B.N. Free radicals, DNA damage, and aging. In Modern Biological Theories of Aging, pp. 113–130 (ed. Warner, H.R., Butler, R.N., Sprott, R.L., & Schneider, E.L.). Raven Press, New York, 1987.

533 Adelman, R., Saul, R.L., & Ames, B.N. Oxidative damage to DNA: relation to species metabolic rate and life span. Proc. Natl. Acad. Sci. USA 85, 2706–2708. 1988.

534 Perez-Campo, R., Lopez-Torres, M., Cadenas, S., Rojas, C., & Barja, G. The rate of free radical production as a determinant of the rate of aging: evidence from the comparative approach. J. Comp. Physiol. B 168, 149–158. 1998.

535 Cooke, M.S., Evans, M.D., Dizdaroglu, M., & Lunec, J. Oxidative DNA damage: mechanisms, mutation, and disease. FASEB J. 17, 1195–1214. 2003.

536 Poulsen, H.E. Oxidative DNA modifications. Exp. Toxicol. Pathol. 57 (suppl 1), 161–169. 2005.

537 Collins, A.R., Cadet, J., Moller, L., Poulsen, H.E., & Vina, J. Are we sure we know how to measure 8-oxo-7,8-dihydroguanine in DNA from human cells? Arch. Biochem. Biophys. 423, 57–65. 2004.

538 Richter, C., Park, J.W., & Ames, B.N. Normal oxidative damage to mitochondrial and nuclear DNA is extensive. Proc. Natl. Acad. Sci. USA 85, 6465–6467. 1988.

539 Vlassara, H. & Palace, M.R. Glycoxidation: the menace of diabetes and aging. Mt Sinai J. Med. 70, 232–241. 2003.

540 Lee, A.T. & Cerami, A. Elevated glucose 6-phosphate levels are associated with plasmid mutations in vivo. Proc. Natl. Acad. Sci. USA 84, 8311–8314. 1987.

541 Cerami, A. Accumulation of advanced glycosylation endproducts on proteins and nucleic acids: role in ageing. Prog. Clin. Biol. Res. 195, 79–90. 1985.

542 Barrows, L.R. & Magee, P.N. Nonenzymatic methylation of DNA by S-adenosylmethionine *in vitro.* Carcinogenesis 3, 349–351. 1982.

543 Rydberg, B. & Lindahl, T. Nonenzymatic methylation of DNA by the intracellular methyl group donor S-adenosyl-L-methionine is a potentially mutagenic reaction. EMBO J. 1, 211–216. 1982.

544 Wogan, G.N., Hecht, S.S., Felton, J.S., Conney, A.H., & Loeb, L.A. Environmental and chemical carcinogenesis. Semin. Cancer Biol. 14, 473–486. 2004.

545 Everson, R.B., Randerath, E., Santella, R.M., Cefalo, R.C., Avitts, T.A., & Randerath, K. Detection of smoking-related covalent DNA adducts in human placenta. Science 231, 54–57. 1986.

546 Randerath, E., Avitts, T.A., Reddy, M.V., Miller, R.H., Everson, R.B., & Randerath, K. Comparative 32P-analysis of cigarette smoke-induced DNA damage in human tissues and mouse skin. Cancer Res. 46, 5869–5877. 1986.

547 Phillips, D.H., Hewer, A., Martin, C.N., Garner, R.C., & King, M.M. Correlation of DNA adduct levels in human lung with cigarette smoking. Nature 336, 790–792. 1988.

548 Baan, R.A., van den Berg, P.T., Steenwinkel, M.J., & van der Wulp, C.J. Detection of benzo[a]pyrene-DNA adducts in cultured cells treated with benzo[a]pyrene diol-epoxide by quantitative immunofluorescence microscopy and 32P-postlabelling; immunofluorescence analysis of benzo[a]pyrene-DNA adducts in bronchial cells from smoking individuals. IARC Sci. Publ. 89, 146–154. 1988.

549 Xue, W. & Warshawsky, D. Metabolic activation of polycyclic and heterocyclic aromatic hydrocarbons and DNA damage: a review. Toxicol. Appl. Pharmacol. 206, 73–93. 2005.

550 Mullaart, E., Buytenhek, M., Brouwer, A., Lohman, P.H., & Vijg, J. Genotoxic effects of intragastrically administered benzo[a]pyrene in rat liver and intestinal cells. Carcinogenesis 10, 393–395. 1989.

551 Cerutti, P.A. Prooxidant states and tumor promotion. Science 227, 375–381. 1985.

552 Cadet, J., Sage, E., & Douki, T. Ultraviolet radiation-mediated damage to cellular DNA. Mutat. Res. 571, 3–17. 2005.

553 Haugen, A., Becher, G., Benestad, C., Vahakangas, K., Trivers, G.E., Newman, M.J. *et al.* Determination of polycyclic aromatic hydrocarbons in the urine, benzo(a)pyrene diol epoxide-DNA adducts in lymphocyte DNA, and antibodies to the adducts in sera from coke oven workers exposed to measured amounts of polycyclic aromatic hydrocarbons in the work atmosphere. Cancer Res. 46, 4178–4183. 1986.

554 Shamsuddin, A.K., Sinopoli, N.T., Hemminki, K., Boesch, R.R., & Harris, C.C. Detection of benzo(a)pyrene:DNA adducts in human white blood cells. Cancer Res. 45, 66–68. 1985.

555 Perera, F.P., Hemminki, K., Young, T.L., Brenner, D., Kelly, G., & Santella, R.M. Detection of polycyclic aromatic hydrocarbon-DNA adducts in white blood cells of foundry workers. Cancer Res. 48, 2288–2291. 1988.

556 Charles, M. UNSCEAR report 2000: sources and effects of ionizing radiation. United Nations Scientific Comittee on the Effects of Atomic Radiation. J. Radiol. Prot. 21, 83–86. 2001.

557 Tracy, B.L., Krewski, D., Chen, J., Zielinski, J.M., Brand, K.P., & Meyerhof, D. Assessment and management of residential radon health risks: a report from the health Canada radon workshop. J. Toxicol. Environ. Health A 69, 735–758. 2006.

558 Lindholm, C., Makelainen, I., Paile, W., Koivistoinen, A., & Salomaa, S. Domestic radon exposure and the frequency of stable or unstable chromosomal aberrations in lymphocytes. Int. J. Radiat. Biol. 75, 921–928. 1999.

559 Lobrich, M., Rief, N., Kuhne, M., Heckmann, M., Fleckenstein, J., Rube, C. *et al.* In vivo formation and repair of DNA double-strand breaks after computed tomography examinations. Proc. Natl. Acad. Sci. USA 102, 8984–8989. 2005.

560 Ahnstrom, G. Techniques to measure DNA single-strand breaks in cells: a review. Int. J. Radiat. Biol. 54, 695–707. 1988.

561 Boerrigter, M.E., Mullaart, E., van der Schans, G.P., & Vijg, J. Quiescent human peripheral blood lymphocytes do not contain a sizable amount of preexistent DNA single-strand breaks. Exp. Cell Res. 180, 569–573. 1989.

562 Jostes, R., Reese, J.A., Cleaver, J.E., Molero, M., & Morgan, W.F. Quiescent human lymphocytes do not contain DNA strand breaks detectable by alkaline elution. Exp. Cell Res. 182, 513–520. 1989.

563 Collins, A.R. The comet assay for DNA damage and repair: principles, applications, and limitations. Mol. Biotechnol. 26, 249–261. 2004.

564 Ono, T., Okada, S., & Sugahara, T. Comparative studies of DNA size in various tissues of mice during the aging process. Exp. Gerontol. 11, 127–132. 1976.

565 Lawson, T. & Stohs, S. Changes in endogenous DNA damage in aging mice in response to butylated hydroxyanisole and oltipraz. Mech. Ageing Dev. 30, 179–185. 1985.

566 Mullaart, E., Boerrigter, M.E., Brouwer, A., Berends, F., & Vijg, J. Age-dependent accumulation of alkali-labile sites in DNA of post-mitotic but not in that of mitotic rat liver cells. Mech. Ageing Dev. 45, 41–49. 1988.

567 Chetsanga, C.J., Boyd, V., Peterson, L., & Rushlow, K. Single-stranded regions in DNA of old mice. Nature 253, 130–131. 1975.

568 Dean, R.G. & Cutler, R.G. Absence of significant age-dependent increase of single-stranded DNA extracted from mouse liver nuclei. Exp. Gerontol. 13, 287–292. 1978.

569 Mori, N. & Goto, S. Estimation of the single stranded region in the nuclear DNA of mouse tissues during aging with special reference to the brain. Arch. Gerontol. Geriatr. 1, 143–150. 1982.

570 Mullaart, E., Boerrigter, M.E., Boer, G.J., & Vijg, J. Spontaneous DNA breaks in the rat brain during development and aging. Mutat. Res. 237, 9–15. 1990.

571 Walker, A.P. & Bachelard, H.S. Studies on DNA damage and repair in the mammalian brain. J. Neurochem. 51, 1394–1399. 1988.

572 Chetsanga, C.J., Tuttle, M., Jacoboni, A., & Johnson, C. Age-associated structural alterations in senescent mouse brain DNA. Biochim. Biophys. Acta 474, 180–187. 1977.

573 Hartnell, J.M., Storrie, M.C., & Mooradian, A.D. The tissue specificity of the age-related changes in alkali-induced DNA-unwinding. Mutat. Res. 219, 187–192. 1989.

574 Singh, N.P., Danner, D.B., Tice, R.R., Brant, L., & Schneider, E.L. DNA damage and repair with age in individual human lymphocytes. Mutat. Res. 237, 123–130. 1990.

575 Singh, N.P., Danner, D.B., Tice, R.R., Pearson, J.D., Brant, L.J., Morrell, C.H. et al. Basal DNA damage in individual human lymphocytes with age. Mutat. Res. 256, 1–6. 1991.

576 Gedik, C.M., Grant, G., Morrice, P.C., Wood, S.G., & Collins, A.R. Effects of age and dietary restriction on oxidative DNA damage, antioxidant protection and DNA repair in rats. Eur. J. Nutr. 44, 263–272. 2005.

577 Rothkamm, K. & Lobrich, M. Evidence for a lack of DNA double-strand break repair in human cells exposed to very low x-ray doses. Proc. Natl. Acad. Sci. USA 100, 5057–5062. 2003.

578 Gupta, R.C., Reddy, M.V., & Randerath, K. 32P-postlabeling analysis of non-radioactive aromatic carcinogen—DNA adducts. Carcinogenesis 3, 1081–1092. 1982.

579 Randerath, K., Liehr, J.G., Gladek, A., & Randerath, E. Age-dependent covalent DNA alterations (I-compounds) in rodent tissues: species, tissue and sex specificities. Mutat. Res. 219, 121–133. 1989.

580 Gupta, K.P., van Golen, K.L., Randerath, E., & Randerath, K. Age-dependent covalent DNA alterations (I-compounds) in rat liver mitochondrial DNA. Mutat. Res. 237, 17–27. 1990.

581 Gaubatz, J.W. Postlabeling analysis of indigenous aromatic DNA adducts in mouse myocardium during aging. Arch. Gerontol. Geriatr. 8, 47–54. 1989.

582 Sokhansanj, B.A. & Wilson, D.M., 3rd. Oxidative DNA damage background estimated by a system model of base excision repair. Free Radic. Biol. Med. 37, 422–427. 2004.

583 Agarwal, S. & Sohal, R.S. DNA oxidative damage and life expectancy in houseflies. Proc. Natl. Acad. Sci. USA 91, 12332–12335. 1994.

584 Mecocci, P., Fano, G., Fulle, S., MacGarvey, U., Shinobu, L., Polidori, M.C. *et al.* Age-dependent increases in oxidative damage to DNA, lipids, and proteins in human skeletal muscle. Free Radic. Biol. Med. 26, 303–308. 1999.

585 Anson, R.M., Senturker, S., Dizdaroglu, M., & Bohr, V.A. Measurement of oxidatively induced base lesions in liver from Wistar rats of different ages. Free Radic. Biol. Med. 27, 456–462. 1999.

586 Hirano, T., Yamaguchi, R., Asami, S., Iwamoto, N., & Kasai, H. 8-Hydroxyguanine levels in nuclear DNA and its repair activity in rat organs associated with age. J. Gerontol. A Biol. Sci. Med. Sci. 51, B303–B307. 1996.

587 Le, X.C., Xing, J.Z., Lee, J., Leadon, S.A., & Weinfeld, M. Inducible repair of thymine glycol detected by an ultrasensitive assay for DNA damage. Science 280, 1066–1069. 1998.

588 Bjorksten, J. & Tenhu, H. The crosslinking theory of aging–added evidence. Exp. Gerontol. 25, 91–95. 1990.

589 Russell, A.P., Dowling, L.E., & Herrmann, R.L. Age-related differences in mouse liver DNA melting and hydroxylapatite fractionation. Gerontologia 16, 159–171. 1970.

590 Kurtz, D.I., Russell, A.P., & Sinex, F.M. Multiple peaks in the derivative melting curve of chromatin from animals of varying age. Mech. Ageing Dev. 3, 37–49. 1974.

591 Izzotti, A., Cartiglia, C., Taningher, M., De Flora, S., & Balansky, R. Age-related increases of 8-hydroxy-2'-deoxyguanosine and DNA-protein crosslinks in mouse organs. Mutat. Res. 446, 215–223. 1999.

592 Muthuswamy, A.D., Vedagiri, K., Ganesan, M., & Chinnakannu, P. Oxidative stress-mediated macromolecular damage and dwindle in antioxidant status in aged rat brain regions: role of l-carnitine and dl-alpha-lipoic acid. Clin. Chim. Acta 368, 84–92. 2006.

593 Haripriya, D., Sangeetha, P., Kanchana, A., Balu, M., & Panneerselvam, C. Modulation of age-associated oxidative DNA damage in rat brain cerebral cortex, striatum and hippocampus by L-carnitine. Exp. Gerontol. 40, 129–135. 2005.

594 Lan, J., Li, W., Zhang, F., Sun, F.Y., Nagayama, T., O'Horo, C. *et al.* Inducible repair of oxidative DNA lesions in the rat brain after transient focal ischemia and reperfusion. J. Cereb. Blood Flow Metab. 23, 1324–1339. 2003.

595 Won, M.H., Kang, T.C., Jeon, G.S., Lee, J.C., Kim, D.Y., Choi, E.M. *et al.* Immunohistochemical detection of oxidative DNA damage induced by ischemia-reperfusion insults in gerbil hippocampus in vivo. Brain Res. 836, 70–78. 1999.

596 Mullaart, E., Boerrigter, M.E., Ravid, R., Swaab, D.F., & Vijg, J. Increased levels of DNA breaks in cerebral cortex of Alzheimer's disease patients. Neurobiol. Aging 11, 169–173. 1990.

597 Wang, J., Xiong, S., Xie, C., Markesbery, W.R., & Lovell, M.A. Increased oxidative damage in nuclear and mitochondrial DNA in Alzheimer's disease. J. Neurochem. 93, 953–962. 2005.

598 Rehman, A., Nourooz-Zadeh, J., Moller, W., Tritschler, H., Pereira, P., & Halliwell, B. Increased oxidative damage to all DNA bases in patients with type II diabetes mellitus. FEBS Lett. 448, 120–122. 1999.

599 Cooke, M.S., Olinski, R., & Evans, M.D. Does measurement of oxidative damage to DNA have clinical significance? Clin. Chim. Acta 365, 30–49. 2006.

600 Lu, T., Pan, Y., Kao, S.Y., Li, C., Kohane, I., Chan, J. *et al.* Gene regulation and DNA damage in the ageing human brain. Nature 429, 883–891. 2004.

601 Yakes, F.M. & van Houten, B. Mitochondrial DNA damage is more extensive and persists longer than nuclear DNA damage in human cells following oxidative stress. Proc. Natl. Acad. Sci. USA. 94, 514–519. 1997.

602 Longo, L.D. Classic pages in obstetrics and gynecology. The chromosome number in man. Joe Hin Tjio and Albert Levan. Hereditas, vol. 42, pp. 1–6, 1956. Am. J. Obstet. Gynecol. 130, 722. 1978.

603 Kurzrock, R., Kantarjian, H.M., Druker, B.J., & Talpaz, M. Philadelphia chromosome-positive leukemias: from basic mechanisms to molecular therapeutics. Ann. Intern. Med. 138, 819–830. 2003.

604 Curtis, H. & Crowley, C. Chromosome aberrations in liver cells in relation to the somatic mutation theory of aging. Radiat. Res. 19, 337–344. 1963.

605 Ramsey, M.J., Moore, D.H., 2nd, Briner, J.F., Lee, D.A., Olsen, L., Senft, J.R. *et al.* The effects of age and lifestyle factors on the accumulation of cytogenetic damage as measured by chromosome painting. Mutat. Res. 338, 95–106. 1995.

606 Tucker, J.D., Spruill, M.D., Ramsey, M.J., Director, A.D., & Nath, J. Frequency of spontaneous chromosome aberrations in mice: effects of age. Mutat. Res. 425, 135–141. 1999.

607 Fenech, M. & Morley, A.A. The effect of donor age on spontaneous and induced micronuclei. Mutat. Res. 148, 99–105. 1985.

608 Rehen, S.K., McConnell, M.J., Kaushal, D., Kingsbury, M.A., Yang, A.H., & Chun, J. Chromosomal variation in neurons of the developing and adult mammalian nervous system. Proc. Natl. Acad. Sci. USA 98, 13361–13366. 2001.

609 Rehen, S.K., Yung, Y.C., McCreight, M.P., Kaushal, D., Yang, A.H., Almeida, B.S. *et al.* Constitutional aneuploidy in the normal human brain. J. Neurosci. 25, 2176–2180. 2005.

610 Kingsbury, M.A., Friedman, B., McConnell, M.J., Rehen, S.K., Yang, A.H., Kaushal, D. *et al.* Aneuploid neurons are functionally active and integrated into brain circuitry. Proc. Natl. Acad. Sci. USA 102, 6143–6147. 2005.

611 Iafrate, A.J., Feuk, L., Rivera, M.N., Listewnik, M.L., Donahoe, P.K., Qi, Y. *et al.* Detection of large-scale variation in the human genome. Nat. Genet. 36, 949–951. 2004.

612 Sebat, J., Lakshmi, B., Troge, J., Alexander, J., Young, J., Lundin, P. *et al.* Large-scale copy number polymorphism in the human genome. Science 305, 525–528. 2004.

613 Feuk, L., Marshall, C.R., Wintle, R.F., & Scherer, S.W. Structural variants: changing the landscape of chromosomes and design of disease studies. Hum. Mol. Genet. 15 (suppl 1), R57–R66. 2006.

614 Reynolds, B.A. & Rietze, R.L. Neural stem cells and neurospheres—re-evaluating the relationship. Nat. Methods 2, 333–336. 2005.

615 Bailey, K.J., Maslov, A.Y., & Pruitt, S.C. Accumulation of mutations and somatic selection in aging neural stem/progenitor cells. Aging Cell 3, 391–397. 2004.

616 Stout, J.T. & Caskey, C.T. HPRT: gene structure, expression, and mutation. Annu. Rev. Genet. 19, 127–148. 1985.

617 Albertini, R.J., Castle, K.L., & Borcherding, W.R. T-cell cloning to detect the mutant 6-thioguanine-resistant lymphocytes present in human peripheral blood. Proc. Natl. Acad. Sci. USA 79, 6617–6621. 1982.

618 Morley, A.A., Trainor, K.J., Seshadri, R., & Ryall, R.G. Measurement of in vivo mutations in human lymphocytes. Nature 302, 155–156. 1983.

619 Turner, D.R., Morley, A.A., Haliandros, M., Kutlaca, R., & Sanderson, B.J. In vivo somatic mutations in human lymphocytes frequently result from major gene alterations. Nature 315, 343–345. 1985.

620 Jones, I.M., Thomas, C.B., Tucker, B., Thompson, C.L., Pleshanov, P., Vorobtsova, I. *et al.* Impact of age and environment on somatic mutation at the hprt gene of T lymphocytes in humans. Mutat. Res. 338, 129–139. 1995.

621 Dempsey, J.L., Pfeiffer, M., & Morley, A.A. Effect of dietary restriction on in vivo somatic mutation in mice. Mutat. Res. 291, 141–145. 1993.

622 Inamizu, T., Kinohara, N., Chang, M.P., & Makinodan, T. Frequency of 6-thioguanine-resistant T cells is inversely related to the declining T-cell activities in aging mice. Proc. Natl. Acad. Sci. USA 83, 2488–2491. 1986.

623 da Cruz, A.D., Curry, J., Curado, M.P., & Glickman, B.W. Monitoring hprt mutant frequency over time in T-lymphocytes of people accidentally exposed to high doses of ionizing radiation. Environ. Mol. Mutagen. 27, 165–175. 1996.

624 Grist, S.A., McCarron, M., Kutlaca, A., Turner, D.R., & Morley, A.A. In vivo human somatic mutation: frequency and spectrum with age. Mutat. Res. 266, 189–196. 1992.

625 Aidoo, A., Mittelstaedt, R.A., Bishop, M.E., Lyn-Cook, L.E., Chen, Y.J., Duffy, P. *et al.* Effect of caloric restriction on Hprt lymphocyte mutation in aging rats. Mutat. Res. 527, 57–66. 2003.

626 Martin, G.M., Ogburn, C.E., Colgin, L.M., Gown, A.M., Edland, S.D., & Monnat, R.J., Jr. Somatic mutations are frequent and increase with age in human kidney epithelial cells. Hum. Mol. Genet. 5, 215–221. 1996.

627 Glazer, P.M., Sarkar, S.N., & Summers, W.C. Detection and analysis of UV-induced mutations in mammalian cell DNA using a lambda phage shuttle vector. Proc. Natl. Acad. Sci. USA 83, 1041–1044. 1986.

628 Wood, W.B. Host specificity of DNA produced by Escherichia coli: bacterial mutations affecting the restriction and modification of DNA. J. Mol. Biol. 16, 118–133. 1966.

629 Gossen, J.A. & Vijg, J. E. coli C: a convenient host strain for rescue of highly methylated DNA. Nucleic Acids Res. 16, 9343. 1988.

630 Gossen, J.A., de Leeuw, W.J., Tan, C.H., Zwarthoff, E.C., Berends, F., Lohman, P.H. *et al.* Efficient rescue of integrated shuttle vectors from transgenic mice: a model for studying mutations in vivo. Proc. Natl. Acad. Sci. USA 86, 7971–7975. 1989.

631 Kohler, S.W., Provost, G.S., Fieck, A., Kretz, P.L., Bullock, W.O., Putman, D.L. *et al.* Analysis of spontaneous and induced mutations in transgenic mice using a lambda ZAP/lacI shuttle vector. Environ. Mol. Mutagen. 18, 316–321. 1991.

632 Gossen, J.A., Molijn, A.C., Douglas, G.R., & Vijg, J. Application of galactose-sensitive E. coli strains as selective hosts for LacZ- plasmids. Nucleic Acids Res. 20, 3254. 1992.

633 Gossen, J.A. & Vijg, J. A selective system for lacZ- phage using a galactose-sensitive E. coli host. Biotechniques 14, 326, 330. 1993.

634 Gossen, J.A., de Leeuw, W.J., Molijn, A.C., & Vijg, J. Plasmid rescue from transgenic mouse DNA using LacI repressor protein conjugated to magnetic beads. Biotechniques 14, 624–629. 1993.

635 Dollé, M.E., Martus, H.J., Gossen, J.A., Boerrigter, M.E., & Vijg, J. Evaluation of a plasmid-based transgenic mouse model for detecting in vivo mutations. Mutagenesis 11, 111–118. 1996.

636 Boerrigter, M.E., Dolle, M.E., Martus, H.J., Gossen, J.A., & Vijg, J. Plasmid-based transgenic mouse model for studying in vivo mutations. Nature 377, 657–659. 1995.

637 Gossen, J.A., de Leeuw, W.J., Verwest, A., Lohman, P.H., & Vijg, J. High somatic mutation frequencies in a LacZ transgene integrated on the mouse X-chromosome. Mutat. Res. 250, 423–429. 1991.

638 Leach, E.G., Gunther, E.J., Yeasky, T.M., Gibson, L.H., Yang-Feng, T.L., & Glazer, P.M. Frequent spontaneous deletions at a shuttle vector locus in transgenic mice. Mutagenesis 11, 49–56. 1996.

639 Lee, A.T., DeSimone, C., Cerami, A., & Bucala, R. Comparative analysis of DNA mutations in lacI transgenic mice with age. FASEB J. 8, 545–550. 1994.

640 Dollé, M.E., Giese, H., Hopkins, C.L., Martus, H.J., Hausdorff, J.M., & Vijg, J. Rapid accumulation of genome rearrangements in liver but not in brain of old mice. Nat. Genet. 17, 431–434. 1997.

641 Dollé, M.E., Snyder, W.K., Gossen, J.A., Lohman, P.H., & Vijg, J. Distinct spectra of somatic mutations accumulated with age in mouse heart and small intestine. Proc. Natl. Acad. Sci. USA 97, 8403–8408. 2000.

642 Giese, H., Snyder, W.K., van Oostrom, C., van Steeg, H., Dolle, M.E., & Vijg, J. Age-related mutation accumulation at a lacZ reporter locus in normal and tumor tissues of Trp53-deficient mice. Mutat. Res. 514, 153–163. 2002.

643 Martin, S.L., Hopkins, C.L., Naumer, A., Dollé, M.E., & Vijg, J. Mutation frequency and type during ageing in mouse seminiferous tubules. Mech. Ageing Dev. 122, 1321–1331. 2001.

644 Ono, T., Ikehata, H., Nakamura, S., Saito, Y., Hosoi, Y., Takai, Y. *et al.* Age-associated increase of spontaneous mutant frequency and molecular nature of mutation in newborn and old lacZ-transgenic mouse. Mutat. Res. 447, 165–177. 2000.

645 Stuart, G.R., Oda, Y., de Boer, J.G., & Glickman, B.W. Mutation frequency and specificity with age in liver, bladder and brain of lacI transgenic mice. Genetics 154, 1291–1300. 2000.

646 Hill, K.A., Halangoda, A., Heinmoeller, P.W., Gonzalez, K., Chitaphan, C., Longmate, J. *et al.* Tissue-specific time courses of spontaneous mutation frequency and deviations in mutation pattern are observed in middle to late adulthood in Big Blue mice. Environ. Mol. Mutagen. 45, 442–454. 2005.

647 Walter, C.A., Intano, G.W., McCarrey, J.R., McMahan, C.A., & Walter, R.B. Mutation frequency declines during spermatogenesis in young mice but increases in old mice. Proc. Natl. Acad. Sci. USA 95, 10015–9. 1998.

648 Dollé, M.E. & Vijg, J. Genome dynamics in aging mice. Genome Res. 12, 1732–1738. 2002.

649 Zhang, X.B., Urlando, C., Tao, K.S., & Heddle, J.A. Factors affecting somatic mutation frequencies in vivo. Mutat. Res. 338, 189–201. 1995.

650 Bodnar, A.G., Ouellette, M., Frolkis, M., Holt, S.E., Chiu, C.P., Morin, G.B. *et al.* Extension of life-span by introduction of telomerase into normal human cells. Science 279, 349–352. 1998.

651 Vulliamy, T., Marrone, A., Szydlo, R., Walne, A., Mason, P.J., & Dokal, I. Disease anticipation is associated with progressive telomere shortening in families with dyskeratosis congenita due to mutations in TERC. Nat. Genet. 36, 447–449. 2004.

652 Lansdorp, P.M., Verwoerd, N.P., van de Rijke, F.M., Dragowska, V., Little, M.T., Dirks, R.W. *et al.* Heterogeneity in telomere length of human chromosomes. Hum. Mol. Genet. 5, 685–691. 1996.

653 Baird, D.M., Rowson, J., Wynford-Thomas, D., & Kipling, D. Extensive allelic variation and ultrashort telomeres in senescent human cells. Nat. Genet. 33, 203–207. 2003.

654 Butler, M.G., Tilburt, J., DeVries, A., Muralidhar, B., Aue, G., Hedges, L. *et al.* Comparison of chromosome telomere integrity in multiple tissues from subjects at different ages. Cancer Genet. Cytogenet. 105, 138–144. 1998.

655 Vaziri, H., Schachter, F., Uchida, I., Wei, L., Zhu, X., Effros, R. *et al.* Loss of telomeric DNA during aging of normal and trisomy 21 human lymphocytes. Am J. Hum. Genet. 52, 661–667. 1993.

656 Iwama, H., Ohyashiki, K., Ohyashiki, J.H., Hayashi, S., Yahata, N., Ando, K. *et al.* Telomeric length and telomerase activity vary with age in peripheral blood cells obtained from normal individuals. Hum. Genet. 102, 397–402. 1998.

657 Cawthon, R.M., Smith, K.R., O'Brien, E., Sivatchenko, A., & Kerber, R.A. Association between telomere length in blood and mortality in people aged 60 years or older. Lancet 361, 393–395. 2003.

658 Martin-Ruiz, C.M., Gussekloo, J., van Heemst, D., von Zglinicki, T., & Westendorp, R.G. Telomere length in white blood cells is not associated with morbidity or mortality in the oldest old: a population-based study. Aging Cell 4, 287–290. 2005.

659 Taylor, R.W. & Turnbull, D.M. Mitochondrial DNA mutations in human disease. Nat. Rev. Genet. 6, 389–402. 2005.

660 Larsen, N.B., Rasmussen, M., & Rasmussen, L.J. Nuclear and mitochondrial DNA repair: similar pathways? Mitochondrion 5, 89–108. 2005.

661 Rossignol, R., Faustin, B., Rocher, C., Malgat, M., Mazat, J.P., & Letellier, T. Mitochondrial threshold effects. Biochem. J. 370, 751–762. 2003.

662 Soong, N.W., Hinton, D.R., Cortopassi, G., & Arnheim, N. Mosaicism for a specific somatic mitochondrial DNA mutation in adult human brain. Nat. Genet. 2, 318–323. 1992.

663 Khrapko, K., Bodyak, N., Thilly, W.G., van Orsouw, N.J., Zhang, X., Coller, H.A. *et al*. Cell-by-cell scanning of whole mitochondrial genomes in aged human heart reveals a significant fraction of myocytes with clonally expanded deletions. Nucleic Acids Res. 27, 2434–2441. 1999.

664 Bodyak, N.D., Nekhaeva, E., Wei, J.Y., & Khrapko, K. Quantification and sequencing of somatic deleted mtDNA in single cells: evidence for partially duplicated mtDNA in aged human tissues. Hum. Mol. Genet. 10, 17–24. 2001.

665 Brierley, E.J., Johnson, M.A., Lightowlers, R.N., James, O.F., & Turnbull, D.M. Role of mitochondrial DNA mutations in human aging: implications for the central nervous system and muscle. Ann. Neurol. 43, 217–223. 1998.

666 Cao, Z., Wanagat, J., McKiernan, S.H., & Aiken, J.M. Mitochondrial DNA deletion mutations are concomitant with ragged red regions of individual, aged muscle fibers: analysis by laser-capture microdissection. Nucleic Acids Res. 29, 4502–4508. 2001.

667 Gokey, N.G., Cao, Z., Pak, J.W., Lee, D., McKiernan, S.H., McKenzie, D. *et al*. Molecular analyses of mtDNA deletion mutations in microdissected skeletal muscle fibers from aged rhesus monkeys. Aging Cell 3, 319–326. 2004.

668 Kraytsberg, Y., Kudryavtseva, E., McKee, A.C., Geula, C., Kowall, N.W., & Khrapko, K. Mitochondrial DNA deletions are abundant and cause functional impairment in aged human substantia nigra neurons. Nat. Genet. 38, 518–520. 2006.

669 Nekhaeva, E., Bodyak, N.D., Kraytsberg, Y., McGrath, S.B., Van Orsouw, N.J., Pluzhnikov, A. *et al*. Clonally expanded mtDNA point mutations are abundant in individual cells of human tissues. Proc. Natl. Acad. Sci. USA 99, 5521–5526. 2002.

670 Taylor, R.W., Barron, M.J., Borthwick, G.M., Gospel, A., Chinnery, P.F., Samuels, D.C. *et al*. Mitochondrial DNA mutations in human colonic crypt stem cells. J. Clin. Invest. 112, 1351–1360. 2003.

671 Britten, R.J. & Davidson, E.H. Gene regulation for higher cells: a theory. Science 165, 349–357. 1969.

672 Kimberland, M.L., Divoky, V., Prchal, J., Schwahn, U., Berger, W., & Kazazian, H.H., Jr. Full-length human L1 insertions retain the capacity for high frequency retrotransposition in cultured cells. Hum. Mol. Genet. 8, 1557–1560. 1999.

673 Bourc'his, D. & Bestor, T.H. Meiotic catastrophe and retrotransposon reactivation in male germ cells lacking Dnmt3L. Nature 431, 96–99. 2004.

674 Haoudi, A., Semmes, O.J., Mason, J.M., & Cannon, R.E. Retrotransposition-competent human LINE-1 induces apoptosis in cancer cells with intact p53. J. Biomed. Biotechnol. 2004, 185–194. 2004.

675 Bouffler, S.D., Bridges, B.A., Cooper, D.N., Dubrova, Y., McMillan, T.J., Thacker, J. *et al*. Assessing radiation-associated mutational risk to the germline: repetitive DNA sequences as mutational targets and biomarkers. Radiat. Res. 165, 249–268. 2006.

676 Jeffreys, A.J., Wilson, V., & Thein, S.L. Individual-specific 'fingerprints' of human DNA. Nature 316, 76–79. 1985.

677 Jeffreys, A.J., Neumann, R., & Wilson, V. Repeat unit sequence variation in minisatellites: a novel source of DNA polymorphism for studying variation and mutation by single molecule analysis. Cell 60, 473–485. 1990.

678 Dubrova, Y.E., Jeffreys, A.J., & Malashenko, A.M. Mouse minisatellite mutations induced by ionizing radiation. Nat. Genet. 5, 92–94. 1993.

679 Krichevsky, S., Pawelec, G., Gural, A., Effros, R.B., Globerson, A., Yehuda, D.B. *et al.* Age related microsatellite instability in T cells from healthy individuals. Exp. Gerontol. 39, 507–515. 2004.

680 Monckton, D.G. Using robots to find needles. Mech. Ageing Dev. 126, 1046–1050. 2005.

681 Coolbaugh-Murphy, M.I., Xu, J., Ramagli, L.S., Brown, B.W., & Siciliano, M.J. Microsatellite instability (MSI) increases with age in normal somatic cells. Mech. Ageing Dev. 126, 1051–1059. 2005.

682 Krontiris, T.G. Minisatellites and human disease. Science 269, 1682–1683. 1995.

683 Johnson, R. & Strehler, B.L. Loss of genes coding for ribosomal RNA in ageing brain cells. Nature 240, 412–414. 1972.

684 Strehler, B.L. Genetic instability as the primary cause of human aging. Exp. Gerontol. 21, 283–319. 1986.

685 Peterson, C.R., Cryar, J.R., & Gaubatz, J.W. Constancy of ribosomal RNA genes during aging of mouse heart cells and during serial passage of WI-38 cells. Arch. Gerontol. Geriatr. 3, 115–125. 1984.

686 Buys, C.H., Osinga, J., & Anders, G.J. Age-dependent variability of ribosomal RNA-gene activity in man as determined from frequencies of silver staining nucleolus organizing regions on metaphase chromosomes of lymphocytes and fibroblasts. Mech. Ageing Dev. 11, 55–75. 1979.

687 Thomas, S. & Mukherjee, A.B. A longitudinal study of human age-related ribosomal RNA gene activity as detected by silver-stained NORs. Mech. Ageing Dev. 92, 101–109. 1996.

688 Olsen, A.K., Bjørtuft, H., Wiger, R., Holme, J., Seeberg, E., Bjoras, M. *et al.* Highly efficient base excision repair (BER) in human and rat male germ cells. Nucleic Acids Res. 29, 1781–1790. 2001.

689 Evans, H.J. Mutation and mutagenesis in inherited and acquired human disease. The first EEMS Frits Sobels Prize Lecture, Noordwijkerhout, The Netherlands, June 1995. Mutat. Res. 351, 89–103. 1996.

690 Kamiguchi, Y., Tateno, H. and Mikamo, K. Chromosomally abnormal gametes as a cause of developmental and congenital anomalies in humans. Cong. Anom. 34, 1–12. 1994.

691 Plachot, M., Junca, A.M., Mandelbaum, J., de Grouchy, J., Salat-Baroux, J., & Cohen, J. Chromosome investigations in early life. II. Human preimplantation embryos. Hum. Reprod. 2, 29–35. 1987.

692 Marchetti, F. & Wyrobek, A.J. Mechanisms and consequences of paternally-transmitted chromosomal abnormalities. Birth Defects Res. C Embryo. Today 75, 112–129. 2005.

693 Haldane, J.B. The rate of spontaneous mutation of a human gene. 1935. J. Genet. 83, 235–244. 2004.

694 Suh, Y. & Vijg, J. Maintaining genetic integrity in aging: a zero sum game. Antioxid. Redox Signal. 8, 559–571. 2006.

695 Wilmut, I., Schnieke, A.E., McWhir, J., Kind, A.J., & Campbell, K.H. Viable offspring derived from fetal and adult mammalian cells. Nature 385, 810–813. 1997.

696 Wakayama, T., Perry, A.C., Zuccotti, M., Johnson, K.R., & Yanagimachi, R. Full-term development of mice from enucleated oocytes injected with cumulus cell nuclei. Nature 394, 369–374. 1998.

697 Wakayama, T. & Yanagimachi, R. Mouse cloning with nucleus donor cells of different age and type. Mol. Reprod. Dev. 58, 376–383. 2001.

698 Tian, X.C., Kubota, C., Enright, B., & Yang, X. Cloning animals by somatic cell nuclear transfer–biological factors. Reprod. Biol. Endocrinol. 1, 98. 2003.

699 Shiga, K., Umeki, H., Shimura, H., Fujita, T., Watanabe, S., & Nagai, T. Growth and fertility of bulls cloned from the somatic cells of an aged and infertile bull. Theriogenology 64, 334–343. 2005.

700 Shiels, P.G., Kind, A.J., Campbell, K.H., Wilmut, I., Waddington, D., Colman, A. *et al.* Analysis of telomere length in Dolly, a sheep derived by nuclear transfer. Cloning 1, 119–125. 1999.

701 Wakayama, T., Shinkai, Y., Tamashiro, K.L., Niida, H., Blanchard, D.C., Blanchard, R.J. *et al.* Cloning of mice to six generations. Nature 407, 318–319. 2000.

702 Gavrilov, L.A., Gavrilova, N.S., Kroutko, V.N., Evdokushkina, G.N., Semyonova, V.G., Gavrilova, A.L. *et al.* Mutation load and human longevity. Mutat. Res. 377, 61–62. 1997.

703 Holliday, R. Mutations and epimutations in mammalian cells. Mutat. Res. 250, 351–363. 1991.

704 Feinberg, A.P., Ohlsson, R., & Henikoff, S. The epigenetic progenitor origin of human cancer. Nat. Rev. Genet. 7, 21–33. 2006.

705 Jones, P.A. & Gonzalgo, M.L. Altered DNA methylation and genome instability: a new pathway to cancer? Proc. Natl. Acad. Sci. USA 94, 2103–2105. 1997.

706 Richardson, B. Impact of aging on DNA methylation. Ageing Res. Rev. 2, 245–261. 2003.

707 Bandyopadhyay, D. & Medrano, E.E. The emerging role of epigenetics in cellular and organismal aging. Exp. Gerontol. 38, 1299–1307. 2003.

708 Wilson, V.L., Smith, R.A., Ma, S., & Cutler, R.G. Genomic 5-methyldeoxycytidine decreases with age. J. Biol. Chem. 262, 9948–9951. 1987.

709 Lengauer, C., Kinzler, K.W., & Vogelstein, B. DNA methylation and genetic instability in colorectal cancer cells. Proc. Natl. Acad. Sci. USA 94, 2545–2550. 1997.

710 Chen, R.Z., Pettersson, U., Beard, C., Jackson-Grusby, L., & Jaenisch, R. DNA hypomethylation leads to elevated mutation rates. Nature 395, 89–93. 1998.

711 Slagboom, P.E., de Leeuw, W.J., & Vijg, J. Messenger RNA levels and methylation patterns of GAPDH and beta-actin genes in rat liver, spleen and brain in relation to aging. Mech. Ageing Dev. 53, 243–257. 1990.

712 Clark, S.J., Harrison, J., Paul, C.L., & Frommer, M. High sensitivity mapping of methylated cytosines. Nucleic Acids Res. 22, 2990–2997. 1994.

713 van Remmen, H., Ward, W.F., Sabia, R.V., & Richardson, A. Gene expression and protein degradation. In Handbook of Physiology—Aging, pp. 171–234 (ed. Masoro, E.). Oxford University Press, New York, 1995.

714 Hatada, I., Hayashizaki, Y., Hirotsune, S., Komatsubara, H., & Mukai, T. A genomic scanning method for higher organisms using restriction sites as landmarks. Proc. Natl. Acad. Sci. USA 88, 9523–9527. 1991.

715 Tra, J.K.T., Lu, Q., Kuick, R., Hanash, S., & Richardson, B. Infrequent occurrence of age-dependent changes in CpG island methylation as detected by restriction landmark genome scanning. Mech. Ageing Dev. 123, 1487–1503. 2002.

716 Fraga, M.F., Ballestar, E., Paz, M.F., Ropero, S., Setien, F., Ballestar, M.L. *et al.* Epigenetic differences arise during the lifetime of monozygotic twins. Proc. Natl. Acad. Sci. USA 102, 10604–10609. 2005.

717 Wareham, K.A., Lyon, M.F., Glenister, P.H., & Williams, E.D. Age related reactivation of an X-linked gene. Nature 327, 725–727. 1987.

718 Bennett-Baker, P.E., Wilkowski, J., & Burke, D.T. Age-associated activation of epigenetically repressed genes in the mouse. Genetics 165, 2055–2062. 2003.

719 Migeon, B.R., Axelman, J., & Beggs, A.H. Effect of ageing on reactivation of the human X-linked HPRT locus. Nature 335, 93–96. 1988.

720 Holliday, R. Ageing. X-chromosome reactivation. Nature 327, 661–662. 1987.

721 Kator, K., Cristofalo, V., Charpentier, R., & Cutler, R.G. Dysdifferentiative nature of aging: passage number dependency of globin gene expression in normal human diploid cells grown in tissue culture. Gerontology 31, 355–361. 1985.

722 Ono, T. & Cutler, R.G. Age-dependent relaxation of gene repression: increase of endogenous murine leukemia virus-related and globin-related RNA in brain and liver of mice. Proc. Natl. Acad. Sci. USA 75, 4431–4435. 1978.

723 Kanungo, M.S. Genes and Aging. Cambridge University Press, New York, 1994.

724 Villeponteau, B. The heterochromatin loss model of aging. Exp. Gerontol. 32, 383–394. 1997.

725 Liu, L., Wylie, R.C., Andrews, L.G., & Tollefsbol, T.O. Aging, cancer and nutrition: the DNA methylation connection. Mech. Ageing Dev. 124, 989–998. 2003.

726 Ames, B.N. DNA damage from micronutrient deficiencies is likely to be a major cause of cancer. Mutat. Res. 475, 7–20. 2001.

727 Mann, D.B., Springer, D.L., & Smerdon, M.J. DNA damage can alter the stability of nucleosomes: effects are dependent on damage type. Proc. Natl. Acad. Sci. USA 94, 2215–2220. 1997.

728 Morley, A.A. The estimation of in vivo mutation rate and frequency from samples of human lymphocytes. Mutat. Res. 357, 167–176. 1996.

729 Shane, B.S., Smith-Dunn, D.L., de Boer, J.G., Glickman, B.W., & Cunningham, M.L. Mutant frequencies and mutation spectra of dimethylnitrosamine (DMN) at the lacI and cII loci in the livers of Big Blue transgenic mice. Mutat. Res. 452, 197–210. 2000.

730 Mahner, M. & Kary, M. What exactly are genomes, genotypes and phenotypes? And what about phenomes? J. Theor. Biol. 186, 55–63. 1997.

731 Rossi, D.J., Bryder, D., Zahn, J.M., Ahlenius, H., Sonu, R., Wagers, A.J. *et al.* Cell intrinsic alterations underlie hematopoietic stem cell aging. Proc. Natl. Acad. Sci. USA 102, 9194–9199. 2005.

732 Kamminga, L.M. & de Haan, G. Cellular memory and hematopoietic stem cell aging. Stem Cells 24, 1143–1149. 2006.

733 Conboy, I.M., Conboy, M.J., Wagers, A.J., Girma, E.R., Weissman, I.L., & Rando, T.A. Rejuvenation of aged progenitor cells by exposure to a young systemic environment. Nature 433, 760–764. 2005.

734 Maisel, A.S., Bhalla, V., & Braunwald, E. Cardiac biomarkers: a contemporary status report. Nat. Clin. Pract. Cardiovasc. Med. 3, 288. 2006.

735 Warner, H.R. Current status of efforts to measure and modulate the biological rate of aging. J. Gerontol. A Biol. Sci. Med. Sci. 59, 692–696. 2004.

736 Duggirala, R., Uttley, M., Williams, K., Arya, R., Blangero, J., & Crawford, M.H. Genetic determination of biological age in the Mennonites of the Midwestern United States. Genet. Epidemiol. 23, 97–109. 2002.

737 Guillozet, A.L., Weintraub, S., Mash, D.C., & Mesulam, M.M. Neurofibrillary tangles, amyloid, and memory in aging and mild cognitive impairment. Arch. Neurol. 60, 729–736. 2003.

738 Yancik, R. & Ries, L.A. Aging and cancer in America. Demographic and epidemiologic perspectives. Hematol. Oncol. Clin. North Am. 14, 17–23. 2000.

739 Macieira-Coelho, A. From Weismann's theory to present day gerontology 1889–2003. Pathol. Biol. (Paris) 51, 550–562. 2003.

740 Forsyth, N.R., Elder, F.F., Shay, J.W., & Wright, W.E. Lagomorphs (rabbits, pikas and hares) do not use telomere-directed replicative aging in vitro. Mech. Ageing Dev. 126, 685–691. 2005.

741 Prowse, K.R. & Greider, C.W. Developmental and tissue-specific regulation of mouse telomerase and telomere length. Proc. Natl. Acad. Sci. USA 92, 4818–4822. 1995.

742 Hemann, M.T. & Greider, C.W. Wild-derived inbred mouse strains have short telomeres. Nucleic Acids Res. 28, 4474–4478. 2000.

743 von Zglinicki, T., Saretzki, G., Docke, W., & Lotze, C. Mild hyperoxia shortens telomeres and inhibits proliferation of fibroblasts: a model for senescence? Exp. Cell Res. 220, 186–193. 1995.

744 Chang, E. & Harley, C.B. Telomere length and replicative aging in human vascular tissues. Proc. Natl. Acad. Sci. USA 92, 11190–11194. 1995.

745 Herbig, U., Ferreira, M., Condel, L., Carey, D., & Sedivy, J.M. Cellular senescence in aging primates. Science 311, 1257. 2006.

746 Effros, R.B. Insights on immunological aging derived from the T lymphocyte cellular senescence model. Exp. Gerontol. 31, 21–27. 1996.

747 Krtolica, A., Parrinello, S., Lockett, S., Desprez, P.Y., & Campisi, J. Senescent fibroblasts promote epithelial cell growth and tumorigenesis: a link between cancer and aging. Proc. Natl. Acad. Sci. USA 98, 12072–12077. 2001.

748 Lawen, A. Apoptosis-an introduction. Bioessays 25, 888–896. 2003.

749 Jacobson, M.D., Weil, M., & Raff, M.C. Programmed cell death in animal development. Cell 88, 347–354. 1997.

750 Warner, H.R., Fernandes, G., & Wang, E. A unifying hypothesis to explain the retardation of aging and tumorigenesis by caloric restriction. J. Gerontol. A Biol. Sci. Med. Sci. 50, B107–B109. 1995.

751 James, S.J. & Muskhelishvili, L. Rates of apoptosis and proliferation vary with caloric intake and may influence incidence of spontaneous hepatoma in C57BL/6 x C3H F1 mice. Cancer Res. 54, 5508–5510. 1994.

752 Balmain, A. Cancer genetics: from Boveri and Mendel to microarrays. Nat. Rev. Cancer 1, 77–82. 2001.

753 Fearon, E.R. & Vogelstein, B. A genetic model for colorectal tumorigenesis. Cell 61, 759–767. 1990.

754 Hahn, W.C., Counter, C.M., Lundberg, A.S., Beijersbergen, R.L., Brooks, M.W., & Weinberg, R.A. Creation of human tumour cells with defined genetic elements. Nature 400, 464–468. 1999.

755 Loeb, L.A., Springgate, C.F., & Battula, N. Errors in DNA replication as a basis of malignant changes. Cancer Res. 34, 2311–2321. 1974.

756 Loeb, L.A., Loeb, K.R., & Anderson, J.P. Multiple mutations and cancer. Proc. Natl. Acad. Sci. USA 100, 776–781. 2003.

757 Hanahan, D. & Weinberg, R.A. The hallmarks of cancer. Cell 100, 57–70. 2000.

758 Turker, M.S. Somatic cell mutations: can they provide a link between aging and cancer? Mech. Ageing Dev. 117, 1–19. 2000.

759 Gates, G.A. & Mills, J.H. Presbycusis. Lancet 366, 1111–1120. 2005.

760 Hutchin, T.P. & Cortopassi, G.A. Mitochondrial defects and hearing loss. Cell. Mol. Life Sci. 57, 1927–1937. 2000.

761 McFadden, S.L., Ding, D., Reaume, A.G., Flood, D.G., & Salvi, R.J. Age-related cochlear hair cell loss is enhanced in mice lacking copper/zinc superoxide dismutase. Neurobiol. Aging 20, 1–8. 1999.

762 Strenk, S.A., Strenk, L.M., & Koretz, J.F. The mechanism of presbyopia. Prog. Retin. Eye Res. 24, 379–393. 2005.

763 Bloemendal, H., de Jong, W., Jaenicke, R., Lubsen, N.H., Slingsby, C., & Tardieu, A. Ageing and vision: structure, stability and function of lens crystallins. Prog. Biophys. Mol. Biol. 86, 407–485. 2004.

764 Schiffman, S.S. Taste and smell losses in normal aging and disease. JAMA 278, 1357–1362. 1997.

765 Yaar, M. & Gilchrest, B.A. Aging of the skin. In Fitzpatrick's Dermatology in General Medicine, pp. 1386–1398 (ed. Freedberg, I.M., Eisen, A.Z., Wolff, K., Austen, K.F., Goldsmith, L.A., & Katz, S.I.). McGraw-Hill, New York, 2003.

766 Goukassian, D.A. & Gilchrest, B.A. The interdependence of skin aging, skin cancer, and DNA repair capacity: a novel perspective with therapeutic implications. Rejuvenation Res. 7, 175–185. 2004.

767 Kosmadaki, M.G. & Gilchrest, B.A. The role of telomeres in skin aging/photoaging. Micron 35, 155–159. 2004.

768 Boukamp, P. Skin aging: a role for telomerase and telomere dynamics? Curr. Mol. Med. 5, 171–177. 2005.

769 Seeman, E. Estrogen, androgen, and the pathogenesis of bone fragility in women and men. Curr. Osteoporos. Rep. 2, 90–96. 2004.

770 Nair, K.S. Aging muscle. Am. J. Clin. Nutr. 81, 953–963. 2005.

771 Chan, G.K. & Duque, G. Age-related bone loss: old bone, new facts. Gerontology 48, 62–71. 2002.

772 Troen, B.R. Molecular mechanisms underlying osteoclast formation and activation. Exp. Gerontol. 38, 605–614. 2003.

773 Leifke, E., Gorenoi, V., Wichers, C., Von Zur Muhlen, A., Von Buren, E., & Brabant, G. Age-related changes of serum sex hormones, insulin-like growth factor-1 and sex-hormone binding globulin levels in men: cross-sectional data from a healthy male cohort. Clin. Endocrinol. (Oxf.) 53, 689–695. 2000.

774 Musaro, A., McCullagh, K., Paul, A., Houghton, L., Dobrowolny, G., Molinaro, M. *et al.* Localized Igf-1 transgene expression sustains hypertrophy and regeneration in senescent skeletal muscle. Nat. Genet. 27, 195–200. 2001.

775 Peters, R. Ageing and the brain. Postgrad. Med. J. 82, 84–88. 2006.

776 Gandhi, S. & Wood, N.W. Molecular pathogenesis of Parkinson's disease. Hum. Mol. Genet. 14, 2749–2755. 2005.

777 Helmuth, L. Detangling Alzheimer's disease. Sci. Aging Knowledge Environ. 2001, oa2. 2001.

778 Hardy, J. Has the amyloid cascade hypothesis for Alzheimer's disease been proved? Curr. Alzheimer Res. 3, 71–73. 2006.

779 Barnham, K.J., Masters, C.L., & Bush, A.I. Neurodegenerative diseases and oxidative stress. Nat. Rev. Drug Discov. 3, 205–214. 2004.

780 Dobson, C.M. Protein folding and its links with human disease. In From Protein Foliding To New Enzymes. Biochemical Society Symposium vol. 68, pp. 1–26. Portland Press, London, 2001.

781 Robbins, J.H., Otsuka, F., Tarone, R.E., Polinsky, R.J., Brumback, R.A., & Nee, L.E. Parkinson's disease and Alzheimer's disease: hypersensitivity to X rays in cultured cell lines. J. Neurol. Neurosurg. Psychiatry 48, 916–923. 1985.

782 Boerrigter, M.E., van Duijn, C.M., Mullaart, E., Eikelenboom, P., van der Togt, C.M., Knook, D.L. *et al.* Decreased DNA repair capacity in familial, but not in sporadic Alzheimer's disease. Neurobiol. Aging 12, 367–370. 1991.

783 Bender, A., Krishnan, K.J., Morris, C.M., Taylor, G.A., Reeve, A.K., Perry, R.H. *et al.* High levels of mitochondrial DNA deletions in substantia nigra neurons in aging and Parkinson disease. Nat. Genet. 38, 515–517. 2006.

784 Lakatta, E.G. Hemodynamic adaptations to stress with advancing age. Acta Med. Scand. Suppl. 711, 39–52. 1986.

785 Hansson, G.K. Inflammation, atherosclerosis, and coronary artery disease. N. Engl. J. Med. 352, 1685–1695. 2005.

786 McCaffrey, T.A., Du, B., Consigli, S., Szabo, P., Bray, P.J., Hartner, L. *et al.* Genomic instability in the type II TGF-beta1 receptor gene in atherosclerotic and restenotic vascular cells. J. Clin. Invest. 100, 2182–2188. 1997.

787 Clark, K.J., Cary, N.R., Grace, A.A., & Metcalfe, J.C. Microsatellite mutation of type II transforming growth factor-beta receptor is rare in atherosclerotic plaques. Arterioscler. Thromb. Vasc. Biol. 21, 555–559. 2001.

788 Andreassi, M.G. & Botto, N. DNA damage as a new emerging risk factor in atherosclerosis. Trends Cardiovasc. Med. 13, 270–275. 2003.

789 Bridges, B.A. International Commission for Protection against Environmental Mutagens and Carcinogens. ICPEMC Topic No. 1. Are somatic mutations involved in atherosclerosis? Mutat. Res. 182, 301–302. 1987.

790 Arvanitis, D.A., Flouris, G.A., & Spandidos, D.A. Genomic rearrangements on VCAM1, SELE, APEG1and AIF1 loci in atherosclerosis. J. Cell. Mol. Med. 9, 153–159. 2005.

791 Schmucker, D.L. Age-related changes in liver structure and function: Implications for disease ? Exp. Gerontol. 40, 650–659. 2005.

792 Handler, J.A. & Brian, W.R. Effect of aging on mixed-function oxidation and conjugation by isolated perfused rat livers. Biochem. Pharmacol. 54, 159–164. 1997.

793 Buemi, M., Nostro, L., Aloisi, C., Cosentini, V., Criseo, M., & Frisina, N. Kidney aging: from phenotype to genetics. Rejuvenation Res. 8, 101–109. 2005.

794 Effros, R.B. Ageing and the immune system. Novartis Found. Symp. 235, 130–139. 2001.

795 Walford, R.L. The Immunological Theory of Aging. Copenhagen, Munksgaard, 1969.

796 Gravekamp, C., Sypniewska, R., & Hoflack, L. The usefulness of mouse breast tumor models for testing and optimization of breast cancer vaccines at old age. Mech. Ageing Dev. 125, 125–127. 2004.

797 Plackett, T.P., Boehmer, E.D., Faunce, D.E., & Kovacs, E.J. Aging and innate immune cells. J. Leukoc. Biol. 76, 291–299. 2004.

798 Finch, C.E. & Crimmins, E.M. Inflammatory exposure and historical changes in human life-spans. Science 305, 1736–1739. 2004.

799 Burnet, F.M. The Clonal Selection Theory of Acquired Immunity. Cambridge University Press, Cambridge, 1957.

800 Jerne, N.K. The somatic generation of immune recognition. Eur. J. Immunol. 1, 1–9. 1971.

801 Maizels, N. Immunoglobulin gene diversification. Annu. Rev. Genet. 39, 23–46. 2005.

802 Castellino, F. & Germain, R.N. Cooperation between CD4+and CD8+T cells: when, where, and how. Annu. Rev. Immunol. 24, 519–540. 2006.

803 Sansoni, P., Cossarizza, A., Brianti, V., Fagnoni, F., Snelli, G., Monti, D. *et al.* Lymphocyte subsets and natural killer cell activity in healthy old people and centenarians. Blood 82, 2767–2773. 1993.

804 George, A.J. & Ritter, M.A. Thymic involution with ageing: obsolescence or good housekeeping? Immunol. Today 17, 267–272. 1996.

805 Taub, D.D. & Longo, D.L. Insights into thymic aging and regeneration. Immunol. Rev. 205, 72–93. 2005.

806 Franceschi, C. Cell proliferation, cell death and aging. Aging (Milano) 1, 3–15. 1989.

807 Kim, E., Lowenson, J.D., MacLaren, D.C., Clarke, S., & Young, S.G. Deficiency of a protein-repair enzyme results in the accumulation of altered proteins, retardation of growth, and fatal seizures in mice. Proc. Natl. Acad. Sci. USA 94, 6132–6137. 1997.

808 Moskovitz, J., Bar-Noy, S., Williams, W.M., Requena, J., Berlett, B.S., & Stadtman, E.R. Methionine sulfoxide reductase (MsrA) is a regulator of antioxidant defense and lifespan in mammals. Proc. Natl. Acad. Sci. USA 98, 12920–12925. 2001.

809 Harrison, D.E. Mouse erythropoietic stem cell lines function normally 100 months: loss related to number of transplantations. Mech. Ageing Dev. 9, 427–433. 1979.

810 Thomas, H. Ageing in plants. Mech. Ageing Dev. 123, 747–753. 2002.

811 Ligthart, G.J., Corberand, J.X., Fournier, C., Galanaud, P., Hijmans, W., Kennes, B. *et al*. Admission criteria for immunogerontological studies in man: the SENIEUR protocol. Mech. Ageing Dev. 28, 47–55. 1984.

812 Miller, R.A. New paradigms for research on aging and late-life illness. Mech. Ageing Dev. 122, 130–132. 2001.

813 Lakatta, E.G. Cardiovascular aging: perspectives from humans to rodents. Am. J. Geriatr. Cardiol. 7, 32–45. 1998.

814 Selkoe, D.J., Bell, D.S., Podlisny, M.B., Price, D.L., & Cork, L.C. Conservation of brain amyloid proteins in aged mammals and humans with Alzheimer's disease. Science 235, 873–877. 1987.

815 Nakayama, H., Uchida, K., & Doi, K. A comparative study of age-related brain pathology–are neurodegenerative diseases present in nonhuman animals? Med. Hypotheses 63, 198–202. 2004.

816 Buffenstein, R. The naked mole-rat: a new long-living model for human aging research. J. Gerontol. A Biol. Sci. Med. Sci. 60, 1369–1377. 2005.

817 Herndon, L.A., Schmeissner, P.J., Dudaronek, J.M., Brown, P.A., Listner, K.M., Sakano, Y. *et al*. Stochastic and genetic factors influence tissue-specific decline in ageing C. elegans. Nature 419, 808–814. 2002.

818 Huang, C., Xiong, C., & Kornfeld, K. Measurements of age-related changes of physiological processes that predict lifespan of Caenorhabditis elegans. Proc. Natl. Acad. Sci. USA 101, 8084–8089. 2004.

819 Klass, M.R. Aging in the nematode Caenorhabditis elegans: major biological and environmental factors influencing life span. Mech. Ageing Dev. 6, 413–429. 1977.

820 Iijima, K., Liu, H.P., Chiang, A.S., Hearn, S.A., Konsolaki, M., & Zhong, Y. Dissecting the pathological effects of human Abeta40 and Abeta42 in Drosophila: a potential model for Alzheimer's disease. Proc. Natl. Acad. Sci. USA 101, 6623–6628. 2004.

821 Paternostro, G., Vignola, C., Bartsch, D.U., Omens, J.H., McCulloch, A.D., & Reed, J.C. Age-associated cardiac dysfunction in Drosophila melanogaster. Circ. Res. 88, 1053–1058. 2001.

822 Melov, S., Lithgow, G.J., Fischer, D.R., Tedesco, P.M., & Johnson, T.E. Increased frequency of deletions in the mitochondrial genome with age of Caenorhabditis elegans. Nucleic Acids Res. 23, 1419–1425. 1995.

823 Yui, R., Ohno, Y., & Matsuura, E.T. Accumulation of deleted mitochondrial DNA in aging Drosophila melanogaster. Genes Genet. Syst. 78, 245–251. 2003.

824 Klass, M., Nguyen, P.N., & Dechavigny, A. Age-correlated changes in the DNA template in the nematode Caenorhabditis elegans. Mech. Ageing Dev. 22, 253–263. 1983.

825 Botstein, D., Chervitz, S.A., & Cherry, J.M. Yeast as a model organism. Science 277, 1259–1260. 1997.

826 Sinclair, D.A., Mills, K., & Guarente, L. Accelerated aging and nucleolar fragmentation in yeast sgs1 mutants. Science 277, 1313–1316. 1997.

827 Lesur, I. & Campbell, J.L. The transcriptome of prematurely aging yeast cells is similar to that of telomerase-deficient cells. Mol. Biol. Cell 15, 1297–1312. 2004.

828 Fabrizio, P., Gattazzo, C., Battistella, L., Wei, M., Cheng, C., McGrew, K. *et al.* Sir2 blocks extreme life-span extension. Cell 123, 655–667. 2005.

829 Gavrilov, L.A. & Gavrilova, N.S. Reliability theory of aging and longevity. In Handbook of the Biology of Aging, 6th edn, pp. 3–42 (eds. Masoro, E.J. & Austad, S.N.). Elsevier Academic Press, San Diego, CA, 2006.

830 Kaern, M., Elston, T.C., Blake, W.J., & Collins, J.J. Stochasticity in gene expression: from theories to phenotypes. Nat. Rev. Genet. 6, 451–464. 2005.

831 Becskei, A., Kaufmann, B.B., & van Oudenaarden, A. Contributions of low molecule number and chromosomal positioning to stochastic gene expression. Nat. Genet. 37, 937–944. 2005.

832 Magee, J.A., Abdulkadir, S.A., & Milbrandt, J. Haploinsufficiency at the Nkx3.1 locus. A paradigm for stochastic, dosage-sensitive gene regulation during tumor initiation. Cancer Cell 3, 273–283. 2003.

833 Klein, C.A., Seidl, S., Petat-Dutter, K., Offner, S., Geigl, J.B., Schmidt-Kittler, O. *et al.* Combined transcriptome and genome analysis of single micrometastatic cells. Nat. Biotechnol. 20, 387–392. 2002.

834 Bahar, R., Hartmann, C.H., Rodriguez, K.A., Denny, A.D., Busuttil, R.A., Dollé, M.E. T. *et al.* Increased cell-to-cell variation in gene expression in aging mouse heart. Nature 441, 1011–1014. 2006.

835 Finch, C.E. & Kirkwood, T.B. Chance, Development, and Aging. Oxford University Press, Oxford, 1999.

836 Fraser, H.B., Hirsh, A.E., Giaever, G., Kumm, J., & Eisen, M.B. Noise minimization in eukaryotic gene expression. PLoS Biol. 2, e137. 2004.

837 Srivastava, S. & Moraes, C.T. Double-strand breaks of mouse muscle mtDNA promote large deletions similar to multiple mtDNA deletions in humans. Hum. Mol. Genet. 14, 893–902. 2005.

838 Suh, Y. & Vijg, J. SNP discovery in associating genetic variation with human disease phenotypes. Mutat. Res. 573, 41–53. 2005.

839 Carlson, C.S., Eberle, M.A., Kruglyak, L., & Nickerson, D.A. Mapping complex disease loci in whole-genome association studies. Nature 429, 446–452. 2004.

840 Risch, N.J. Searching for genetic determinants in the new millennium. Nature 405, 847–856. 2000.

841 International HapMap Consortium. A haplotype map of the human genome. Nature 437, 1299–1320. 2005.

842 Cardon, L.R. & Abecasis, G.R. Using haplotype blocks to map human complex trait loci. Trends Genet. 19, 135–140. 2003.

843 Terwilliger, J.D., Haghighi, F., Hiekkalinna, T.S., & Goring, H.H. A bias-ed assessment of the use of SNPs in human complex traits. Curr. Opin. Genet. Dev. 12, 726–734. 2002.

844 Service, R.F. Gene sequencing. The race for the $1000 genome. Science 311, 1544–1546. 2006.

845 Geiger-Thornsberry, G.L., & Mackay, T.F. Quantitative trait loci affecting natural variation in Drosophila longevity. Mech. Ageing Dev. 125, 179–189. 2004.

846 Service, P.M. How good are quantitative complementation tests? Sci. Aging Knowledge Environ. 12, pe13. 2004.

847 Jackson, A.U., Galecki, A.T., Burke, D.T., & Miller, R.A. Mouse loci associated with life span exhibit sex-specific and epistatic effects. J. Gerontol. A Biol. Sci. Med. Sci. 57, B9–B15. 2002.

848 Finch, C.E. & Tanzi, R.E. Genetics of aging. Science 278, 407–411. 1997.

849 Perls, T.T., Bubrick, E., Wager, C.G., Vijg, J., & Kruglyak, L. Siblings of centenarians live longer. Lancet 351, 1560. 1998.

850 Perls, T.T., Wilmoth, J., Levenson, R., Drinkwater, M., Cohen, M., Bogan, H. *et al.* Life-long sustained mortality advantage of siblings of centenarians. Proc. Natl. Acad. Sci. USA 99, 8442–8447. 2002.

■ INDEX

Note: Page numbers in *italics* refer to Figures and Tables, whilst those in **bold** refer to Glossary entries.